MW00761216

Lecture Notes in Physics

Springer
Berlin
Heidelberg
New York
Barcelona
Hong Kong
London
Milan
Paris
Singapore
Tokyo

Physics and Astronomy | **ONLINE LIBRARY**

http://www.springer.de/phys/

The Editorial Policy for Proceedings

The series Lecture Notes in Physics reports new developments in physical research and teaching – quickly, informally, and at a high level. The proceedings to be considered for publication in this series should be limited to only a few areas of research, and these should be closely related to each other. The contributions should be of a high standard and should avoid lengthy redraftings of papers already published or about to be published elsewhere. As a whole, the proceedings should aim for a balanced presentation of the theme of the conference including a description of the techniques used and enough motivation for a broad readership. It should not be assumed that the published proceedings must reflect the conference in its entirety. (A listing or abstracts of papers presented at the meeting but not included in the proceedings could be added as an appendix.)

When applying for publication in the series Lecture Notes in Physics the volume's editor(s) should submit sufficient material to enable the series editors and their referees to make a fairly accurate evaluation (e.g. a complete list of speakers and titles of papers to be presented and abstracts). If, based on this information, the proceedings are (tentatively) accepted, the volume's editor(s), whose name(s) will appear on the title pages, should select the papers suitable for publication and have them refereed (as for a journal) when appropriate. As a rule discussions will not be accepted. The series editors and Springer-Verlag will normally not interfere with the detailed editing except in fairly obvious cases or on technical matters.

Final acceptance is expressed by the series editor in charge, in consultation with Springer-Verlag only after receiving the complete manuscript. It might help to send a copy of the authors' manuscripts in advance to the editor in charge to discuss possible revisions with him. As a general rule, the series editor will confirm his tentative acceptance if the final manuscript corresponds to the original concept discussed, if the quality of the contribution meets the requirements of the series, and if the final size of the manuscript does not greatly exceed the number of pages originally agreed upon. The manuscript should be forwarded to Springer-Verlag shortly after the meeting. In cases of extreme delay (more than six months after the conference) the series editors will check once more the timeliness of the papers. Therefore, the volume's editor(s) should establish strict deadlines, or collect the articles during the conference and have them revised on the spot. If a delay is unavoidable, one should encourage the authors to update their contributions if appropriate. The editors of proceedings are strongly advised to inform contributors about these points at an early stage.

The final manuscript should contain a table of contents and an informative introduction accessible also to readers not particularly familiar with the topic of the conference. The contributions should be in English. The volume's editor(s) should check the contributions for the correct use of language. At Springer-Verlag only the prefaces will be checked by a copy-editor for language and style. Grave linguistic or technical shortcomings may lead to the rejection of contributions by the series editors. A conference report should not exceed a total of 500 pages. Keeping the size within this bound should be achieved by a stricter selection of articles and not by imposing an upper limit to the length of the individual papers. Editors receive jointly 30 complimentary copies of their book. They are entitled to purchase further copies of their book at a reduced rate. As a rule no reprints of individual contributions can be supplied. No royalty is paid on Lecture Notes in Physics volumes. Commitment to publish is made by letter of interest rather than by signing a formal contract. Springer-Verlag secures the copyright for each volume.

The Production Process

The books are hardbound, and the publisher will select quality paper appropriate to the needs of the author(s). Publication time is about ten weeks. More than twenty years of experience guarantee authors the best possible service. To reach the goal of rapid publication at a low price the technique of photographic reproduction from a camera-ready manuscript was chosen. This process shifts the main responsibility for the technical quality considerably from the publisher to the authors. We therefore urge all authors and editors of proceedings to observe very carefully the essentials for the preparation of camera-ready manuscripts, which we will supply on request. This applies especially to the quality of figures and halftones submitted for publication. In addition, it might be useful to look at some of the volumes already published. As a special service, we offer free of charge LaTeX and TeX macro packages to format the text according to Springer-Verlag's quality requirements. We strongly recommend that you make use of this offer, since the result will be a book of considerably improved technical quality. To avoid mistakes and time-consuming correspondence during the production period the conference editors should request special instructions from the publisher well before the beginning of the conference. Manuscripts not meeting the technical standard of the series will have to be returned for improvement.

For further information please contact Springer-Verlag, Physics Editorial Department II, Tiergartenstrasse 17, D-69121 Heidelberg, Germany

Series homepage – http://www.springer.de/phys/books/lnpp

J. Klamut B. W. Veal B. M. Dabrowski
P. W. Klamut M. Kazimierski (Eds.)

New Developments in High Temperature Superconductivity

Proceedings of the 2nd Polish-US Conference
Held at Wrocław and Karpacz, Poland,
17-21 August 1998

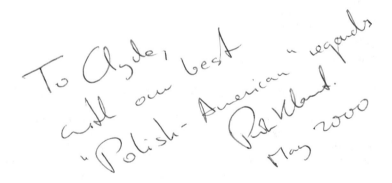

 Springer

Editors

Jan Klamut
Institute of Low Temperature and Structure Research, PAS, and
International Laboratory of High Magnetic Fields and Low Temparatures
95 Gajowicka Str., 53-529 Wrocław, Poland

Boyd W. Veal
Argonne National Laboratory, 9700 S. Cass Ave.
Argonne, IL 60439, USA

Bogdan M. Dabrowski
Department of Physics, Northern Illinois University
DeKalb, IL 60115, USA

Piotr W. Klamut
Maciej Kazimierski
Institute of Low Temperature and Structure Research, PAS
P.O.Box 1410, 50-950 Wrocław, Poland

Library of Congress Cataloging-in-Publication Data applied for.

Die Deutsche Bibliothek - CIP-Einheitsaufnahme

New developments in high temperature superconductivity : proceedings
of the 2nd Polish US conference, held at Wroclaw and Karpacz, Poland,
17 - 21 August 1998 / J. Klamut ... (ed.). - Berlin ; Heidelberg ; New
York ; Barcelona ; Hong Kong ; London ; Milan ; Paris ; Singapore ;
Tokyo : Springer, 2000
  (Lecture notes in physics ; 545)
  ISBN 3-540-67188-9

ISSN 0075-8450
ISBN 3-540-67188-9 Springer-Verlag Berlin Heidelberg New York

Springer-Verlag is a company in the BertelsmannSpringer publishing group.
© Springer-Verlag Berlin Heidelberg 2000
Printed in Germany

The use of general descriptive names, registered names, trademarks, etc. in this publication
does not imply, even in the absence of a specific statement, that such names are exempt
from the relevant protective laws and regulations and therefore free for general use.

Typesetting: Camera-ready by the authors/editors
Cover design: *design & production*, Heidelberg

Printed on acid-free paper
SPIN: 10720759   55/3144/du - 5 4 3 2 1 0

Preface

This volume contains the proceedings of The Second Polish–US Conference on High Temperature Superconductivity which was held August 18–21, 1998 in Karpacz, Poland. The conference followed The First Polish–US Conference on High Temperature Superconductivity organized in 1995, proceedings of which were published by Springer–Verlag in 1996 (Recent Developments in High Temperature Superconductivity, Lecture Notes in Physics 475).

High Temperature Superconductivity (HTSC) in complex copper oxides has become a household name after twelve years of intense research following its discovery in 1986 by J.G. Bednorz and K.A. Müller. Because of the rapid growth of the HTSC field, there is a need for periodic summary and condensation both for scientists working in the field and, especially, for young researchers entering the field of oxide materials. Following the First Conference, it was recognized that an extended format of lectures perfectly satisfied that need, providing adequate time for experts from the international community to fully introduce and develop complex ideas. Thus, the format of the Second Conference brought together by cooperating scientists from the Institute of Low Temperature and Structure Research of the Polish Academy of Science at Wrocław, Northern Illinois University, and Argonne National Laboratory remained mostly unchanged. Again, we were delighted to receive enthusiastic responses from distinguished US and Polish scientists who were invited to participate. The focused sessions on microscopic description, physical properties, materials, crystal chemistry, and applications of HTSC provided forums for intense discussion of common research topics for US and Polish scientists. The Conference also provided a base for personal scientific interactions, especially important for young Polish researchers. The high level of scientific presentations, the high altitude, and perfect weather all contributed to a particularly pleasant atmosphere for the meeting.

The Conference included approximately 80 participants who contributed 21 invited lecturers (10 US and 11 Polish speakers), as well as 44 posters describing the current status of research on HTSC in Poland. The articles presented in this book span the field from the theoretical investigations of the pairing mechanism to the experiments relating to new materials applications. In an effort to present the most current status of HTSC, the texts were updated just prior to publication (fall of 1999). The editors anticipate that the book will become a valuable resource not only for the advanced reader, but also for a larger readership seeking reviews of current problems in HTSC.

We would like to express our sincere thanks to the scientific staff of the Institute of Low Temperature and Structural Research for their superior organization work. We want particularly to acknowledge the leading abilities of S. Gołąb in administration of the Conference and the cooperative spirit of the team, comprising also D. Włosewicz, A.J. Zaleski, and A. Baszczuk, M. Matusiak, H. Misiorek, T. Plackowski, A. Sikora, Cz. Sułkowski.

The Conference was organized under the patronage of the Physics Committee of the Polish Academy of Sciences, and the U.S. Science and Technology Center for Superconductivity. On behalf of all the participants, we would like to express our sincere gratitude to the U.S. National Science Foundation and Polish State Committee for Scientific Research for their financial support which made this conference possible. We thank all the authors for their contributions.

Wrocław, De Kalb, and Argonne,   *The Editors*
January 2000

Contents

Muon Spin Rotation Studies of Doping in High-T_c Superconductors

J. I. Budnick[1] and Ch. Niedermayer[2]

[1] Department of Physics and Institute for Materials Science,
University of Connecticut, Storrs, CT 06269, USA
[2] Fakultät für Physik, Universität Konstanz,
D–78457 Konstanz, Germany

Abstract. Muon spin rotation (µSR) studies on high temperature superconducting (HTS) cuprates will be reviewed. After an introduction to the technique, studies of the superfluid density will be described and the universal variation of the superfluid density n_s as a function of [hole concentration] p will be discussed. Important exceptions will be noted, such as the $YBa_2Cu_3O_{7-\delta}$ system, where, besides the intrinsically superconducting CuO_2 planes, an interlayer may be metallized (here the CuO chains) which consequently contributes to a significant enhancement in superfluid density and associated improvement in technologically interesting properties such as flux pinning and critical current density. A common phase diagram of the antiferromagnetic correlations for Sr doped La_2CuO_4 and Ca doped $YBa_2Cu_3O_6$ and, in particular the coexistence of strong electronic magnetism within the superconducting state, will be discussed in terms of µSR experiments in zero external magnetic field.

1 Introduction

The technique of muon spin rotation or relaxation (µSR) is a powerful tool for studying the internal distribution of magnetic fields within solids [1]. In the context of high temperature superconducting (HTS) cuprates µSR experiments have provided important contributions to a better understanding of the physics of the vortex state, superfluid density and the complex interplay between magnetism and superconductivity that distinguishes the HTS cuprates. These materials are strongly anisotropic due to their distinctive structure which comprises quasi–two–dimensional CuO_2 planes separated by insulating interlayers. The electronic correlations causing magnetism and superconductivity originate primarily in the CuO_2 planes. One of the most striking features of the HTS cuprates is the strong dependence of these electronic correlations on the hole concentration per CuO_2 plane, p, resulting in a generic p–dependent phase behavior as summarized in Fig. 1. At very low doping these materials are antiferromagnetic (AF) insulators but with increasing p the AF correlations are weakened and the Néel temperature, T_N, falls rather sharply to zero. At a critical concentration there occurs an insulator–metal transition coinciding with the onset of superconductivity.

The critical temperature $T_c(p)$ follows a universal, approximately parabolic p–dependence, that can be conveniently expressed as [2]

$$T_c = T_{c\,max}[1 - 82.6(p - 0.16)^2]\,,\qquad\qquad(1)$$

where the maximum is reached for an optimum doping of $p \approx 0.16$ holes per CuO_2 plane. This p–dependence appears to be common to the HTS cuprates and all that varies between the different HTS compounds is the magnitude of the optimal value, $T_{c\,max}$. As also shown in the Figure, at intermediate doping levels extending from the Néel State and well into the superconducting domain, short–ranged AF fluctuations survive and at low temperatures freeze into a disordered spin–glass state (annotated SG) which coexists with superconductivity.

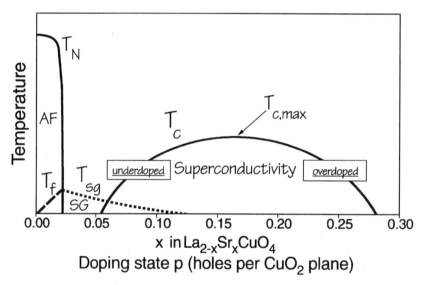

Fig. 1. Schematic representation of the doping–dependent phase diagram for HTS cuprates. AF and SG denote the antiferromagnetic and spin glass phases

After an introduction to the technique, studies on the superfluid density will be described and the universal variation of the superfluid density n_s as a function of p will be presented. Important exceptions will be discussed, such as the $YBa_2Cu_3O_{7-\delta}$ system, where, besides the intrinsically superconducting CuO_2 planes, an interlayer may be metallized (here the CuO chains) which consequently contributes to a significant enhancement in superfluid density and associated improvement in technologically interesting properties such as flux pinning and critical current density.

2   The μSR Technique

The basic idea of a μSR experiment is very similar to that of the NMR technique. Positive muons are incorporated as local probes of interstitial regions of the sample to be studied. The spin of the muon and the related magnetic moment act as a sensitive probe for the local magnetic field through its precession in the field with a frequency of $\omega_\mu = \gamma_\mu B_{loc}$, where $\gamma_\mu = 851.4\,\mathrm{MHz/T}$ is the gyromagnetic ratio of the muon and B_{loc} is the local field. The polarization of the muon is conveniently determined by the fact that when a muon decays (half life 2.2 μs) the resulting positron is emitted preferentially in the direction of the instantaneous polarization. A schematic diagram of the standard μSR experiment is shown in Fig. 2. A beam of 100% spin–polarized muons is directed onto the cuprate sample which, for studies of the vortex state and superfluid density, is mounted in a magnetic field of strength H transverse to the polarization of the muon spin. The sample may be a sintered polycrystalline body, an oriented single crystal, a mosaic of single crystals or even a compact of powder. The injected muons thermalize rapidly without any significant loss in polarization and come to rest at locations in the sample which are random on a scale of the London penetration depth, λ (100–300 nm) but at distinct sites in the crystallographic unit cell, forming a muoxyl bond with apical or chain oxygens [3]. The muon spin precesses about the local field which may be modulated due to flux vortices in the presence of the field but may, in zero field, arise from magnetic ordering or local moments. In the transverse–field experiment, the precession frequency is randomly distributed due to the random distribution of local fields and so the muons dephase and progressively lose their polarization. The time–resolved polarization signal is thus oscillatory with decaying amplitude. Its depolarization rate provides a measure for the inhomogeneity of the magnetic field in the vortex state and hence for the magnetic penetration depth. By Fourier transformation one can obtain the frequency–resolved signal which in the case of single–crystalline materials exhibits the characteristic features of the vortex structure, i.e. a tail towards high frequencies (fields) caused by the vortex cores, a peak at the field of the saddle point between two vortices and a sharp cut-off on the low field side. For polycrystalline samples the distribution of precession frequencies is almost symmetrical and of approximately Gaussian shape. In this case, the Gaussian depolarization rate, σ, may be found from the second moment $\langle \Delta\omega^2 \rangle$ of the frequency distribution or more conveniently it may be more directly determined from the envelope of the oscillatory time–resolved polarization, as given by:

$$P(t) \propto \exp[-\frac{1}{2}\sigma^2 t^2] \,. \tag{2}$$

The key factor is that σ, being a measure of the field distribution, is proportional to λ_{eff}^{-2} where λ_{eff} is an effective magnetic penetration depth related to the in–plane and out–of–plane penetration depths λ_{ab} and λ_c by

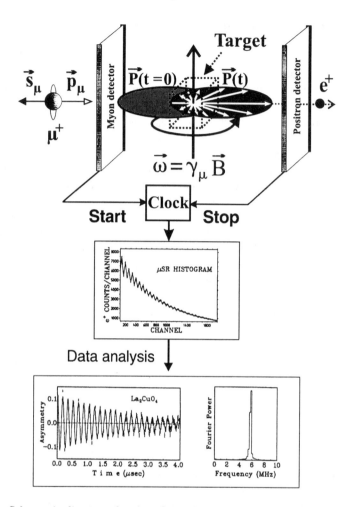

Fig. 2. Schematic diagram showing the main components of the μSR technique

the degree of anisotropy and field orientation. As such it is a measure of the superfluid density, n_s, as will be seen.

Barford and Gunn [4] have analysed the situation for highly anisotropic systems such as the HTS cuprates and find for a polycrystalline sample that $\lambda_{\text{eff}} = 1.23\,\lambda_{ab}$ provided that the anisotropy $\gamma = \lambda_c/\lambda_{ab} > 5$, which is generally satisfied. They deduce:

$$\sigma\,[\mu\text{s}^{-1}] = 7.086 \times 10^4\,\lambda_{ab}^{-2}\,[\text{nm}] \; ; \tag{3}$$

$$= 2.75 \times 10^9\,n_s/m_{ab}^*\,[\text{cm}^{-3}\text{kg}^{-1}] \; . \tag{4}$$

Here n_s is the superfluid density expressed as the density of *single* quasi-particles which contribute to the condensate, and m_{ab}^* is the electronic effective mass for ab plane transport. We stress that for an ideal homogeneous superconductor in the absence of pairbreaking the low–temperature super-fluid density is expected to be equal to the carrier concentration.

3   Superconducting Condensate Density

The early µSR studies on the cuprates appeared largely confined to under-doped and near optimally doped samples. This led to the remarkable obser-vation by Uemura and coworkers [5,6] that, for a wide range of cuprates, T_c and the low temperature depolarization rate are linearly related:

$$T_c \propto \sigma_o \propto \lambda_o^{-2} \propto n_s(T=0)/m^* \,, \tag{5}$$

where λ and m are to be understood as λ_{ab} and m_{ab}. This relationship is shown in Fig. 3. This universal linearity was taken at the time to be evid-ence of the superconducting transition being a Bose–Einstein condensation of preformed real–space pairs [5] in which:

$$T_c - E_F = \pi \hbar^2 n/m^* \tag{6}$$

for a 2D free–electron gas. While this view continues to maintain some sup-port it is not upheld by heat capacity measurements. Dissociation of pairs at elevated temperatures would lead to a clear increase in entropy/T which is not evident [7]. Several alternative models have been shown to lead to this key $T_c \propto \sigma_o$ relationship (see [8] and references therein).

For quite some time the expectation was that the superfluid density $n_s(T=0)$ should continue to increase with growing carrier concentration in the overdoped regime. Experimentally it was found that n_s was strongly de-pressed in the overdoped region so that T_c versus $n_s(T=0)$ follows a reentrant loop as shown in Fig. 4. This behavior was first demonstrated for $Tl_2Ba_2CuO_{6+\delta}$ [9–11] and later reproduced for $Yb_{0.7}Ca_{0.3}Ba_{1.6}Sr_{0.4}Cu_3O_{7-\delta}$ [12], a system, in which the hole concentration can be varied throughout the entire range from heavily underdoped to heavily overdoped. This behavior was modeled in terms of intrinsic pairbreaking progressively developing on the overdoped side [9,10] and providing an increasing density of normal–state carriers. This view was underscored by heat capacity measurements on the same samples of Tl-2201 [13] which showed the low–temperature linear coefficient of the electronic heat capacity, γ_o, progressively rising with overdoping from zero towards the normal–state value, γ_n, well above T_c. This increasing density of low–energy excitations is strongly suggestive of pair breaking. The suppressed condensate density was confirmed in Tl-1212, La-214 and in Ca–substituted $RBa_2Cu_3O_{7-\delta}$ with $R = Y$ and Yb [14] and may be considered to be a generic effect amongst the cuprates. More recent infrared reflectivity measurements on overdoped Bi-2212 confirm the reduction in λ_{ab}^{-2} and at the same time show a dramatic filling of the gap in $\sigma(\omega)$ with overdoping and an increase in elastic scattering [15].

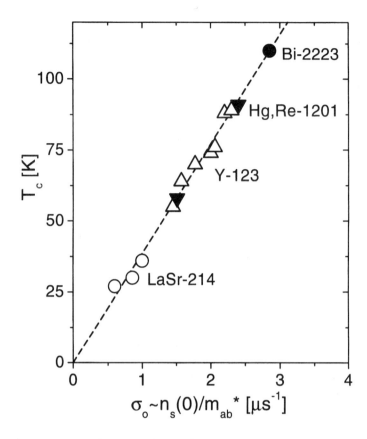

Fig. 3. T_c plotted as a function of the μSR depolarization rate for underdoped cuprates showing the universal linear relation $T_c \sim \sigma_0 \sim n_s/m^*$

In contrast, $YBa_2Cu_3O_{7-\delta}$ develops a broad plateau as $\delta \to 0$ (open circles in Fig. 4). An important structural aspect of this high-T_c–super-conductor is the presence of linear CuO–chains in addition to the CuO_2–planes. We have argued, that upon full oxygenation the chains become metallic and n_s rapidly increases due to the additional condensate density induced there [14,16]. To further elucidate this chain condensation we studied a series of $Y_{1-x}Ca_xBa_2Cu_3O_{7-\delta}$ with $x = 0.03, 0.06, 0.13, 0.2$ and $Yb_{0.7}Ca_{0.3}Ba_{1.6}Sr_{0.4}Cu_3O_{7-\delta}$. In this 1-2-3 system, hole doping of the CuO_2 planes is not only achieved by oxygenation of the CuO–chains but also by replacing Y^{3+} by Ca^{2+}. With increasing Ca–content the complete oxygenation of the CuO–chains is thus shifted towards the overdoped regime. While for $x = 0$ the final filling of the CuO–chains coincides with optimum doping in the planes ($T_{c\,max}$) and one therefore observes the well known "plateau" in σ_0 versus T_c, the "plateau" disappears gradually with increasing Ca–content

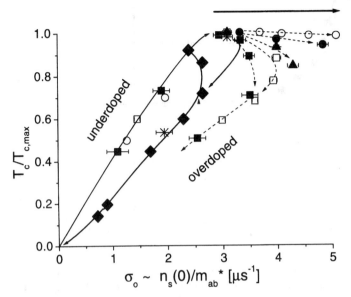

Fig. 4. $T_c/T_{c\,max}$ plotted as a function of the µSR depolarization rate extending from the under– to overdoped region showing the generic re–entrant loop behavior on the overdoped side. The "plane + chain" samples of $Y_{1-x}Ca_xBa_2Cu_3O_{7-\delta}$ are represented by (O) for $x = 0$, (●) – $x = 0.03$, (▲) – $x = 0.13$, (□) – $x = 0.2$, and (■) for $Yb_{0.7}Ca_{0.3}Ba_{1.6}Sr_{0.4}Cu_3O_{7-\delta}$. The broad plateau occurs for $\delta < 0.15$. Here σ_o doubles as the chains become fully oxygenated. "Plane–only" samples are (♦) $Tl_2Ba_2CuO_{6+\delta}$, (✳) $Y_{1-x}Ca_xBa_2Cu_3O_{6.2}Br_z$ ($x = 0$ and $x = 0.2$) and the highly deoxygenated "plane + chain" samples ($\delta > 0.3$)

and is completely absent for $x = 0.2$ and 0.3. In these compounds the planes are already overdoped when the additional chain–condensate is formed.

To separate the chain and plane contributions to the superconducting condensate density, we performed µSR–experiments on samples with optimally doped CuO_2–planes but with different degrees of chain filling. The results for σ_o versus oxygen deficiency δ are shown in Fig. 5. For all samples with $\delta > 0.3$ we observe the same value of $\sigma_o \approx 3.0(1)\,µs^{-1}$, which therefore represents the contribution of the CuO_2 planes to the condensate density. The additional contribution due to the appearance of superconductivity in the chains is only observed for samples with lower oxygen deficiencies ($\delta < 0.25$). The solid line is a fit by a model where the additional chain condensate is mobile just along the CuO–chains and rapidly destroyed by pair–breaking due to oxygen vacancies, which we assume to be randomly distributed. We deduce a chain–coherence length of $\xi_0^{ch} = 5.6\,nm$ and a pronounced in–plane anisotropy with $\lambda_a = 150\,nm$ and λ_b as low as $80\,nm$. Such an in–plane anisotropy for λ_b and λ_a is confirmed by microwave experiments [17].

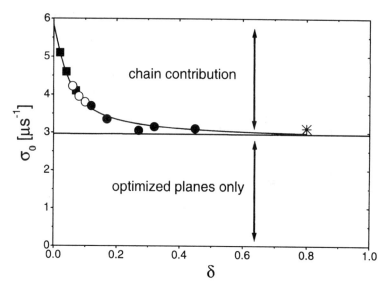

Fig. 5. The low temperature depolarization rate σ_0 for optimally doped $Y_{1-x}Ca_xBa_2Cu_3O_{7-\delta}$ plotted as a function of the oxygen deficiency, δ. (■) slightly overdoped $YBa_2Cu_3O_{7-\delta}$ for $\delta = 0.02$, 0.04 and 0.07; (●) optimized samples for $x = 0$, 0.03, 0.06, 0.13 and 0.20 and (*) $YBa_2Cu_3O_{6.2}Br_z$ with $T_c = 92\,K$. *Open circles* represent data for $TmBa_2Cu_3O_{7-\delta}$. The *solid line* is a fit by a model where the chain condensate is mobile just along the CuO–chains and suppressed by pair-breaking effects that are caused by randomly distributed oxygen vacancies

Cu–NQR experiments provide information on the average length n of the ordered chain–fragments [18]. For (Y,Gd,Tm)-123 Lütgemeier *et al.* observe a sharp increase in n when $\delta < 0.25$, most probably because the oxygen vacancies tend to form clusters. Comparing their data with our µSR–results (open circles in Fig. 5) we find that the increase in n is correlated to the growth of the chain–condensate. Rather short chain fragments (due to a random distribution of the O–vacancies) are found in case of $NdBa_2Cu_3O_{7-\delta}$. Consequently, the observed σ_0–value of $2.85\,\mu s^{-1}$ for $NdBa_2Cu_3O_{6.94}$ nearly equals that of a slightly underdoped "plane–only" sample. These results indicate that the chain–condensate is extremely sensitive on any disruption of the long range order in the CuO–chains.

4 Antiferromagnetism and Superconductivity in HTS Cuprates

Whenever the insulating composition of a given class of high T_c superconductors is chemically stable, it generally exhibits long–range AF order that is rapidly destroyed by small amounts of doped carriers. Short–range 2D AF correlations, however, persist into the superconducting regime. It is therefore of great importance to study the evolution of magnetism as more holes are doped into the CuO_2 planes and to explore the interplay between short–range magnetic order and superconductivity.

The zero–field (ZF) μSR technique is especially suited for such studies since the positive muon is an extremely sensitive local probe able to detect internal magnetic fields as small as 0.1 mT and covering a time window from 10^{-6} s to about 10^{-10} s. Another advantage is the sensitivity of the muon probe to extremely short–ranged magnetic correlations. Systematic studies so far focused on the La-214 [19,20] and Y-123 [21,22] systems. In Y-123 the phase diagram has to be drawn versus oxygen content and a reliable determination of p is difficult due to the rather complicated charge transfer from the CuO chains to the CuO_2 planes. The $Y_{1-x}Ca_xBa_2Cu_3O_6$ system, i.e. with $\delta = 1.0$, avoids this complication, because hole doping is achieved by the substitution of Y^{3+} by Ca^{2+}. This allows one to directly control the hole concentration in the CuO_2 planes in a quantitative manner and $p = x/2$.

Representative ZF-μSR time spectra are shown in Fig. 6. At low temperature and for $p \leq 0.08$, the time evolution of the muon spin polarization is well described by the Ansatz:

$$G_z(t) = \frac{2}{3}\cos(\gamma_\mu B_\mu t + \Phi)\exp\left[-\frac{1}{2}(\gamma_\mu \Delta B_\mu t)^2\right] + \frac{1}{3}\exp[-\lambda t] , \quad (7)$$

where $\gamma_\mu = 851.4$ MHz/T is the gyromagnetic ratio of the muon, B_μ – the average internal magnetic field at the muon site and ΔB – its rms deviation. The two terms arise from the random orientation of the local magnetic field in a polycrystalline sample, which on average points parallel (perpendicular) to the muon spin direction with probability 1/3 (2/3) [1]. In analogy to NMR the dynamic spin lattice relaxation rate $\lambda = 1/T_1$ is given by

$$\frac{1}{T_1} = \gamma_\mu^2 \langle B_t^2 \rangle \frac{\tau_c}{1 + (\omega_\mu \tau_c)^2} . \quad (8)$$

A slowing down of magnetic fluctuations typically causes a maximum of $1/T_1$ at $\omega_\mu \tau_c \approx 1$, where ω_μ is the μ^+ Zeeman frequency, $\langle B_t^2 \rangle$ – the mean of the square of the fluctuating transverse field components and τ_c – their average correlation time. A precessing 2/3 component indicates static magnetic order on the time scale of the μSR technique ($\tau_c < 10^{-6}$ s). For $p > 0.08$ no oscillations were observed and the 2/3 part of $G_z(t)$ was better represented by an exponential relaxation $\exp[-\Lambda t]$ (see Fig. 6c), which may indicate either a very strongly disordered static field distribution or rapid fluctuations.

For only lightly doped systems the Cu^{2+} spins and those of the holes order independently. As an example we discuss the data on $Y_{0.94}Ca_{0.06}Ba_2Cu_3O_{6.02}$ which are displayed in Fig. 7. Well below the 3D Néel temperature of $T_N \approx$

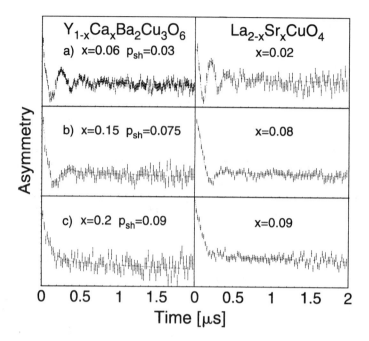

Fig. 6. ZF-μSR spectra obtained at low temperatures ($T < 1\,\mathrm{K}$) for various degrees of hole doping in $Y_{1-x}Ca_xBa_2Cu_3O_{6.02(1)}$ and $La_{2-x}Sr_xCuO_4$. *Dotted curves* are the fit to the data using (7)

170 K a second magnetic transition occurs at a temperature $T_f \approx 25\,\mathrm{K}$. This is evident from the peak in the longitudinal relaxation rate $1/T_1$ and the upturn of the muon spin precession frequency. A corresponding transition within the AF state has been reported recently from La-NQR [23] and μSR studies [24] on La,Sr-214 where $T_f = (815\,\mathrm{K}) \cdot p$ has been obtained for $p < 0.02$. This transition was ascribed to a freezing of the spins of the doped holes into a spin glass state which is superimposed on the preexisting 3D AF long–range order of the Cu^{2+} spins. Interestingly, we find that the spin freezing temperature T_f exhibits the same linear dependence on the planar hole content for Y,Ca-123 and La,Sr-214 (see Fig. 8). According to the model of Gooding *et al.* [25], in which $k_B T_f \approx J_{eff} p$, this implies that the effective in plane exchange coupling constant, J_{eff}, is identical for both systems and that the freezing of the spin degrees of freedom is a property of the hole dynamics within a single plane. The Néel–state, however, persists to higher hole content in Y,Ca-123 ($0 < p < 0.035$) as compared to La,Sr-214 ($p < 0.02$). This suggests that the bilayer coupling makes the 3D AF–state more robust to the presence of doped holes. A similar result was reported from a ^{89}Y NMR study of T_N in Y,Ca-123 [26].

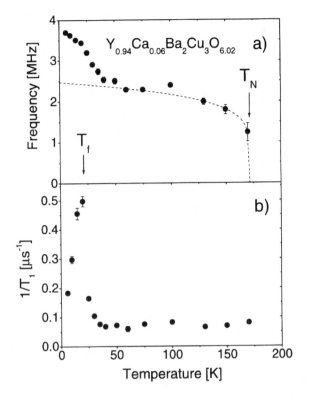

Fig. 7. ZF-μSR results on $Y_{0.94}Ca_{0.06}Ba_2Cu_3O_{6.02(1)}$ plotted as a function of temperature. **a)** The muon spin precession frequency and **b)** the longitudinal relaxation rate $1/T_1$. The *dashed line* in a) represents a fit of the data with a power law $(1 - T/T_N)^\beta$ with $\beta = 0.2$

Only a single magnetic transition into a short range AF correlated spin–glass like state is observed for $p > 0.02$ in La,Sr-214 and $p > 0.035$ in Y,Ca--123. This transition is characterized by a slowing down of the AF fluctuations towards a glass transition which is defined by the maximum in $1/T_1$ (corresponding to a correlation time of the spin fluctuations of about 10^{-7} s). The spin–glass character of this magnetic state has been demonstrated recently for $La_{1.96}Sr_{0.04}CuO_4$ where the susceptibility exhibits irreversible and remanent behavior and obeys scaling laws [27]. T_g is significantly higher due to bilayer interactions in Y,Ca-123 than in La,Sr-214.

It is notable that spin–glass behavior is observed for both compounds up to about $p = 0.11$, the same point where static susceptibility, spin susceptibility and heat capacity show the superconducting gap rapidly fills in. This also coincides with the 60 K plateau for a wide range of Ca– and La–substituted Y-123 samples and all of this features have been attributed to dynamic phase separation which freezes out at low temperature [28].

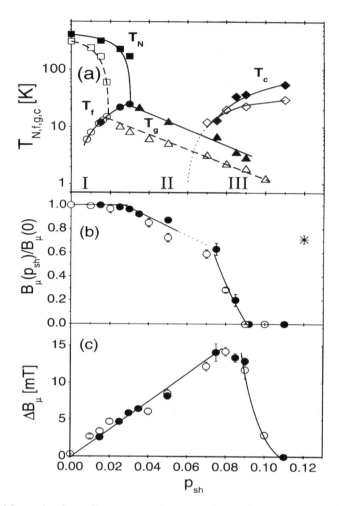

Fig. 8. Magnetic phase diagram as a function of the hole concentration per CuO_2 sheet for $La_{2-x}Sr_xCuO_4$ (*open symbols*) and $Y_{1-x}Ca_xBa_2Cu_3O_{6.02}$ (*full symbols*). **a)** In regime I two transitions are observed. The Néel temperatures T_N (*squares*), at which the Cu^{2+} spins order into a 3D AF state and a freezing transition of the spins of the doped holes at $T_f = (815\,K) \cdot P_{sh}$ (*circles*, including data from [24]). T_g indicates a transition into a spin–glass like state (*up triangles*, regime II) with strong magnetic correlations which coexist with superconductivity in regime III. *Diamonds* represent the superconducting transition temperatures. **b)** Doping dependence of the normalized average internal magnetic field at the muon site. The *star* at $p_{sh} = 0.12$ represents the data for $La_{1.58}Nd_{0.3}Sr_{0.12}CuO_4$. **c)** rms deviation ΔB. Data in b) and c) are for $T < 1\,K$

In the picture of electronic phase separation the underdoped cuprates evolve on doping into a phase with hole poor antiferromagnetic islands separated by grain boundaries of a hole rich "metallic" phase [29]. It is tempting to identify the antiferromagnetic island phase with domain size $L(p) \sim p^{-1/2}$ ($p^{-1/2}$ is the average distance between two holes) with the spin glass phase. In this case the spin glass transition temperature should be given by $T_g \sim J_{xy}(\xi_0/a)^2 \sim L^2 \sim 1/p$ in qualitative agreement with the experiment.

The spin glass regime extends far into the SC state (regime III). For strongly underdoped SC samples with $0.06 < p_{sh} < 0.01$ we still observe a freezing of the spin degrees of freedom. Except for the somewhat smaller ordering temperature the signature of the transition is the same as for the non–SC samples in regime II. From the amplitude of the rapidly damped muon spin polarization (see Fig. 8 b,c) we can obtain information about the volume fraction of the magnetically correlated regions. We find that all the muons stopped inside the sample experience a non zero local magnetic field, which implies that the magnetic order persists throughout the entire volume of the superconducting sample. The magnetic ground state may still be inhomogeneous but the size of the non magnetic hole rich regions must be smaller than the typical length scale (about 2 nm) of the μSR experiment. By decoupling experiments in a longitudinal field we have confirmed the static nature of the magnetic ground state [30]. From transverse field measurements we find that the flux line lattice which is formed below $T_c > T_g$ extends throughout the entire volume of the sample [14]. Note, that these results are markedly different from the μSR results that have been obtained on the "superoxide" $La_2CuO_{4.13}$ where long range oxygen diffusion leads to macroscopic phase separation with an average domain size of about 300 nm [31]. In this compound finite size effects are negligible and the hole poor phase (40% of the volume fraction) displays a temperature dependence and absolute values of the internal fields identical to those in stoichiometric La_2CuO_4. Simultaneously, a flux line lattice forms only within the hole rich regions which account for the remaining 60% of the volume [32]. As described above, our present μSR results are fundamentally different and indicate a microscopic coexistence of the AF and SC order parameter. We want to stress that identical magnetic behavior is observed for both the single layer system La,Sr-214 and the bilayer compound Y,Ca-123.

The consistency of our results suggests that the coexistence of SC and AF order is an intrinsic property of the CuO_2 planes and not an artifact of chemical or structural impurities. Our data show that the strength of the AF correlation is determined solely by the hole content of the CuO_2 planes and does not depend on the concentration of dopant atoms. For a given hole content the number of dopant atoms (Ca^{2+} or Sr^{2+}) is twice the number in Y,Ca-123 compared to La,Sr-214.

In contrast to T_g which evolves rather smoothly, the internal magnetic field at the muon site exhibits a strong change for $p_{sh} \approx 0.06 - 0.08$ as one enters the SC regime. The change in slope is rather significant and indicates a distinct change in the ground state properties of the CuO_2 planes. From the µSR experiment alone we cannot decide whether it is the competition between the AF and the SC order parameter or an underlying change of the electronic properties of the CuO_2 planes which causes the suppression of the internal field. Further experiments will be required in order to clarify if the SC order parameter is affected by the static AF correlation.

Notably, the AF correlation is fully restored at $p_{sh} \approx 1/8$. A depression of T_c at this hole concentration at first appeared to be uniquely present in $La_{2-x}Ba_xCuO_4$ [33], but recent studies on $La_{2-x}Sr_xCuO_4$ [34] have shown the presence of a shallow cusp at the same doping level and this behavior may also be related to the 60 K plateau in Y-123 [35]. Detailed µSR studies [36] by Luke et al. and Kumagai et al. show that at this doping level static magnetic order is restored at temperatures below 35 K. Tranquada et al. [37] showed that the static order in $La_{1.6-0.125}Nd_{0.4}Ba_{0.125}CuO_4$ comprised a spatial separation of the spin and charge into AF stripes three lattice spacings wide (hole poor) separated by antiphase domain boundaries of one lattice dimension where the doped holes reside on every second site. If we consider a picture in which a stripe phase were to be established through connectivity of the hole doped regions already existing in region II, the averaged internal magnetic field is expected to be 75% of the value of the undoped compound. Interestingly, this value is observed for both the SC compound $La_{1.93}Sr_{0.07}CuO_4$ and the non–SC static stripe phase compound $La_{1.58}Nd_{0.3}Sr_{0.12}CuO_4$.

In summary we have presented a magnetic phase diagram for the single layer system $La_{2-x}Sr_xCuO_4$ and the bilayer compound $Y_{1-x}Ca_xBa_2Cu_3O_6$. We observe a common phase diagram as a function of hole doping per CuO_2 plane. In the 3D AF regime at low doping concentrations, well below T_N, we observe a freezing of the spin degrees of freedom of the doped holes at a temperature T_f, which increases linearly with the number of doped holes in both systems suggesting that the hole dynamics in a single plane is responsible for the observed behavior. Due to plane to plane correlations this regime extends to higher hole concentrations in the bilayer system. For higher doping levels (regime II), we observe a single magnetic transition into a spin glass like state with T_g(Y,Ca-123)$>$ T_g(La,Sr-214) and extending well into the superconducting regime III. The evolution of the internal magnetic field with doping is understood on the basis of a microscopic phase segregation of the doped holes into hole rich and hole poor regions. We observe a microscopic coexistence of superconductivity and frozen antiferromagnetic correlations at low temperatures for underdoped samples.

5   Conclusions

As an internal probe of the local fields in HTS cuprates the µSR technique has proved to be a powerful tool for revealing many generic features in the magnetic phase behavior of these materials and the systematic changes in superfluid density with doping and substitution. Many of these were quite new results with important implications for the origins and physics of superconductivity while, for others, μSR played a complementary role. In all cases the strength of the technique has been in investigating a systematic series of samples with progressive doping of one sort or another where spurious impurity phases and grain boundary effects do not affect the intrinsic response. The list is impressive: penetration depths, proximity–induced superconductivity, temperature– and scattering–dependent superfluid density, the symmetry of the order parameter, local moments and coexisting magnetism and superconductivity.

Acknowledgement

We are grateful to our many collaborators: C. Bernhard, T. Blasius, A. Golnik, A. Weldinger, E.J. Ansaldo, J.L. Tallon, A. Moodenbaugh. The experiments described herein have been performed at the *Paul Scherrer* Institute, Villigen, Switzerland and TRIUMF, Vancouver. We thank these institutions and their support staff for continuing assistance. Funding assistance from the BMFB the DOE and the DFG is gratefully acknowledged.

References

1. For a review of the μSR technique and application prior to HTSC see: Schenck A. (1985) Muon Spin Rotation Spectroscopy, Adam Hilger, Bristol
2. Tallon J.L. et al. (1995) Phys Rev B 51:12 911
3. Dawson W.K. et al. (1988) J Appl Phys 64:5803
4. Barford W., Gunn J.M.F. (1988) Physica C 156:515
5. Uemura Y.J. et al. (1989) Phys Rev Lett 62:2317
6. Uemura Y.J. et al. (1991) Phys Rev Lett 66:2665
7. Loram J.W. et al. (1993) Phys Rev Lett 71:1740
8. Emery V.J., Kivelsen S.A. (1995) Nature 374:434
9. Niedermayer Ch. et al. (1993) Phys Rev Lett 71:1764
10. Niedermayer Ch. et al. (1994) Phys Rev Lett 72:2502
11. Uemura Y.J. et al. (1993) Nature 364:605
12. Bernhard C. et al. (1994) Physica C 226:250
13. Wade J.M. et al. (1994) J Superconduct 7:261
14. Bernhard C. et al. (1995) Phys Rev B 52:10 488
15. Prenninger M. (1996) In: Bozoric I., van der Marel D. (Eds.) Spectroscopic Studies of Superconductors, SPIE, Bellingham
16. Tallon J.L. et. al. (1995) Phys Rev Lett 74:1008
17. Basov D.N. et al. (1995) Phys Rev Lett 74:598
18. Lütgemeier H. (1994) In: Sigmund E., Maller K.A. (Eds.) Phase Separation in Cuprate Superconductors, Springer, Berlin, 225
19. Budnick J.I. et al. (1988) Europhys Lett 5:65
20. Weldinger A. et al. (1989) Phys Rev Lett 62:102

21. Nishida N. et al. (1988) J Phys Soc Jpn 57:597
22. Weldinger A. et al. (1990) Hyperf Interact 63:147
23. Chou F.C. et al. (1993) Phys Rev Lett 71:2323
24. Borsa F. et al. (1995) Phys Rev B 52:7334
25. Gooding R.J. et al. (1994) Phys Rev B49:6067
26. Casalta H., Alloul H., Marucco J.-F. (1993) Physica C 204:331
27. Chou F.C. et al. (1995) Phys Rev Lett 75:2204
28. Tallon J.L., Flower N.E., Williams G.V.M. (1999) to be published
29. Emery V.J., Kievelson S.A. (1993) Physica C 209:597
30. Niedermayer Ch., Forgan E.M., Gluckler H., et al. (1999)
    Phys Rev Lett 83:3932–3935
31. Jorgensen J.D. et al. (1988) Phys Rev B 38:11 337
32. Ansaldo E.J. et al. (1989) Phys Rev B 40:2555
33. Moodenbaugh A.R. et al. (1988) Phys Rev B 38:4596
34. Radaelli P.G. et al. (1994) Phys Rev B 49:4163
35. Tallon J.L. et al. (1997) Physica C 282-287:236
36. Luke G.M. et al. (1991) Physica C 185–189 1175;
    Kumagai K. et al. (1991) Physica C 185–189:913
37. Tranquada J.M. et al. (1995) Nature 375:561;
    Tranquada J.M. et al. (1997) Phys Rev Lett 78:338

Dynamic Signatures of Driven Vortex Motion

G. W. Crabtree[1], D. Lopez[1], W. K. Kwok[1], A. M. Petrean[1,2],
R. J. Olsson[1,3], H. Safar[4], and L. M. Paulius[1,2]

[1] Materials Science Division, Argonne National Laboratory,
Argonne, IL 60439, USA
[2] Department of Physics, Western Michigan University,
Kalamazoo, MI 49008, USA
[3] Department of Physics, Michigan State University,
East Lansing, MI 48824, USA
[4] Department of Physics, University of Illinois at Chicago,
Chicago, IL 60680, USA

Abstract. We probe the dynamic nature of driven vortex motion in superconductors with a new type of transport experiment. An inhomogeneous Lorentz driving force is applied to the sample, inducing vortex velocity gradients that distinguish the hydrodynamic motion of the vortex liquid from the elastic and plastic motion of the vortex solid. We observe elastic depinning of the vortex lattice at the critical current, and shear induced plastic slip of the lattice at high Lorentz force gradients.

1 Introduction

The melting of the vortex solid to a liquid is one of the most basic phenomena in high temperature superconductivity. This fundamental phase change from order to disorder induces novel behavior in the structural, electrodynamic, and thermal properties of superconductors. The phase transition itself has many fascinating aspects, including first and second order thermodynamic character, upper and lower critical points, an associated peak effect in the critical current, strong angular dependence, and sensitivity to pinning disorder. (For reviews, see Refs. 1–5.) Strong interest in the nature of melting has stimulated creative experiments on the resistivity, [6–12] magnetization, [13–17] and heat capacity [18–23] at the transition. These experiments provide valuable information on the thermodynamic properties of melting, and on the dramatic differences in pinning strengths and critical currents of the solid and liquid vortex phases.

There are equally interesting *dynamic* properties associated with the vortex phases. In the liquid state, vortices are free to move past each other in response to variations in the driving and pinning forces they experience. The resulting motion is hydrodynamic, where the velocity profile changes continuously on the scale of the intervortex distance. Hydrodynamic motion can be described by differential equations which provide a complete picture of the

---

[1] Present Address: Lucent Technologies, Murray Hill, NJ 07964, USA

velocity distribution if the driving force and boundary conditions are known [24–26]. In the solid vortex states, however, elastic bonds between vortices prevent vortices from moving past each other [27]. If the bonds do not break, the vortex system moves as a single elastic object with one average velocity for all vortices. If shear forces are large enough to break the elastic bonds, planes or surfaces of plastic slip are introduced where the velocity profile changes abruptly. The elastic character of the moving solid is the fundamental feature distinguishing it from the hydrodynamic motion of the liquid. The elastic shear modulus of the solid *resists* an applied shear stress by constraining neighboring vortices to move at the same velocity, while the liquid *accommodates* an applied shear by allowing neighboring vortices to move at different velocities.

Conventional transport experiments where the vortex system is driven with a spatially uniform Lorentz force are only weakly sensitive to the differences between liquid and solid motion. In the absence of pinning, the liquid and solid move with uniform velocity and their differing response to shear forces plays no role in the motion. Pinning disorder introduces shear forces and velocity profiles that in principle could distinguish liquids from solids. However, the velocity differences are averaged out on the scale of the pinning disorder, typically well below the spatial resolution of transport experiments.

In this article, we describe a new class of transport experiments where the vortex system is driven with an inhomogeneous Lorentz force [28]. The response of the vortex system to the gradient in the applied driving force reveals the fundamental signatures of liquid and solid motion. We demonstrate a qualitative change in the character of the motion at the vortex melting transition, and we show explicitly that the moving vortex lattice maintains dynamic coherence over distances of order the sample size. We use a high driving force gradient to introduce plastic slip surfaces into the moving lattice and we see the number of slip surfaces increase with the strength of the applied shear stress.

2   Controlled Gradient Transport

Our experiments are carried out in the disk geometry illustrated in the inset of Fig. 1. Current is injected at the center of the disk and removed at the circumference through gold electrodes evaporated onto the sample. In this geometry the current flows in the radial direction, with inhomogeneous magnitude $J(r) = I/(2\pi r t)$, where t is the sample thickness. The Lorentz force due to this current drives vortices azimuthally in circular orbits around the disk. Vortices in these orbits never encounter the sample surface, ensuring that the measured transport properties reflect only bulk vortex behavior unaffected by entry and exit effects, Bean Livingston barriers, or surface pinning. The inhomogeneous current distribution produces a Lorentz force gradient varying as $1/r^2$. This gradient introduces a spatially varying shear stress which probes the dynamic response of the liquid and solid phases.

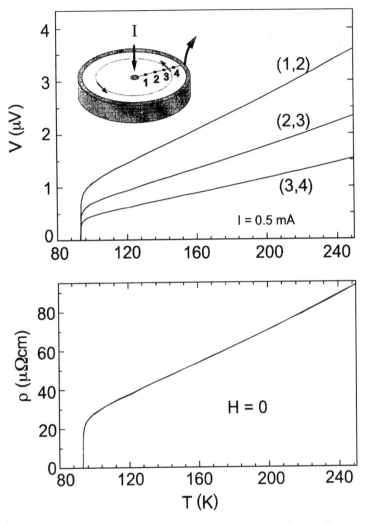

Fig. 1. Upper panel: the measured voltages in the normal metallic state for the three sets of voltage taps indicated in the inset. Lower panel: The measured voltages scaled by (1) of the text. The collapse to a single curve verifies the $1/r$ dependence of the applied current density

Our experiments were carried out on untwinned $YBa_2Cu_3O_x$ crystals carefully polished into disk geometries, of diameter approximately $700\,\mu m$ and thickness $10\,\mu m$. Voltage leads were placed along a radius of the sample at intervals of approximately $60\,\mu m$ from the center current electrode and each other, as indicated in the inset of Fig. 1. Transport measurements were made with standard ac and dc methods. Further details of the experiment may be found in Ref. 28.

3   Normal Metallic State

The upper panel of Fig. 1 shows the voltage measured between pairs of voltage leads as a function of temperature at zero field in the normal state below 240 K. The voltages decrease with increasing radius, reflecting the decreasing current density in the disk geometry. In the normal state with Ohmic resistivity $\rho = E/J$, the voltage as a function of radius is given by

$$V_{n,n+1} = \int_{r_n}^{r_{n+1}} E(r)\, \mathrm{d}r = \frac{\rho I}{2\pi t}\, \ln\left[\frac{r_{n+1}}{r_n}\right] . \tag{1}$$

The voltages scaled by the logarithmic term on the right should collapse onto a single line characteristic of the resistivity. The lower panel of Fig. 1 shows the scaled voltages. Their collapse to a single line confirms that the current density falls off like $1/r$ as expected for the disk geometry.

4   Vortex Liquid State

In the superconducting state in finite field, the transport voltage arises from the motion of vortices driven by the Lorentz force. Here the electric field is not directly related to the current density as in the normal state. Instead, the electric field is governed by the vortex velocity through

$$\boldsymbol{E} = \boldsymbol{B} \times \boldsymbol{v}/c . \tag{2}$$

The vortex velocity, in turn, is determined by the balance between the Lorentz driving force and the pinning forces opposing the motion. In the liquid state, the velocity is given by the solution to the hydrodynamic equation

$$-\gamma \boldsymbol{v} + \eta \nabla_\perp^2 \boldsymbol{v} + \boldsymbol{F}_{\mathrm{L}} = 0 , \tag{3}$$

where η is the shear viscosity and γ the dynamic friction opposing the vortex motion. The solution of the hydrodynamic equation for the disk geometry is given by Marchetti and Nelson [26]. They discuss a method of measuring the dynamic correlation length from the velocity profile in the vicinity of a no–slip boundary condition as might be imposed by the sample edge or an extended pinning structure. Here we consider the free response of the vortex liquid far from any boundary conditions. In this region, a $1/r$ driving current induces a $1/r$ velocity profile. Physically this means that the velocity of the liquid follows the local driving force. The liquid fully accommodates the applied shear force by adjusting its velocity accordingly. This fundamental property of hydrodynamic motion can be directly observed in our experiment.

The electric field induced by the vortex motion in the liquid is $E(r) = Bv(r)/c = (B\varphi/c^2)J(r) = \rho_f J(r)$, where ρ_f is the flux flow resistivity. This is the same form as for the normal state, with the Ohmic resistivity replaced by the flux flow resistivity. Thus hydrodynamic scaling follows (1) with ρ replaced by ρ_f.

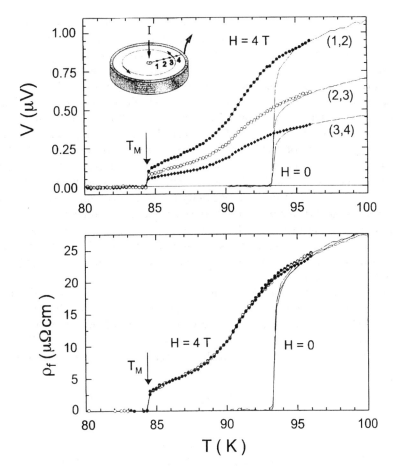

Fig. 2. Upper panel: the measured voltages of the taps shown in the inset in the vortex liquid state as a function of temperature in a magnetic field of 4 T. Lower panel: the measured voltages scaled according to (1) of the text with the resistivity ρ replaced by the flux flow resistivity ρ_f. The scaling of the three curves is a characteristic feature of hydrodynamic flow

The upper panel of Fig. 2 shows the measured voltages in the vortex liquid state, as a function of temperature for applied fields of zero and 4 T.

The voltages decrease with radius, indicating that the vortices circulate at lower velocity as the radius increases. The lower panel of Fig. 2 shows the same voltages scaled according to (1) and expressed as flux flow resistivity. The three curves collapse to a single curve, demonstrating the long range $1/r$ velocity profile characteristic of hydrodynamic motion of the liquid.

5   Vortex Lattice State

In the vortex lattice, the shear viscosity of the liquid is replaced by shear elasticity. The elastic bonds between neighboring vortices require them to move with the same velocity as long as the bonds remain intact. Velocity changes can occur, however, by plastic shear where the elastic bonds between neighboring vortices are broken. Thus the smooth variation of the velocity on the scale of intervortex distances is eliminated and the motion is no longer hydrodynamic. The change from hydrodynamic flow to solid–like flow appears in the experiment as a failure of the hydrodynamic scaling. This is shown in the upper panel of Fig. 3, where the measured voltages are shown in the vicinity of the melting point. Just below the melting point T_M there is a break in the curves (marked by an arrow in Fig. 3) indicating a slowdown in the vortex velocity as the shear elasticity of the lattice enhances its pinning effectiveness. Below this break the three curves begin to approach each other, eventually crossing in pairs. At lower temperatures, the order of the curves is reversed, with the outer vortices traveling faster than the inner ones. This order is opposite to that of the liquid phase and it cannot be described by hydrodynamic motion.

The nature of the motion in the lattice phase is revealed in the lower panel of Fig. 3. Here the measured voltages are scaled according to elastic rotation. If the lattice rotates elastically, the velocity of any vortex is linearly proportional to its radius and the rate of rotation, $v(r) = \omega r$. Then the electric field is $E(r) = B\omega r/c$, and the voltage between taps is given by

$$V_{n,n+1} = \int_{r_n}^{r_{n+1}} E(r)\,\mathrm{d}r = \frac{B\omega}{2c}\left(r_{n+1}^2 - r_n^2\right). \qquad (4)$$

The lower panel of Fig. 3 shows the voltages scaled by (4) and plotted as the angular velocity ω. At the break in the curves just below T_M, the angular velocity of the lattice sampled by the three sets of voltage taps begins to equalize. This marks the termination of hydrodynamic flow with its $1/r$ velocity scaling, and the onset of plastic flow. At lower temperature, the three curves collapse to a single curve. Here the lattice rotates elastically with a single angular velocity. The degree of dynamic correlation in this regime is remarkable – the dynamic correlation length exceeds the distance between the innermost and outermost voltage taps. The velocity correlation is macroscopic, of order the size of the sample.

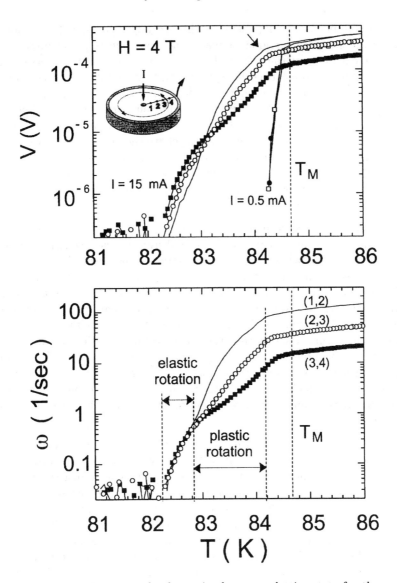

Fig. 3. Upper panel: measured voltages in the vortex lattice state for three pairs of taps indicated in the inset of the lower panel. Results for two transport currents are shown. 0.5 mA is below the critical current of the lattice, while 15 mA is above the critical current and drives the lattice into motion. T_M indicates the melting temperature and the arrow indicates the loss of hydrodynamic motion on cooling. Lower panel: the measured voltages scaled according to (4) of the text and expressed as rotational velocity. The collapse to a single curve at low temperature indicates elastic rotation of the lattice with a single angular velocity

In the temperature range between the hydrodynamic and elastic motion, there is an interesting region of plastic motion reflecting the influence of the shear elasticity of the lattice. Here there is a competition between the shear modulus of the lattice favoring a constant angular velocity and the shear component of the Lorentz force favoring faster rotation of the inner vortices over the outer ones. Neither has sufficient strength to dominate the other, with the result that the lattice moves at short range in elastic flow interrupted by plastic slip surfaces which relieve the applied shear stress over large distances. As the temperature decreases and the rotation slows down, the shear elasticity of the lattice gradually overcomes the applied shear stress, and the dynamic correlation length grows until the entire lattice rotates as a single elastic object.

The nature of plastic motion in the rotating lattice can be seen by following the rotational velocities as measured by the inner, center, and outer sets of voltage taps as the driving shear stress is increased. Figure 4 shows $I - V$ curves for the three pairs of voltage taps plotted as angular velocity at two temperatures in an applied magnetic field of 5 T. Below the critical current, the lattice is stationary. At the critical current, vortices depin and begin to move in circular orbits. The depinning process itself is interesting in this crystal. Figure 4 shows that the lattice rotates elastically immediately on depinning, as indicated by the single angular velocity sampled by all three pairs of voltage taps. Conventional transport experiments cannot distinguish between this kind of elastic depinning and the more familiar plastic depinning where the weakly pinned vortices begin to move while the strongly pinned ones remain at rest.

In elastic rotation the lattice is under shear stress since the driving force distribution varying as $1/r$ is out of balance with the velocity distribution varying as r. This shear stress is accommodated by elastic distortions in the rotating lattice. As the driving current increases, the shear stress also increases, until it exceeds the elastic limit of the shear modulus. At this point the shear stress is relieved by plastic slip in the lattice at a fixed radius, and the entire lattice no longer rotates with a single angular velocity. This effect can be seen in Fig. 4, where the angular velocities sampled by the three voltage taps separate as indicated by arrows. The inner section of the lattice is the first to shear, as the angular velocity sampled by the inner taps rises above the others. The faster rotation of the inner section is consistent with the larger driving force there, and it partially relieves the shear stress in the rotating lattice. At this point plastic slip has occurred only in the innermost section, with the outer two sections rotating elastically with a common angular velocity. As the gradient of the driving force continues to increase, plastic slip interrupts the coherent motion of the center and outermost sections, as indicated by the second arrow in Fig. 4. These data reveal the onset and evolution of plastic motion under increasing shear stress. The lattice separates into concentric rings, each rotating with successively slower angular velocity

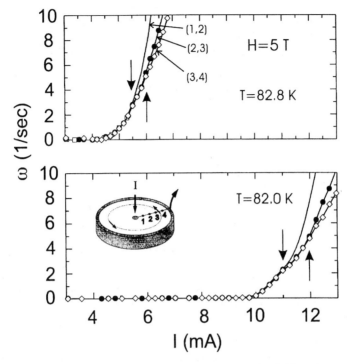

Fig. 4. $I - V$ characteristics for the vortex lattice at the field and temperatures indicated. The solid line indicates the voltage between the inner taps, the filled circles that between the center taps, and the open squares that between the outer taps. Arrows indicate the onset of plastic slip

as the radius increases. As the shear stress increases, more fault lines are introduced into the lattice, reducing the radial dynamic correlation length at each step.

The transformation of elastic flow to plastic flow by the introduction of slip lines can be visualized on the driving force–temperature phase diagram of Fig. 5. Here the regions of pinned solid, elastic rotation, and shear induced slip at 5 T as a function of temperature and driving force are shown. The critical current as a function of temperature defines the boundary between the pinned solid and elastic rotation. At higher driving force the first observable shear induced slip occurs between the inner set of taps and the center and outer sets of taps, indicated by the first arrow in Fig. 4. The middle line in Fig. 5 shows the temperatures and driving forces where this slip occurs at 5 T. The highest lying line indicates the onset of slip between the center and outer sets of voltage taps. The critical currents for depinning and the onset of plastic shear all approach zero at the melting temperature. Here the shear modulus of the lattice disappears, the motion becomes hydrodynamic, and the velocity adjusts to the driving force gradient on the intervortex length scale.

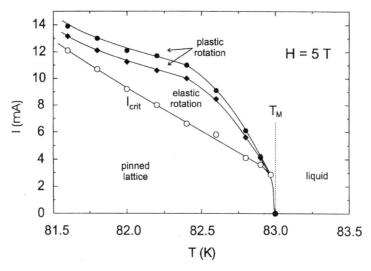

Fig. 5. Phase diagram showing regions of the pinned lattice, elastic rotation, plastic rotation, and hydrodynamic motion as a function of driving force and temperature. The middle and upper lines correspond the to the first and second arrows in Fig. 4, respectively

One interesting question is the nature of the transition from solid to liquid in the presence of strong shear induced slip. As the applied shear stress increases the lattice adjusts by increasing the radial density of plastic slip lines. The physical limit for this process is one slip line per radial intervortex spacing, allowing velocity changes on the same length scale as in hydrodynamic motion. The difference between this high density plastic slip and hydrodynamic flow is an interesting basic question.

6   Summary

We describe a new type of transport experiment where an inhomogeneous current density is applied to the sample to produce a gradient in the Lorentz driving force. The vortex velocity profiles induced by the driving force gradient reveal fundamental features of the dynamics of the driven motion. The measured velocity profiles distinguish hydrodynamic motion of the vortex liquid from elastic and plastic motion of the solid. Elastic depinning of the vortex lattice at the critical current is observed, implying that the elastic bonds between vortices dominate the random pinning forces in our crystal. By controlling the driving force and its gradient, we regulate the number of shear–induced plastic slip lines in the moving lattice. The various pinned, plastic, and elastic dynamic states of the vortex lattice and hydrodynamic flow of the liquid are visualized on a driving force–temperature phase diagram.

The controlled gradient transport experiment described above offers new opportunities to explore the basic features of the vortex phases in superconductors. We have examined the responses of the vortex liquid and lattice to a gradient in the driving force, finding characteristic signatures of the underlying viscous or elastic natures of these phases. The response of the disordered vortex states such as the vortex and Bose glasses, and of vortex liquid states that are strongly pinned or entangled promises new insights into their fundamental static and dynamic character.

Acknowledgement

This work was supported by the US Department of Energy, Office of Basic Energy Sciences–Materials Sciences, under contract #W–31–ENG–38 (GWC, WKK) and the NSF Science and Technology Center for Superconductivity under contract #DMR91–20000 (DL).

References

1. Crabtree G.W., Kwok W.K., Welp U., Lopez D., Fendrich J.A. (1999) Vortex Melting and the Liquid State in YBa_2Cu3O_x. In: Kossowsky R., Bose S., Pan V., Durusoy Z., (Eds.) Physics and Materials Science of Vortex States, Flux Pinning, and Dynamics (Proc. of the NATO ASI in Kusadasi, Turkey, 1998) Kluwer, Dordrecht, 357

2. Crabtree George W., Nelson David R. (1997) Vortex Physics in High Temperature Superconductors, Phys Today 50(4):38–45

3. Crabtree G.W., Welp U., Kwok W.K., Fendrich J.A., Veal B.W. (1997) The Equilibrium Vortex Melting Transition in $YBa_2Cu_3O_7$. J Alloys Compd 250:609–614

4. Crabtree G.W., Kwok W.K., Welp U., Fendrich J.A., Veal B.W. (1996) Static and Dynamic Vortex Transitions in Clean $YBa_2Cu_3O_7$. J Low Temp Phys 105:1073–1082

5. Blatter G., Feigel'man M.V., Geshkenbein V.B., Larkin A.I., Vinokur V.M. (1994) Vortices in High Temperature Superconductors, Rev Mod Phys 66:1125–1388

6. Safar H., Gammel P.L., Huse D.A., Bishop D.J., Rice J.P., Ginsberg D.M. (1992) Experimental Evidence for a First Order Vortex Lattice Melting Transition in Untwinned, Single Crystal YBCO. Phys Rev Lett 69:824–828

7. Kwok W.K., Fleshler S., Welp U., Vinokur V.M., Downey J., Crabtree G.W., Miller M.M. (1992) Vortex Lattice Melting in Untwinned and Twinned Single Crystals of YBCO. Phys Rev Lett 69:3370

8. Safar H. Gammel P.L., Huse D.A., Bishop D.J., Lee W.C., Giapintzakis J., Ginsberg D.M. (1993) Experimental Evidence for a Multicritical Point in the Magnetic Phase Diagram for the Mixed State of Clean, Untwinned YBCO. Phys Rev Lett 70:3800–3803

9. Kwok W.K., Fendrich J., Fleshler S., Welp U., Downey J., Crabtree G.W. (1994) Vortex Liquid Disorder and the First Order Melting Transition in YBCO. Phys Rev Lett 72:1092–1096

10. Fendrich J.A., Kwok W.K., Giapintzakis J., van der Beek C.J., Vinokur V.M., Fleshler S., Welp U., Viswanathan H.K., Crabtree G.W. (1995) Vortex Liquid State in an Electron Irradiated Untwinned YBCO Crystal. Phys Rev Lett 74:1210–1214

11. Lopez D., Krusin-Elbaum L., Safar H., Righi E., de la Cruz F., Grigera S., Feild C., Kwok W.K., Paulius L., Crabtree G.W. (1998) Pinned Vortex Liquid above the Critical Point of the First Order Melting Transition: A Consequence of Point–Like Disorder. Phys Rev Lett 80:1070–1073

12. Kwok W.K., Fendrich J.A., Vinokur V.M., Koshelev A.E., Crabtree G.W. (1996) Vortex Shear Modulus and Lattice Melting in Twin Boundary Channels of $YBa_2Cu_3O_x$. Phys Rev Lett 76:4596

13. Pastoriza H., Goffman M.F., Arribére A., de la Cruz F. (1994) First Order Phase Transition at the Irreversibility Line of $Bi_2Sr_2CaCu_2O_8$. Phys Rev Lett 72:2951–2954

14. Zeldov E., Majer D., Konczykowski M., Geshkenbein V.B., Vinokur V.M. (1995) Thermodynamic Observation of First Order Vortex–Lattice Melting Transition in BSCCO, Nature 375:373–376

15. Liang R., Bonn D.A., Hardy W. (1996) Discontinuity of Reversible Magnetization in Untwinned YBCO Single Crystals at the First Order Vortex Melting Transition. Phys Rev Lett 76:835–839

16. Welp U., Fendrich J.A., Kwok W.K., Crabtree G.W., Veal B.W. (1996) Thermodynamic Evidence for a Flux Line Lattice Melting Transition in YBCO. Phys Rev Lett 76:4809–4813

17. Fendrich J.A., Welp U., Kwok W.K., Koshelev A.E., Crabtree G.W., Veal B.W. (1996) Static and Dynamic Vortex Phases in YBCO. Phys Rev Lett 77:2073–2077

18. Schilling A., Fisher R.A., Phillips N.E., Welp U., Kwok W.K., Crabtree G.W. (1997) Anisotropic Latent Heat of Vortex–Lattice Melting in Untwinned YBCO. Phys Rev Lett 78:4833–4838

19. Roulin M., Junod A., Walker E. (1998) Observation of Second Order Transitions Below T_c in the Specific Heat of YBCO: Case for the Melting of a Vortex Glass. Physica C 296:137–152

20. Roulin M., Junod A., Erb A., Walker E. (1998) Calorimetric Transitions in the Melting Line of the Vortex System as a Function of Oxygen Deficiency in High Purity YBCO. Phys Rev Lett 80:1722–1726

21. Bouquet F., Marcenat C., Calemczuk R., Erb A., Junod A., Roulin M., Welp U., Kwok W.K., Crabtree G.W., Phillips N.E., Fisher R.A., Schilling A. (1999) Calorimetric Evidence of the Transitions Between the Different Vortex States in YBCO. In: Kossowsky R., Bose S., Pan V., Durusoy Z. (Eds.) Physics and Materials Science of Vortex States, Flux Pinning, and Dynamics. Proc. of the NATO ASI in Kusadasi, Turkey, 1998. Kluwer, Dordrecht, 743

22. Roulin M., Revaz B., Junod A., Erb A, Walker E. (1999) High Resolution Specific Heat Experiments on the Vortex Melting Line in MBa_2Cu3O_x (M = Y, Dy, and Eu) Crystals: Observation of First and Second Order Transitions up to 16 T. In: Kossowsky R., Bose S., Pan V., Durusoy Z. (Eds.) Physics and Materials Science of Vortex States, Flux Pinning, and Dynamics (Proc. of the NATO ASI in Kusadasi, Turkey, 1998) Kluwer, Dordrecht, 489

23. Revaz B., Junod A., Erb A. (1998) Specific Heat Peaks Observed up to 16 T on the Melting Line of Vortex Matter in DyBCO. Phys Rev B 58:11 153–11 156

24. Marchetti M.C., Nelson D.R. (1990) Hydrodynamics of Flux Liquids.
    Phys Rev B 42:9938
25. Marchetti M.C., Nelson D.R. (1991) Dynamics of Flux Line Liquids in
    High-T_c Superconductors. Physica C 174:40–62
26. Marchetti M.C., Nelson D.R. (1999) Patterned Geometries and
    Hydrodynamics at the Vortex Bose Glass Transition.
    Phys Rev B 59:13 624–13 627
27. Crabtree G.W., Leaf G.K., Kaper H.G., Vinokur V.M., Koshelev A.E.,
    Braun D.W., Levine D.M. (1996) Driven Motion of Vortices in
    Superconductors. In: Klamut J. et al. (Eds.) Recent Developments in High
    Temperature Superconductivity (Proc. of the First Polish–US Conference on
    Superconductors, Duszniki Zdrój, Poland, Sept 10–14, 1995) Lecture Notes in
    Physics Vol 475. Springer, Berlin Heidelberg, 303–315
28. Lopez D., Kwok W.K., Safar H., Olsson R.J., Petrean A.M., Paulius L.,
    Crabtree G.W. (1999) Spatially Resolved Dynamic Correlation in the Vortex
    State of High Temperature Superconductors. Phys Rev Lett 82:1277

Improved Y-123 Materials by Chemical Substitutions

B. Dabrowski[1], K. Rogacki[1,2], P. W. Klamut[1], Z. Bukowski[1],
O. Chmaissem[1], and J. D. Jorgensen[3]

[1] Physics Department, Northern Illinois University, De Kalb, IL 60115, USA;
[2] Institute of Low Temperature and Structure Research, Polish Ac. Sci.,
PL–50-950 Wrocław, Poland;
[3] Materials Science Division, Argonne National Laboratory,
Argonne, IL 60439, USA

Abstract. By using the knowledge accumulated during last decade about the relationship between the structural features and the superconducting properties we have attempted to design an improved Y123 material that would display an increased superconducting transition temperature, T_c, and enhanced critical current, J_c, and irreversibility fields, B_{irr}. Chemical substitutions on various crystallographic sites were made to reduce the buckling of the CuO_2–planes. Substitutions of Sr for Ba and transition elements for copper were made to shorten the distance between the double–CuO_2–planes while preserving the "metallicity" of the blocking layer. The synthesis and annealing conditions were optimized to obtain the highest T_c. Improved critical currents and irreversibility fields were found for $(Nd_{1-x}La_x)123$ and $YBaSrCu_{3-y}Mo_yO_{7+d}$ compounds. The existence of these materials indicates that it is possible to further enhance superconducting properties of $YBa_2Cu_3O_7$ materials by clever design of the crystal chemistry.

1 Introduction

Since its discovery in 1987, the $YBa_2Cu_3O_7$ (Y123) high temperature superconductor has been the subject of vigorous research of fundamental physical properties and a prominent candidate material for power and magnet applications. Interest for applications has emerged because the parameters of practical importance, the critical current, J_c, and the irreversibility fields, B_{irr}, remain quite large in intense magnetic fields at 77 K for Y123 [1]. These attractive intrinsic parameters can be further enhanced by the complex melt–texturing processing that generates a grain–aligned material with finely dispersed extrinsic flux pinning centers [2].

For several years we have studied the chemical and structural features that control the superconducting T_c, concluding that the T_c's of optimally doped materials with the same number of the CuO_2–planes are proportional to an easily measured parameter, the distance between the planar Cu and the apical O atoms, d(apical) (see Fig. 1 for the definition of the relevant structural parameters) [3]. The distance d(apical) is inversely related to the buckling of the CuO_2–planes; so, as a consequence, the compounds with the longest

d(apical) and flat CuO_2–planes have the highest T_c's [4]. Y123 has a short d(apical) and relatively heavily buckled double–CuO_2–planes; therefore, it would be expected that the $T_c = 93\,K$ of Y123 could be increased if d(apical) could be increased by chemical substitutions.

Recently, we have also investigated how intrinsic superconducting parameters J_c, and B_{irr} depend on the structural characteristics for compounds with multiple–CuO_2–planes. In particular, we focused interest on the structural features and physical properties of the blocking layer in the crystal structure [5]. By comparing J_c and B_{irr} for several families of HTSC, we have concluded that the intrinsic flux pinning is controlled to a large extent by the conductivity of the blocking layer. The width of the intermediate–region (the layer between the multiple–CuO_2–planes consisting of the blocking layer, d(block), and twice the distance between the planar Cu and apical O) is less significant. The number of adjacent CuO_2–planes is of least importance. Thus, high J_c and B_{irr} of Y123 arise mostly from a strong intrinsic flux pinning that originates from a short and metallic blocking layer that provides effective coupling between the successive double–CuO_2–planes.

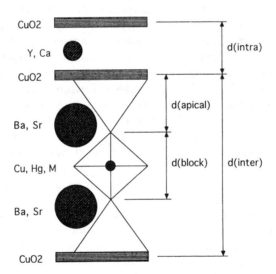

Fig. 1. Schematic structure of the double CuO_2–plane compounds with a single metal–oxygen layer in the intermediate region [3]

During the last 10 years numerous chemical substitutions were attempted to increase T_c and improve the superconducting properties of Y123. While some substitutions of large rare earth's (Nd, Sm, Eu) for Y were found to increase the onset of T_c to 96 K [6], all substitutions for Cu produced inferior materials [7]. The increased T_c's together with the recently found improved J_c and B_{irr} of Nd123 [8] led to intense current research on $RE123$ materials.

We have studied several substitutions on distinct crystallographic sites in the Nd123 compound: Sr, La, and Ca for Ba, La and Ca for Nd, and Mo^{6+}, V^{5+}, Ti^{4+}, Fe^{3+}, and Co^{2+} for Cu. None of these substitutions lead to an increased T_c. Superconducting parameters were investigated by magnetic ac susceptibility and dc magnetization measurements for materials with the highest T_c's for each substitution. Improved irreversibility fields were observed only for $(Nd_{1-x}La_x)123$ with $x = 0.1 - 0.25$.

We have performed a similar exploration of the T_c's for substituted $REBa_{2-x}La_xCu_3O_7$ materials. Several synthesis and annealing methods were attempted to control the preferential site substitution of La for Ba and to tune the oxygen content to achieve the highest T_c. We have found that the T_c of most of the RE123 can be increased to 94–95 K. To achieve these increased T_c's, moderate amounts of the substituted La ($x \approx 0.1$) were required for the smaller ionic–size RE = Y and Yb. For the larger ionic–size RE = Gd and Sm the increase of T_c was attained at the lower substitution levels. Materials with the highest T_c's are currently being studied to determine J_c and B_{irr}.

Recently, we have utilized the acquired knowledge of the relationship between structural and superconducting properties to design the Y123 materials with a shorter intermediate region and conserved magnitude of d(apical) [9]. Simultaneously we have attempted to preserve the "metallicity" of the blocking layer by substituting small ionic–size and multivalent elements for Cu. We have investigated the Y123 materials substituted with a smaller Sr in place of the larger Ba in the intermediate region of the structure and with transition elements substituted for Cu. It was found that during synthesis in air several transition elements preferentially substitute for Cu in the blocking layer (on the Cu–chain site). In particular, it was found that Mo shows a wide range of solubility and that the Mo–substituted materials are easy to synthesize and can be annealed at elevated oxygen pressure to increase T_c. The synthesis conditions and compositions were optimized to obtain the highest T_c's for the double–substituted $YBa_{2-x}Sr_xCu_{3-y}Mo_yO_z$ compounds. The $x = 2$ compound can be synthesized only for $y \approx 0.2$ and is tetragonal with $T_c < 40$ K. Using annealing at high oxygen pressure, the oxygen content was increased to $z \approx 7.30$. The T_c's were increased to ≈ 77 K by the concurrent substitution of small amounts of Ca for Y. For the $YBaSrCu_{3-y}Mo_yO_z$ ($x = 1$) compounds, the highest transitions, $T_c \approx 86$ K, were obtained for $y = 0.05 - 0.10$ after high oxygen pressure anneals. The $x = 1$ materials are orthorhombic for $y < 0.1$ and tetragonal for $y > 0.1$. The scaled irreversibility fields are comparable to pure Y123. The structural studies indicate that the enhanced flux pinning properties may arise from a shorter width of the intermediate region and from the randomly distributed defects, dimmers of corner shared $Mo–O_6$ octahedra in the blocking layer that may act as the intrinsic flux–pinning centers. The Mo–substituted $YBaSrCu_{3-y}Mo_yO_z$ material is the first example of an improved superconducting 123 compound by substitution of a transition element for Cu. It provides a hint that similar improvements can be also obtained for $YBa_2Cu_3O_7$.

2   Synthesis and Experimental Details

Polycrystalline samples were synthesized from a stoichiometric mixture of oxides and carbonates. Samples were fired several times in air or controlled oxygen pressures for several days at increasing temperatures, checked for phase purity, annealed under several oxygen pressures and temperatures, and checked for T_c. The high pressure anneals were done for 12 hours in 20% O_2 in argon at a total pressure of 3 kbar (600 atm. O_2 pressure) at $950 - 1100\,^\circ$C or in pure oxygen (250–300 atm. O_2) at $400 - 900\,^\circ$C followed by slow cooling (0.2 deg/min.) to room temperature. Sample homogeneity was checked by powder X-ray diffraction. Neutron powder diffraction data were obtained using the Special Environment Powder Diffractometer (SEPD) at Argonne National Laboratory's Intense Pulsed Neutron Source (IPNS). The grain sizes and shapes were examined by a Hitachi Scanning Electron Microscope (SEM). Susceptibility and magnetization measurements were performed with a Quantum Design PPMS system. Resistivity was measured using a standard four–lead dc method.

3   Results and Discussion

3.1   Substituted Nd123 Compounds

Among the $RE123$ materials, the Nd123 compound has the highest T_c onset ≈ 96 K. Formation of the stoichiometric Nd123 compound requires synthesis at a reduced pressure of oxygen near $1000\,^\circ$C to suppress the substitution of Nd for Ba that may lower T_c. Numerous studies have established that synthesis at $0.1 - 1\%$ O_2 followed by a prolonged annealing in pure O_2 at $\approx 300\,^\circ$C is optimal for obtaining stoichiometric Nd123 with the highest T_c. The choice of the oxygen pressure and temperature is important because in addition to controlling T_c it may also affect the inter–grain coupling. To establish the most favorable synthesis conditions for both pure and substituted Nd123 materials we have synthesized several samples at various oxygen pressures and temperatures. Figure 2 shows the resistivity of several single–phase samples of Nd123 that were prepared in 1% O_2 around $1000\,^\circ$C and then annealed in O_2 at $\approx 300\,^\circ$C. All samples showed T_c's of 94–95 K but the width of the superconducting transition and the magnitude of the resistivity varied considerably. The sharpest superconducting transition, the largest resistivity ratio, $R(300)/R(100)$, and the lowest magnitude of the resistivity were obtained for a sample fired at $1000\,^\circ$C, indicating that the best inter–grain coupling was obtained for that sample. Figure 3 shows the real, χ', and imaginary, χ'', components of the ac susceptibility for the solid pieces of the samples fired at 980, 1000, and 1030 °C. The inter– and intra–grain contributions to the susceptibility are clearly separated into the higher-T_c intra–grain and lower-T_c inter–grain components, confirming that the synthesis in 1% O_2 at 1000 °C is optimal for obtaining high T_c and good transport properties that are only weakly affected by the grain boundaries.

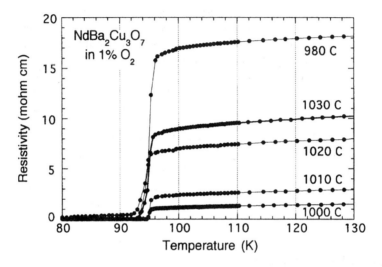

Fig. 2. Resistivity as a function of temperature for the Nd123 samples prepared in 1% O_2 at several temperatures (980 – 1030 °C) and then annealed in pure O_2 at about 300 °C

Synthesis of substituted Nd123 compositions was carried out using similar conditions of the temperature and oxygen pressure. We have studied several substitutions on the Ba and Nd sites: Sr(5–15%), La(0–5%), and Ca(0–10%) for Ba, and La(0–50%) and Ca(0–2.5%) for Nd. All these compositions were single–phase according to the X-ray diffraction; however, none of them led to an increase the onset of T_c above 96 K. Figure 4 shows the normalized resistance near the superconducting transition for selected compositions and for pure Nd123 and Y123. With an exception of the $(Nd_{0.75}La_{0.25})$123 material, all substituted samples show sharp transitions and lower superconducting T_c's than Nd123. The onset of T_c for the $(Nd_{0.75}La_{0.25})$123 material is slightly higher than for the pure Nd123, but the broad two–step transition suggests the presence of two superconducting phases.

Superconducting parameters were investigated by ac susceptibility, χ, and dc magnetization, M, measurements for materials with the highest T_c's for each substitution. Solid pieces as well as powdered samples with masses of about 100 mg were used for both χ and M measurements. For solid samples, measured at zero dc field, a 1 Oe ac field was large enough to separate

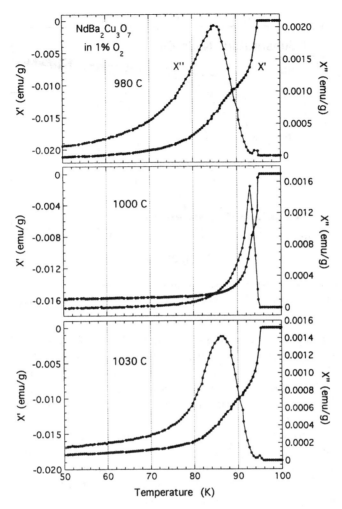

Fig. 3. Real, χ', and imaginary, χ'', components of the ac susceptibility for the solid pieces of the Nd123 samples prepared at 980, 1000 and 1030 °C

the higher-T_c intra–grain and lower-T_c inter–grain components. To remove the inter–grain contribution from the diamagnetic signal, solid pieces were powdered and measured again. Calculation of the intra–grain persistent critical currents was then done from the magnetization loops. The irreversibility field B_{irr} was obtained from dc magnetization by measuring the fields at which the magnetic hysteresis for the $M(B)$ loops disappears at a constant temperature using a criterion $\Delta 4\pi M = 0.05$ G. Irreversibility lines are shown in Fig. 5 for several Nd123 substituted compounds. The highest irreversibility lines and the largest slopes dB_{irr}/dT, that indicate the strongest flux pinning, were obtained for $(Nd_{0.75}La_{0.25})123$ samples. The $B_{irr}(T)$ line for

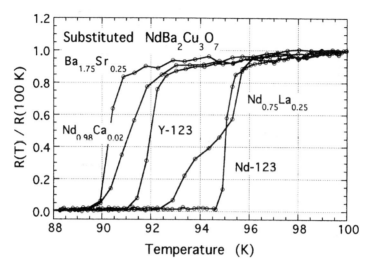

Fig. 4. Normalized resistance near the superconducting transition temperature for pure Y123 and Nd123, and for several substituted Nd123 compounds

that sample is placed about $0.5 - 1$ T above the line for the Nd123 sample. Similar improved irreversibility fields were observed for the $(Nd_{1-x}La_x)123$ samples with $x = 0.1 - 0.25$. This is probably the result of extrinsic flux pinning, possibly originating from the phase separation. It is noteworthy that the $(Nd_{0.98}Ca_{0.02})123$ material has a $B_{irr}(T)$ line similar to that of Nd123 despite its lower T_c.

Substitutions of a wide range of transition elements, that occur in multiple oxidation states, Mo^{6+}, V^{5+}, Ti^{4+}, Fe^{3+}, and Co^{2+}, for Cu produced materials containing visible amounts of second phases. Alteration of the synthesis conditions of oxygen pressure did not improve the purity of the samples. After annealing in oxygen, the T_c's of these materials were markedly lower than 96 K and transition widths were very broad. We did not measure the critical superconducting properties for these samples.

3.2   The $REBa_{2-x}La_xCu_3O_7$ Materials

An increase of T_c by substitution of La for Ba in $YBa_{2-x}La_xCu_3O_7$ was observed shortly after the discovery of Y123 [10]. We have observed that further increases of T_c can be produced by high pressure oxygen anneals of these materials. Recently, we have extended these findings to $REBa_{2-x}La_xCu_3O_7$ compounds, where RE = Yb, Gd, Sm, and Nd, with $x = 0, 0.02, 0.04, 0.06, 0.08, 0.10$. Several synthetic and annealing methods were attempted to control the preferential site substitution of La for Ba and to tune the oxygen content to achieve the highest T_c. Synthesis in air at temperatures closed to an appearance of the liquid phase, at $940 - 1000\,^\circ$C, depending on the RE, led

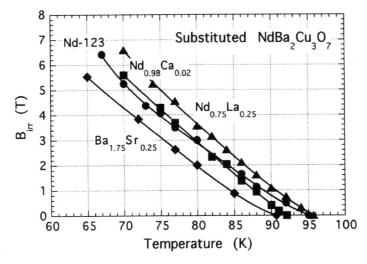

Fig. 5. Irreversibility lines for pure and substituted Nd123 compounds

to the formation of single–phase materials. Annealing at high oxygen pressure usually causes an increase of T_c. We have found that the T_c of most of the $RE123$ materials can be increased to $94 - 95\,\mathrm{K}$. Figure 6 shows the resistive transition close to T_c for the materials with the highest superconducting transitions for each series of $RE\mathrm{Ba}_{2-x}\mathrm{La}_x\mathrm{Cu}_3\mathrm{O}_7$. For the smaller ionic–size $RE = \mathrm{Y}$ and Yb, increased T_c's were observed for moderate amounts of substituted La ($x \approx 0.1$). For the larger ionic–size $RE = \mathrm{Gd}$ and Sm, an increase of T_c was attained at lower substitution levels, 0.08 and 0.02, respectively. No increase of T_c was obtained for $RE = \mathrm{Nd}$. The materials with the highest T_c's are currently being studied for J_c and B_{irr}.

3.3   The $\mathrm{YBa}_{2-x}\mathrm{Sr}_x\mathrm{Cu}_{3-y}\mathrm{Mo}_y\mathrm{O}_z$ Compounds

The layered Y123 materials, that consist of alternating layers of CuO–chains and double–CuO$_2$–planes, offer an opportunity to modify their structural and physical properties by carrying out substitutions of small ionic–size atoms on the Cu–chain site without disturbing the perfect arrangement of the Cu and O atoms in the superconducting planes. Control of the preferential site substitution of the transition and post–transition elements on the Cu–chain site can be accomplished by a choice of the proper synthesis conditions and by the simultaneous substitutions of large ionic–size atoms for Ba or Y. Recently, we have explored these opportunities to design Y123 materials with a shorter width of the intermediate region while attempting to preserve both the magnitude of d(apical) and the "metallicity" of the blocking layer. We have investigated Y123 materials substituted with a smaller Sr in place of the larger Ba in the intermediate region of the structure and with transition

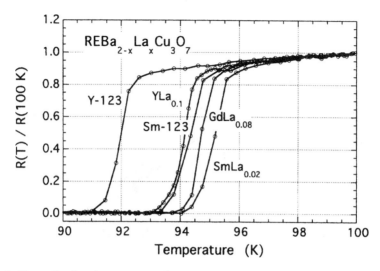

Fig. 6. Normalized resistance near the superconducting transition temperature for pure Y123 and Sm123, and for several $REBa_{2-x}La_xCu_3O_7$ compositions

elements substituted for Cu. It was found that the presence of Sr on the Ba–site enhances preferential substitution of transition elements on the Cu–chain site during synthesis in air [3].

Sr–substituted Y123 forms single–phase only to $x \approx 1$ under normal synthesis conditions in air [11]. For completely substituted compounds ($x = 2$) it was found that the Y123 structure can be stabilized by the addition of the high oxidation–state transition elements like Mo, W, and Re at low substitution levels, $y \approx 0.2$ [3]. Superconducting compounds with $T_c \approx 75\,K$ can then be obtained after annealing at high oxygen pressure that increases the oxygen content to $z \approx 7.30$.

The Mo–substituted materials are easy to synthesize in air, and further annealing in oxygen or at elevated oxygen pressures increases T_c. Recently we have optimized the synthesis conditions and compositions to obtain the highest T_c's for the $YBaSrCu_{3-y}Mo_yO_{7+d}$ compounds. All compositions studied, $y = 0 - 0.20$, were single phase when synthesized in air at $\approx 960\,°C$. High pressure anneals in oxygen at temperatures around $650\,°C$ followed by slow cooling were optimal to increase the oxygen content and to obtain the highest T_c. The structural and superconducting properties were studied in detail for the $y = 0, 0.05, 0.10$ and 0.20 compositions [9].

The structures of these materials were studied by neutron powder diffraction. As for the $x = 2$ compounds, the Mo ion was found exclusively on the Cu–chain site. The resulting $Cu_{1-y}Mo_yO_z$ blocking layer, situated between the two (Ba,Sr) layers, contains more than one oxygen atom per Cu/Mo. Because the chain oxygen atoms observe multiple arraignments of the near–neighbor Sr and Ba atoms and the Cu and Mo atoms, it was difficult to determine their precise locations. However, the dominant positions were refined with reasonable accuracy showing that the oxygen atoms occupy two independent sites $(1/2\ y\ 0)$ and $(x\ 1/2\ 0)$ in the orthorhombic $y = 0.05$ compound with the site occupancies 0.59 ± 0.01 and 0.48 ± 0.01, respectively. The site occupancy difference gives rise to the orthorhombic distortion. For the $y = 0.1$ and 0.2 samples, the occupancy of these sites were equal as required by the tetragonal structures.

The total oxygen contents that were obtained from the refined oxygen occupancies were 7.07, 7.16, and 7.30 for $y = 0.05$, 0.1, and 0.2, respectively. The ratio of the nominal Mo to the amount of excess oxygen in the blocking layer (i.e. 0.07, 0.16, and 0.30) was very close to 2:3 in all samples, indicating that this ratio could arise from the presence of randomly distributed dimmers of the corner shared MoO_6 octahedra in a similar fashion to that observed for $YSr_2Cu_{3-y}M_yO_z$ $(M = Mo, W)$ [3].

The highest $T_c \approx 86\,K$, was observed for the samples with $y = 0.05 - 0.1$ annealed at high oxygen pressure. For samples annealed at 1 atm. O_2, the highest $T_c \approx 82\,K$ was seen for $y = 0 - 0.05$. Figure 7 shows normalized resistance for three single–phase samples which were annealed in oxygen at 1 atm. (Fig. 7a) and high pressure (Fig. 7b). Superconducting transitions for optimally doped $YBaSrCu_{3-y}Mo_yO_{7+d}$ samples are very sharp, indicating good quality of the material. Clearly, T_c is a function of both the doping level y and the oxygen content. The sample with $y = 0$ does not change T_c significantly after the high oxygen pressure treatment. T_c's for samples with $y = 0.05 - 0.10$ are higher than for $YBaSrCu_3O_{7+d}$; i.e. this is the first example of improved superconducting properties of the Y123 material for which a transition element was substituted for copper.

By comparing the superconducting T_c's for three optimally doped Y123 compounds, $YBa_2Cu_3O_{6.94}$, $YBaSrCu_{2.9}Mo_{0.1}O_{7.16}$, and $YSr_2Cu_{2.8}Mo_{0.2}O_{7.3}$, a systematic decrease of $T_c = 93, 86$, and $77\,K$ can be observed that is consistent with the shortening of the apical Cu–O bond–length, $d(\text{apical}) = 2.303$, 2.260, and $2.207\,\text{Å}$, respectively. The $YBaSrCu_{2.9}Mo_{0.1}O_{7.16}$ compound has an intermediate magnitude of $d(\text{apical})$ and, thus, an intermediate T_c. The thickness of the intermediate region, $d(\text{inter}) = 8.228\,\text{Å}$, is also intermediate between values observed for $YBa_2Cu_3O_7$ $(8.307\,\text{Å})$ and $YSr_2Cu_{2.8}Mo_{0.2}O_{7.3}$ $(8.098\,\text{Å})$. However, neither the width of the blocking layer, $d(\text{block})$, nor the intra–CuO_2–plane distance, $d(\text{intra})$, vary in a systematic way with T_c for these compounds.

The irreversibility fields B_{irr}^{ac} and B_{irr} were obtained from ac susceptibility, $\chi(T)$, and dc magnetization measurements, respectively. The irreversibility lines $B_{irr}^{ac}(T)$ were derived form the $\chi(T)$ curves measured at constant dc fields. Each irreversibility point $(B_{irr}^{ac}, T_{irr}^{ac})$ was obtained by measuring the temperature at which the imaginary part of the ac susceptibility, $\chi''(T)$, begins to differ from zero. These $B_{irr}^{ac}(T)$ lines are the upper bound for the

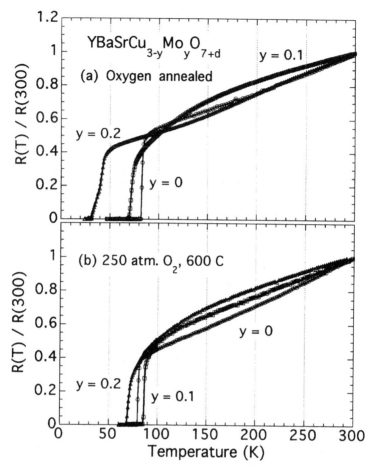

Fig. 7. Normalized resistance as a function of temperature for the YBaSrCu$_{3-y}$Mo$_y$O$_{7+\delta}$ samples with $y = 0$, 0.1 and 0.2, as prepared **(a)** and after the high oxygen pressure anneal **(b)**

$B_{irr}(T)$ lines that are usually used but may occasionally depend on the measurement sensitivity. The pinning properties can be better exhibited when the B_{irr} fields are plotted as a function of reduced temperature $t = T/T_c$ as presented in Fig. 8 for YBaSrCu$_{3-y}$Mo$_y$O$_{7+d}$ with compositions $y = 0$, and 0.05. The data for solid and powder samples shows that the highest irreversibility lines and the largest slopes of dB_{irr}/dT, that indicate the strongest flux pinning, were obtained for the $y = 0.05$ sample. The $B_{irr}(t)$ line for the $y = 0.05$ sample is placed about 1 T above the line for the $y = 0$ sample and is very close to the line obtained for pure polycrystalline Y123 [12]. In Fig. 8 the $B_{irr}(t)$ lines for polycrystalline Hg1201 and single crystals of Bi2212, LSCO and Y123 are also presented. Single crystals were measured in fields parallel to the c–axis, i.e. in the direction of the weakest intrinsic flux pinning. Therefore, results for single crystals are expected to show lower irreversibility lines than the B_{irr}'s that were obtained for polycrystalline materials. It would be desirable to obtain single crystals of the Mo–substituted YBaSrCuO$_7$ materials for comparison with pure Y123.

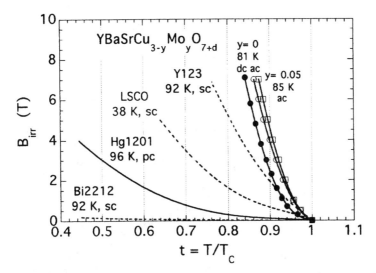

Fig. 8. Irreversibility fields, obtained by dc and ac techniques (closed and open symbols, respectively), versus reduced temperature for the YBaSrCu$_{3-y}$Mo$_y$O$_{7+\delta}$ samples with $y = 0$ and 0.05. For comparison, results for Hg1201 polycrystalline material [13] (pc; *solid line*), and for Bi2212, LSCO and Y123 single crystals [14] (sc; *broken lines*) are also presented. The data for single crystals is for B parallel to the c-axis. The superconducting transition temperature is displayed for each compound

The powder samples were used to measure the intra–grain critical persistent–current density, J_c. To determine J_c, the magnetization loops, $M(B)$, were measured at constant temperature and the Bean formula, $J_c(\mathrm{A/cm}^2) = k\Delta M/w$, was used, where M is in emu/cm^3, $w = (3-4) \times 10^{-4}$ cm is a scaling length, that in our case is the grain diameter, and $k = 40$ is a shape coefficient for roughly spherical grains. The Bean formula was applied only in the range of magnetic fields where M was weakly field dependent, i.e. above the first peak observed for the $M(B)$ curves ≈ 0.5 T. At 5 K, J_c is about 10^6 A/cm^2 and is almost independent of B for all compositions. Similar values of J_c for all compositions arise because at low temperatures the coherence length in the c-axis direction (3–4 Å) is comparable to the d(block) distance and the blocking region between the superconducting CuO$_2$–planes may act as a strong intrinsic pinning center. However, differences in $J_c(B)$ among pure and Mo–substituted samples arise at higher temperatures where the coherence length increases and other types of pinning centers with more extended dimensions may act more effectively. Figure 9 shows J_c as a function of B for $y = 0, 0.05$, and 0.10 at temperatures 70 and 77 K. At 70 K the $y = 0$ and 0.05 compositions have similar values of $J_c(B)$ for $B = 0.5 - 2.5$ T but for higher fields the $y = 0.05$ sample shows a second peak effect, a positive deviation from the approximately exponential decrease of $J_c(B)$ with increasing B. At 77 K this behavior is more clearly visible and, thus, the critical currents above 70 K are much larger for the $y = 0.05$ sample than for $y = 0$. $J_c(B)$ for the $y = 0.1$ sample displays features similar to those observed for $y = 0.05$ but both the pinning force at low fields and the second peak effect are weaker than for $y = 0.05$.

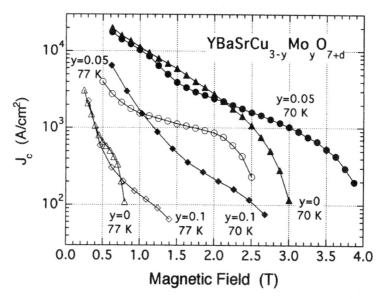

Fig. 9. Critical persistent currents versus magnetic field at 70 K (*closed symbols*) and 77 K (*open symbols*) for the $YBaSrCu_{3-y}Mo_yO_{7+\delta}$ samples with $y = 0$ (*triangles*), 0.05 (*circles*) and 0.1 (*diamonds*)

The second peak effect, present for both Mo–substituted samples, can be explained in terms of the temperature dependence of the superconducting coherence length. The MoO_6 dimers, introducing extended distortions, perturb locally superconductivity in the CuO_2–planes. At elevated temperatures, the isolated perturbation with a dimension of 2–3 unit cells in the ab–plane begins to be comparable to the coherence length (that increases with increasing temperature) and, therefore, may act as an additional pinning center.

4 Conclusion

By using chemical substitutions on various crystallographic sites we have attempted to enhance intrinsic critical currents and irreversibility fields of Y123 material. Substitutions were designed to increase the distance between the planar Cu and the apical O atoms and to shorten the distance between the double CuO_2–planes while preserving the "metallicity" of the blocking layer. The synthesis and annealing conditions were optimized to obtain the highest T_c for each substitution. Improved critical superconducting properties were found for $(Nd_{1-x}La_x)$-123 and $YBaSrCu_{3-y}Mo_yO_{7+d}$ compounds. Increased T_c's were observed for the substituted $REBa_{2-x}La_xCu_3O_7$ compounds.

Improved irreversibility fields observed for the $(Nd_{1-x}La_x)$-123 samples with $x = 0.1-0.25$ may be caused by the extrinsic flux pinning possibly originating from the phase separation. For the Mo–substituted $YBaSrCu_{3-y}Mo_yO_{7+d}$ compounds, T_c's for $y = 0.05 - 0.10$ are higher than for $YBaSrCu_3O_7$, i.e. this is the first example of an increased T_c of the Y123 material by substitution of transition element for copper. The scaled irreversibility fields are comparable to pure Y-123. The MoO_6 dimers are likely to be the intrinsic pinning centers by forming the randomly distributed local distortions perturbing locally superconductivity in the CuO_2–planes. An existence of these improved materials indicates that it may be possible to further enhance intrinsic superconducting properties of Y123 materials by clever design of the crystal chemistry.

Acknowledgment

This work was supported at NIU by the National Science Foundation Science and Technology Center for Superconductivity under grant #DMR 91–20000, at ANL by the U.S. Department of Energy, BES – Materials Sciences under contract No. W–31–109–ENG–38 (JDJ), and at PAS by the TMR Network under contract #ERBFMRX–CT98–0189 (KR).

References

1. Hettinger J.D., Swanson A.G., Skocpol W.J., Brooks J.S., Graybeal J.M., Mankiewicz P.M., Howard R.E., Straughn B.L., Burkhardt E.G. (1989) Phys Rev Lett 62:2044
2. Martinez B., Obradors X., Gou A., Gomis V., Pinol S., Fontcuberta J., Van Tol H. (1996) Phys Rev B 53:2797
3. Dabrowski B., Rogacki K., Koenitzer J.W., Poeppelmeier K.R., Jorgensen J.D. (1997) Physica C 277:24
4. Jorgensen J.D., Hinks D.G., Chmaissem O., Argyriou D.N., Mitchell J.F., Dabrowski B. (1996) In: Klamut J. et al.(Eds.) Recent Developments in High Temperature Superconductivity, Lecture Notes in Physics vol 475, Springer, Berlin Heidelberg, 1–15

5. Klamut P.W., Rogacki K., Dabrowski B. (1998) Solid State Commun 108:425
6. Murphy D.W., Sunshine S., Van Dover R.B., Cava R.J, Batlog B., Zahurak S.M., Schneemeyer L.F. (1987) Phys Rev Lett 58:1888
7. Tarascon J.M., Barboux P., Miceli P.F., Greene L.H., Hull G.W., Eibschutz M., Sunshine S.A. (1988) Phys Rev B 37:7458
8. Murakami M., Sakai N., Higuchi T., Yoo S.I. (1996) Supercond Sci Technol 9:1015
9. Rogacki K., Dabrowski B., Chmaissem O., Jorgensen J.D. (1999) to be published
10. Dabrowski B., Hinks D.G. (1988) In: Kwok H.S., Shaw D.T. (Eds.) Superconductivity and Its Applications, Proc. of the 2nd Ann. Conf. on Superconductivity and Its Applications (Buffalo, 1988). Elsevier, Amsterdam, 141–145
11. Veal B.W., Kwok W.K., Umezawa A., Crabtree G.W., Jorgensen J.D., Downey J.W., Nowicki L.J., Mitchell A.W., Paulikas A.P., Sowers C.H. (1989) Appl Phys Lett 51:279
12. Wisniewski A., Schalk R.M., Weber H.W., Reissner M., Steiner W., Gorecka J. (1991) Physica C 185–189:2211
13. Welp U., Crabtree G.W., Wagner J.L., Hinks D.G., Radaelli P.G., Jorgensen J.D., Mitchell J.F. (1993) Appl Phys Lett 63:693
14. Kishio K. (1996) In: Deutscher G., Revcolevschi A. (Eds.) Coherence in Superconductors, World Sci., Singapore, 212–225

High Pressure Crystal Growth and Properties of Hg–Based Superconductors and One–Dimensional $A_{1-x}CuO_2$ (A = Sr, Ca, Ba) Cuprates

J. Karpiński[1], G. I. Meijer[1], H. Schwer[1], R. Moliński[1], E. Kopnin[1],
M. Angst[1], A. Wiśniewski[2], R. Puźniak[2], J. Hofer[3], and C. Rossel[4]

[1] Laboratorium für Festkörperphysik d. ETH, 8093-Zürich, Switzerland
[2] Institute of Physics, Polish Academy of Sciences, 02-668 Warsaw, Poland
[3] Physik Institut der Universität Zürich, 8057-Zürich, Switzerland
[4] IBM Research Division, 8803-Rüschlikon, Switzerland

Abstract. High gas pressure up to 11 kbar has been applied for the synthesis of cuprates. Both single crystals and polycrystalline samples of Hg–based superconductors and quasi–one–dimensional $A_{1-x}CuO_2$ (A = Sr, Ca, Ba) compounds have been obtained. The influence of substitutions and oxygen content on the magnetic flux–pinning properties of single crystals of Hg–based superconductors has been investigated. The irreversibility field of almost optimally doped, unsubstituted $HgBa_2Ca_2Cu_3O_{8+\delta}$ crystal (T_c = 130 K) is about two–three times larger than the one of underdoped crystal (T_c = 120 K). As a result of Re substitution for Hg a significant improvement of the irreversibility line position for a $Hg_{0.77}Re_{0.23}Ba_2Ca_2Cu_3O_{8+\delta}$ crystal is observed only at low temperatures (below 80 K). Neutron irradiation of $HgBa_2Ca_2Cu_3O_{8+\delta}$ crystal enhances the flux pinning, while also leading to a decrease of the effective mass anisotropy. The magnetic properties of infinite–chain cuprates $Ca_{0.83}CuO_2$, $Sr_{0.73}CuO_2$ and $Ba_{0.66}CuO_2$ have been studied. Susceptibility measurements give some evidence for a singlet ground state. However, all these compounds order antiferromagnetically at $T < 10$ K. Specific heat, elastic neutron scattering of polycrystalline material and magnetic torque on single crystals have been measured. These measurements give clear evidence that the AF ordered state is of long–range 3D character.

1 Introduction

Single crystals and polycrystalline samples of cuprates have been obtained at high gas pressure up to 11 kbar. High hydrostatic pressure can change the properties of materials during crystallization or annealing due to:

a) structural transitions at high pressure,
b) reduced evaporation of volatile components,
c) enhanced pressure activity (fugacity) of gaseous components of the system.

Examples of a) are the spin ladder $Sr_{n-1}Cu_{n+1}O_{2n}$ or infinite layer $Ca_xA_{1-x}CuO_2$ compounds [1,2]. Examples of b) are the $HgBa_2Ca_{n-1}Cu_nO_{2n+2+\delta}$ (Hg–12$(n-1)n$) compounds which melt incongruently and have high decomposition pressures of HgO, Hg, and O_2 [3,4,5]. Examples of c) are the $A_{1-x}CuO_2$ (where $A = Sr, Ca, Ba$) infinite chain compounds [4,5,6].

The position of the irreversibility line (IL) on the $H - T$ phase diagram is often used to describe the applicability of high-T_c superconductors for technical purposes. One of the most important factors limiting possible applications of Hg–12$(n-1)n$ compounds is a relatively low position of the IL. There are several ways which enable improvement of the IL position:

1. Changing intrinsic parameters
   (a) reducing the "blocking layer" (BL) thickness, which should improve coupling between CuO_2 planes, in the case of Hg–12$(n-1)n$ compounds it may be achieved by substitution of smaller Sr atoms for Ba,
   (b) increasing the electrical conductivity in the BL, which could lead to proximity–induced superconductivity in the BL, in the case of Hg–12$(n-1)n$ compounds it may be achieved by substitution of Re for Hg.
   (c) decreasing anisotropy of the compound by increasing oxygen doping.
2. Introduction of new effective pinning centers
   (a) creation of planar defects, in the case of Hg–12$(n-1)n$, e.g. by substitution of Re or Pb for Hg
   (b) creation of defects by irradiation.

We check the effectiveness of all these methods investigating the flux pinning in single crystals of Hg–based superconductors.

All high-T_c superconductors found so far contain two–dimensional (2D) CuO_2 planes. The ground state of the insulating parent compounds shows antiferromagnetic ordering. When these compounds are doped with holes, the antiferromagnetic ordering disappears and superconductivity appears. A rather interesting approach is to modify these planes in order to make part of the structure responsible for superconductivity more one–dimensional (1D) and to investigate the influence of such a modification on the magnetic and/or superconducting properties of a given compound. The electronic properties of one–dimensional materials are usually very different from those observed in higher dimensional systems. A possible way to modify CuO_2 planes is by an application of a high pressure.

In the case of the Sr–Ca–Ba–Cu–O system, the modification of Cu–O planes can be achieved in two ways:

1. By the synthesis under high hydrostatic pressure. Depending on the composition, either the infinite layer structure with CuO_2 corner sharing planes, or the spin ladder structure with CuO_2 corner sharing chains alternating with edge sharing chains can be stabilized.

2. By the synthesis at high oxygen pressure, which results in the quasi one–dimensional structure of edge sharing chains of CuO_4 squares.

Using high oxygen pressure we have synthesized three members of the one–dimensional $A_{1-x}CuO_2$ family: $Ca_{0.83}CuO_2$, $Sr_{0.73}CuO_2$ and $Ba_{0.66}CuO_2$. Structures of these compounds contain square planar CuO_4 units forming edge–sharing infinite–chains. Sr, Ca or Ba atoms are incorporated between the CuO_2 layers. The Cu ions are coupled by 90° Cu–O–Cu bonds. The formal valence of Cu is +2.34 and +2.54 for $Ca_{0.83}CuO_2$ and $Sr_{0.73}CuO_2$ respectively. Since electrical resistivity reveals that the materials are insulating, the holes are considered to be localized. The susceptibility measurements show typical AF behavior, but very peculiar magnetic transitions are observed at $T < 10\,K$ for all three compounds. Specific heat, elastic neutron scattering of polycrystalline material and magnetic torque on single crystals have been measured. These investigations give clear evidence that the AF ordered state is of long–range 3D character.

2 Crystal Growth of Hg–Based Superconductors

Hg–12$(n-1)n$ compounds have high partial decomposition pressure of volatile components Hg, HgO and O_2. These compounds melt peritectically and the pressure in a closed container above inerting temperature reaches a value corresponding to the decomposition pressure of HgO. This is above 100 bar. In order to prevent the decomposition during synthesis and crystal growth we have applied high Ar gas pressure of about 10 kbar [3,4,5]. Because Ar gas has a density of more than 1.5 g/cm^3 at this pressure, the diffusion of volatile components is strongly suppressed. In order to decrease melting temperature, flux growth from solution can be applied. Unfortunately, a typical solvent for the crystal growth of high T_c superconductors (30% $BaCuO_2$ − 70% CuO) has the eutectic melting temperature at pressure $P = 10$ kbar very close to the Hg–12$(n-1)n$ compound for $n > 1$. The addition of 10% of AgO to this solvent decreases the melting temperature by about 10 °C. Such a solvent has been used for most of the processes of crystal growth unless the substitution of Sr for Ba was required. In this case the excess of Ba in the flux prevents the substitution of Sr for Ba. Therefore, in order to grow crystals with partial substitution of Sr for Ba, we had to melt stoichiometric precursors without eutectic additives. An alternative solvent is PbO, but this leads to the substitution of Pb for Hg. The temperature ramp of the crystal growth process contained:

1. Heating up to the temperature of 950 °C, where Hg–12$(n-1)n$ is synthesized as a solid phase.
2. Heating above melting temperature, which is between 1000 and 1130 °C and depends on the composition of the precursor. SrO or ReO containing precursors have higher melting temperatures than unsubstituted ones.

3. Keeping the melt at constant temperature for 1–2 hours.

4. Slow cooling at a rate $1 - 3°/h$.

Using this route we were able to obtain crystals of Hg–12$(n-1)n$ for $n = 1 - 7$, partially substituted with Re, Sr and Pb.

3   Investigation of the Irreversibility Field of Hg-1223 Crystals as a Function of Substitutions and Oxygen Content

Hg-1223 compounds exhibit the highest T_c, but magnetic flux–pinning properties of these materials are not as good as those of some other high-T_c superconductors (e.g. YBa$_2$Cu$_3$O$_{7-x}$). Several authors recently reported significant improvement of the flux pinning in ceramic samples when Hg is partially replaced by Re or Pb and Ba by Sr [7]. However, due to the high anisotropy of HTSC and unavoidable impurity phases in ceramic samples, only measurements on single crystals are reliable for the determination of the true intrinsic properties of these compounds. We have investigated the influence of substitutions and oxygen content on the magnetic flux–pinning properties of single crystals of Hg–based superconductors.

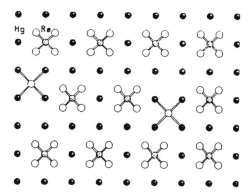

Fig. 1. (Hg,Re)O layer in Hg$_x$Re$_{1-x}$Ba$_2$Ca$_{n-1}$Cu$_n$O$_{2n+2+\delta}$ single crystals

Rhenium can partially substitute Hg. It is coordinated octahedrally by oxygen and forms short bonds leading to a decrease of the lattice parameters (Fig. 1). The maximum substitutions of Re for Hg is 25% because higher Re concentrations would cause short O–O distances which are not possible. Flux pinning in Hg$_{0.77}$Re$_{0.23}$Ba$_2$Ca$_2$Cu$_3$O$_{8+\delta}$ single crystals ($T_c = 130$ K, Fig. 2) and in HgBa$_2$Ca$_2$Cu$_3$O$_{8+\delta}$ crystals with T_c equal to 130 and 120 K was compared (Fig. 3) [8]. At higher temperatures the irreversibility field of an almost optimally doped, unsubstituted ($T_c = 130$ K) crystal is about

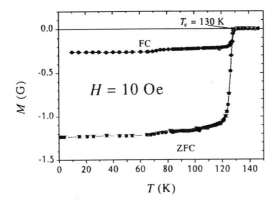

Fig. 2. Magnetization measurement (ZFC/FC) of the $Hg_{0.77}Re_{0.23}Ba_2Ca_2Cu_3O_{8+\delta}$ single crystal

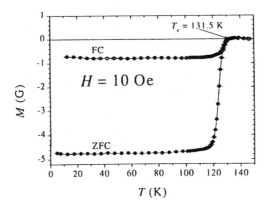

Fig. 3. Magnetization measurement (ZFC/FC) of the $HgBa_2Ca_2Cu_3O_{8+\delta}$ single crystal

two–three times larger than of an underdoped ($T_c = 120$ K) crystal (Fig. 4). The anisotropy γ decreases from 70 for the underdoped crystal ($T_c = 120$ K) to 44 for the almost optimally doped one ($T_c = 130$ K). In some investigated crystals, Re atoms substitute at the Hg site of $HgBa_2Ca_2Cu_3O_{8+\delta}$ and alter the defect structure of the host layer by pulling in four new oxygen atoms at the $(0.34, 0.34, 0)$ position to form an octahedron around Re. As a result, significant improvement of the irreversibility line position at the $H - T$ phase diagram for $Hg_{0.77}Re_{0.23}Ba_2Ca_2Cu_3O_{8+\delta}$ crystal is observed only at low temperatures (below 80 K). However, magnetization measurements performed at temperatures above 100 K show only insignificant change in the irreversibility line position, as compared to that of the parent compound with $T_c = 130$ K. This indicates insignificant influence of new pinning centers, introduced by Re substitution, in the temperature range where thermal activation energy becomes significant in comparison with pinning energy. Furthermore, the ob-

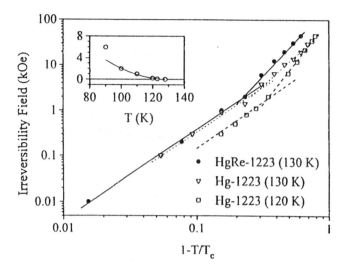

Fig. 4. Irreversibility field of $Hg_{0.77}Re_{0.23}Ba_2Ca_2Cu_3O_{8+\delta}$, underdoped $HgBa_2Ca_2Cu_3O_{8+\delta}$ with $T_c = 120\,K$, almost optimally doped Hg-1223 with $T_c = 130\,K$ single crystals determined in a magnetic field parallel to the c–axis are presented as a function of $(1 - T/T_c)$. The inset presents irreversibility field of $Hg_{0.77}Re_{0.23}Ba_2Ca_2Cu_3O_{8+\delta}$ as a function of T, in higher temperature region

tained data indicates that no improvement of the coupling between superconducting layers was achieved as a result of the expected increase of electrical conductivity in the blocking layer. Since only a small change in the c–axis lattice constant was observed in Re substituted crystals, we can conclude that a larger shortening of the blocking layer in the material is necessary to improve the irreversibility line position in $HgBa_2Ca_2Cu_3O_{8+\delta}$ crystals at high temperatures.

Following this idea, we started recently crystal growth experiments with the compound $Hg_{1-x}Pb_xBa_{2-y}Sr_yCa_2Cu_3O_{8+\delta}$. As a result, we obtained crystals with partial substitution of Pb for Hg and Sr for Ba. These crystals have T_c up to $116\,K$ and preliminary measurements of the irreversibility field show that it is higher than those of unsubstituted or Re substituted crystals.

4   Influence of Neutron Irradiation on Irreversibility Line of Hg-1223

Our studies showed significant influence of the new defect structure on the properties of neutron irradiated $HgBa_2Ca_2Cu_3O_{8+x}$ single crystals (Fig. 5). Neutron irradiation enhances the flux pinning, as expected, but also leads to a decrease of the effective mass anisotropy γ [9]. Torque measurements, which represent the most unambiguous way to measure anisotropy directly, show that the anisotropy decreases down to $\gamma \approx 41$ after irradiation ($\gamma \approx 60$ in the unirradiated state). In the irradiated state we also observed an apparent change of the slope of the IL at low temperatures. This indicates that, in the vicinity of the IL defect, cascades are able to correlate several pancake vortices and "force" them to behave in a 3D rather than a 2D manner. After irradiation we observed a significant shift of the IL in the whole temperature range. The shape of the hysteresis loop is changed – the fishtail effect, observed before irradiation, disappears. The critical current is enhanced at higher temperatures more than one order of magnitude and it's decrease with increasing temperature is slower.

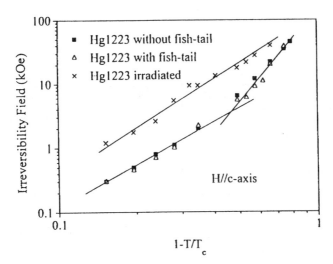

Fig. 5. Comparison of the irreversibility line of $HgBa_2Ca_2Cu_3O_{8+x}$ single crystals ($T_c = 120\,\mathrm{K}$) before and after neutron irradiation presented on the logarithmic scale of $(1 - T/T_c)$. The existence of the fishtail in one crystal does not improve the IL

5   Effect of Oxygen Doping on the Anisotropy γ and the Penetration Depth λ_{ab} of Hg-1201 Crystals

The effective mass anisotropy $\gamma = (m_c^*/m_{ab}^*)^{1/2} = \lambda_{ab}/\lambda_c$ (m_i^* is the effective mass and λ_i penetration depth) is a measure of the effective coupling between adjacent blocks of CuO_2 planes. The anisotropy and the penetration depth λ_{ab} depend strongly on the oxygen doping level. Due to the large anisotropy of superconducting compounds, the γ parameter can by precisely determined only on single crystals. The orientation of grains in aligned powder samples is not enough for this purpose.

The superconducting parameters γ and λ_{ab} have been determined as a function of doping using a miniaturized torque magnetometer [10]. Torque magnetometry is a very powerful technique to determine basic superconducting parameters of high-T_c materials. The reversible torque is a measure of the equilibrium intrinsic properties of the sample, whereas the irreversible torque is due to the pinning of vortices and thus related to the sample microstructure. Using the continuous anisotropic London model, it is possible to derive the superconducting parameters from a reversible angular dependent measurement where the magnetic torque is recorded as a function of the angle between applied field (with fixed amplitude) and the CuO_2 plane of the sample. Figures 6a and b show γ as a function of T_c and oxygen content δ, respectively. The anisotropy γ, as a function of T_c, changes more for underdoped samples, whereas for overdoped samples this dependence is weaker.

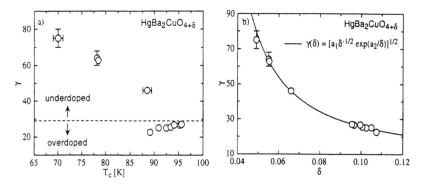

Fig. 6. a) Anisotropy γ for Hg-1201 single crystals as a function of critical temperature T_c. The dashed line represents the value repolled for an optimally doped single crystal ($\gamma = 29$). It corresponds to the boundary between the underdoped and overdoped regime, **b)** γ as a function of doping δ

Anisotropy γ monotonically decreases with increasing δ. This is an important fact, because the decreasing of the anisotropy increases the irreversibility field. These measurements are, to our knowledge, the first systematic measurements of superconducting parameters as a function of the oxygen doping of Hg–based superconductors performed on single crystals. Figure 7 shows the dependence of the in–plane penetration depth λ_{ab} on the oxygen doping δ. It is decreasing with increasing δ in the underdoped regime. The increase of λ_{ab} in the overdoped regime is significant, as tested with different fitting functions. A parabolic fit (dotted line) yields a minimum $\lambda_{ab} \approx 140$ nm for $\delta_{min} = -0.08$. A similar doping dependence of λ_{ab} in the underdoped regime was also observed in all other high-T_c cuprates [11].

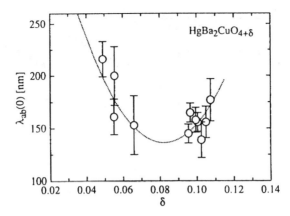

Fig. 7. In plane penetration depth λ_{ab} of Hg-1201 single crystals as a function of doping δ

6   Magnetic Properties of (Hg,Re)-1256 Single Crystals

Re substitution on the Hg site makes the growth of crystals, with the number of CuO_2–planes higher than three, much easier. Hence, it was possible to grow HgRe-1256 crystals of appreciable quality for the first time. Magnetization measurements on single crystals of $Hg_{0.75}Re_{0.22}Ba_2Ca_5Cu_6O_{15}$ grown with a gas–phase high pressure technique at 10 kbar, were done by SQUID magnetometry [12]. In the investigated crystals, Re atoms substitute at the Hg site and alter the blocking layer by pulling in four new oxygen atoms, which form an octahedron around Re (Fig. 1). A crystal with a transition temperature $T_c = 92\,K$ and a transition width of about 15 K was chosen for more detailed investigation. X-ray structure refinement revealed 3% of an infinite layer minority phase. Hysteresis loops and ZFC/FC–curves were recorded on the crystal with $T_c = 92\,K$ in magnetic fields up to 50 kOe. The asymmetrical form of the $M(H)$ loops and the field dependence of the magnetic hysteresis indicate that the irreversible magnetic behavior is dominated by surface barrier effects [13,14] at high and intermediate temperatures. Bulk pinning significantly contributes to the magnetic hysteresis only at low temperatures.

Magnetization data measured in intermediate fields and temperatures, where the magnetic hysteresis is small, was used to determine the penetration depth λ with the London model of the reversible magnetization [15]. While the temperature dependence of λ is consistent with the predictions of the BCS theory, with $\lambda(0) \simeq 220\,nm$, and the fit for λ is not very sensitive to the data, the H_{c2} data fits give rather different values for bulk or surface pinning. While an analysis based on bulk pinning gives H_{c2} values much lower than observed in other cuprate superconductors, an analysis based on the magnetic hysteresis caused by surface barriers gives more reasonable values

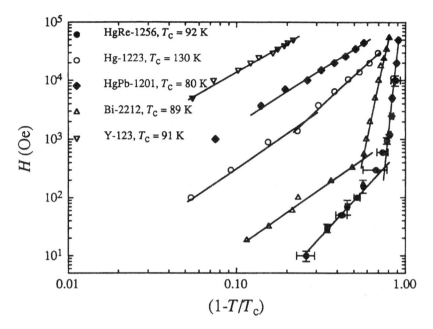

Fig. 8. Irreversibility line of (Hg,Re)-1256, compared with the irreversibility lines of other materials. HgRe-1256 has the lowest irreversibility field of all compared materials. For HgPb-1201, Hg-1223 and HgRe-1256, note the shift of the irreversibility line to lower values with the number of copper oxide planes increasing

ues of H_{c2}. An upward curvature of H_{c2} is observed, consistent with a bipolaronic model of superconductivity. However, the fits for H_{c2} are not very accurate, as it enters the fitting formula in a logarithm. The irreversibility line of (Hg,Re)-1256 was determined from the hysteresis loops $M(H)$ and from zero–field cooling and field cooling $M(T)$ measurements (Fig. 8). It was found to be controlled by surface barrier effects [14] in the whole temperature range investigated. At high temperatures and low fields, the vortex structure for the magnetic field $H \parallel c$–axis can be well described by three–dimensional (3D) flux lines behavior. In higher fields and at lower temperatures a crossover to a two–dimensional (2D) "pancake" vortex structure is observed for layered material. To the first approximation, this crossover occurs at a field $B_{cr} \approx \varphi_0 /(d\gamma)^2$, where φ_0 is the flux quantum, d is the distance between two adjacent superconducting layers, and r is the effective mass anisotropy of the charge carriers [16,17]. The temperature dependence of the irreversibility field is significantly different for the 2D and the 3D region. This can be observed well in Fig. 8. The crossover field is located where the fits to the points of the irreversibility line in high and low fields cross. Thus, it is possible to estimate the anisotropy by measuring the irreversibility line in the geometry $H \parallel c$ only, although the accuracy of such an estimation is in practice very limited. In the case of the measured (Hg,Re)-1256 crystal, the anisotropy, as determined from the crossover field, was estimated to be $\gamma = 70^{+30}_{-15}$. This result is in accordance with the general trend of the Hg–$12(n-1)n$ series, but it should be pointed out that the anisotropy is very sensitive to the oxygen doping level, needing further investigation.

The anisotropy also directs the position of the irreversibility line [16,18]. A lower crossover field (higher anisotropy) generally corresponds to a low irreversibility line, as can be seen in Fig. 8. Direct comparison of the irreversibility line of HgRe-1256 with the irreversibility lines of other members of the Hg–$12(n-1)n$ family, and of Bi-2212 and Y-123 is also presented in Fig. 8. It can be seen clearly that for Hg–$12(n-1)n$ materials, the irreversibility field is shifted to lower values with an increasing number n of the CuO_2 planes. The irreversibility line of HgRe-1256 is the lowest one of those in the graph, even lower than the one of Bi-2212. Thus, while HgRe-1256 is interesting physically to determine the influence of the number of layers on superconducting parameters, it is not suitable for applications.

7 Structure Analysis of HgRe-1256 and HgRe-1267 Single Crystals

After single crystal growth and structure analysis of the $n = 1 - 5$ and $n = \infty$ members of the homologous series $HgBa_2Ca_{n-1}Cu_nO_{2n+2+\delta}$ in the last years, we added two further members to the Hg based family of high-T_c cuprate superconductors. Single crystals of $Hg_{0.75}Re_{0.22}Ba_2Ca_5Cu_6O_{14+x}$ and of $Hg_{1-x}Re_xBa_2Ca_6Cu_7O_{16+x}$ have been grown with a gas–phase high pressure technique at $10\,kbar$ [19]. They crystallize in space group $P4/mmm$ with lattice parameters $a = 3.8542(3)\,Å$, $c = 25.162(3)\,Å$ (HgRe-1256) and $a = 3.8514(5)\,Å$, $c = 28.388(7)\,Å$ (HgRe-1267). They represent the $n = 6, 7$ members of the layered cuprate superconductors of the type $HgBa_2Ca_{n-1}Cu_nO_{2n+2+\delta}$. X-ray single crystal structure analysis showed that Re doping decreases the size of the rock–salt block. Rhenium substitutes mercury up to 25% and forms short bonds of $1.86\,Å$ to oxygen atoms. They coordinate Re octahedrally and are shifted off their ideal positions to $(0.34, 0.34, 0)$. Substitution of Hg by the smaller Re atoms decreases the size and the bond lengths in the rock–salt block of the structure. The perovskite unit is not affected by this process and has dimensions and bond lengths comparable to other Hg–$12(n-1)n$ structures. With increasing n, the crystals contain increasing amounts of stacking faults. They are included in the structure analysis and are refined to 6.1(3)% in HgRe-1256 and to 9.6(9)% in HgRe-1267. The crystals contain stacking faults which are included in the refinements and which introduce additional electron densities close to the position of the O_{excess} atom at $(1/2, 1/2, 0)$. In the case of HgRe-1256 this effect could be corrected, but in HgRe-1267 the stacking faults are so numerous that a refinement of the oxygen atoms in the basal plane was not possible.

In order to prove the validity of the stacking fault model, we simulated the Fourier maps. Calculation of theoretical (difference) electron density maps proved that observable maxima in the AF maps are in fact caused by stacking faults.

The transition temperature onsets of HgRe-1256 and of HgRe-1267 crystals are about 100 K. T_c onset value of 100 K for 1267 crystal is higher then expected, but we can explain it with the influence of stacking faults.

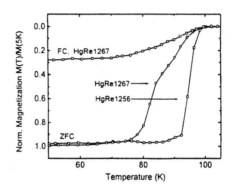

Fig. 9. Magnetization curves of HgRe-1256 and of HgRe-1267 single crystals

8   Single Crystal Structure Analysis of the 1223/1234 Intergrowth Phase HgRe-2457 $Hg_{1.44}Re_{0.5}Ba_4Ca_5Cu_7O_{20}$

With a gas-phase high pressure technique we grew a $Hg_{1.44}Re_{0.5}Ba_4Ca_5Cu_7O_{20}$ single crystal at 10 kbar Ar pressure and at maximum temperatures of 1025 °C [20]. The transition temperature of this HgRe-2457 crystal is 83 K.

HgRe-2457 is a new phase in the Hg–Re–Ba–Ca–Cu–O system and is based on the Hg–12$(n - 1)n$ family (Fig. 10). It has a supercell period of $c = 34.377$ Å and consists of an alternating stacking of 1223 and 1234 units in c–direction. Until now, intergrowth structures in the Hg based series have only been observed on a submicron scale or as a defect. They can either form complex polytypes or simple intergrowth structures like 1212-1223.

X-ray single crystal structure analysis showed that rhenium substitutes mercury to the maximum possible content of 25%. It is bonded to O(7) which is displaced off its ideal positions to $(0.363, 0.363, 0.22688)$. Re is co-ordinated by an almost perfect oxygen octahedron with bond lengths between 1.98 Å and 2.00 Å. No excess oxygen is found at $(1/2, 1/2, z)$ in the (Hg,Re)O plane because Hg is substituted to the maximum level of 25% by ReO_4 and there is no more space left for additional excess oxygen. All bond lengths are very close to those in HgRe-1256 single crystals, but slightly different than those in ceramic HgRe-1223 and HgRe-1234 samples. After refining the ideal 2457 model, strong difference electron density peaks have been observed at $(0, 0, 0.32)$ (Hg'), $(1/2, 1/2, 0.24)$ (Ba1'), $(1/2, 1/2, 0.39)$ (Ba2') and at $(0, 0, 0.19)$ (Cu'). This feature can be explained by a disturbance of the stacking sequence by double 3 or 4 layer units, like, e.g. ...–3–4–3–4–3–3–4–3–4–... A further peak at $(0, 0, 0.13)$ could be attributed to the presence of triple faults in the stacking sequence like ...–3–4–3–3–3–4–3–4–... The observation of this rather complex 3–4 layer intergrowth structure makes it plausible that at least crystals of simpler intergrowth phases like 2413 (1201-1212) or 2435 (1212-1223) could exist.

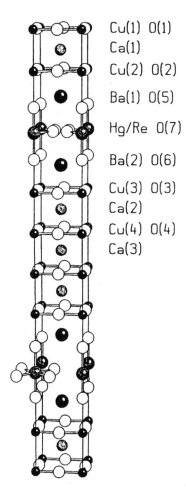

Cu(1) O(1)
Ca(1)
Cu(2) O(2)
Ba(1) O(5)
Hg/Re O(7)
Ba(2) O(6)
Cu(3) O(3)
Ca(2)
Cu(4) O(4)
Ca(3)

Fig. 10. Structure of new phase $Hg_{1.44}Re_{0.5}Ba_4Ca_5Cu_7O_{20}$ ($T_c = 83\,K$)

9   High–Pressure Synthesis of $Sr_{0.73}CuO_2$, $Ca_{0.83}CuO_2$ and $Ba_{0.66}CuO_2$

Synthesis of $Sr_{0.73}CuO_2$ samples have been performed in a double chamber high pressure system [4,6]. A ceramic crucible containing the sample under a high pressure of oxygen is enclosed in a chamber having an internal furnace working under argon atmosphere. The Ar and O_2 pressures are equal. We have investigated several mixtures of $SrCuO_2$ and CuO corresponding to $CuO:SrO$ ratios: 1, 1.174, 1.333, 1.352, 1.4, 1.5, 1.714, 2.333. The synthesis have been done at $1800 < P_{O_2} < 2200$ bar at $950 < T < 1800\,°C$ for 10 to 48 h. As a result, two main phases have been obtained in the samples: $Sr_{0.73}CuO_2$ and $Sr_{14}Cu_{24}O_{41}$. At $1170 - 1180\,°C$ partial melting of the $Sr_{0.73}CuO_2$ occurred and by slow cooling ($5\,°C/h$) small crystals could be grown. These crystals were used for the structure analysis. Crystal growth experiments from the melt with an excess of CuO ($Sr_{0.73}CuO_2 + CuO$) led always to the synthesis of the $Sr_{14}Cu_{24}O_{41}$ compound. $Ca_{0.83}CuO_2$ and $Ba_{0.66}CuO_2$ compounds have also been synthesized at high oxygen pressure, but for these compounds several hundred bar oxygen pressure was sufficient.

10   Magnetic Properties of Quasi–One–Dimensional Cuprates

One–dimensional (ID) spin $S = 1/2$ systems with antiferromagnetic (AF) interactions have attracted considerable interest due to the quantum mechanical nature of their ground state. In the absence of interchain coupling, a nonmagnetic singlet ground state is formed. When a small interchain interaction is present, a long–range magnetic order can be formed. Particularly interesting is the case where doping leads to a competition between the quantum mechanical singlet ground state and the classical long–range ordered state. For example, in the spin–Peierls material $CuGeO_3$, a Néel ordered state is formed by substituting Cu with Ni ($S = 1$) or nonmagnetic Zn ($S = 0$) [21]. In both cases, for a certain doping level, the spin–Peierls singlet ground state seems to coexist with a Néel ordered state. Two further examples of quasi–1D cuprates where doping is realized by introducing holes are $Ca_{0.83}CuO_2$ [22] and $Sr_{0.73}CuO_2$ [23]. These compounds contain CuO_2 infinite chains separated by Ca or Sr layers [6]. The Cu ions are coupled by $90°$ Cu–O–Cu bonds (Fig. 11). The formal valence of Cu is $+2.34$ and $+2.54$ for $Ca_{0.83}CuO_2$ and $Sr_{0.73}CuO_2$ respectively, i.e., there are inherently 0.34 and 0.54 Zhang–Rice singlets/Cu present. The susceptibility of both compounds is consistent with a dimerized Heisenberg chain [22,23] (Fig. 12). This gives an indication of a singlet ground state with a spin gap to the triplet states. However, susceptibility and electron paramagnetic resonance (EPR) measurements indicated that $Ca_{0.83}CuO_2$ and $Sr_{0.73}CuO_2$ exhibit AF order at $T \simeq 10\,\mathrm{K}$. It was speculated that this order might coexist with a singlet ground state like in doped $CuGeO_3$ [23].

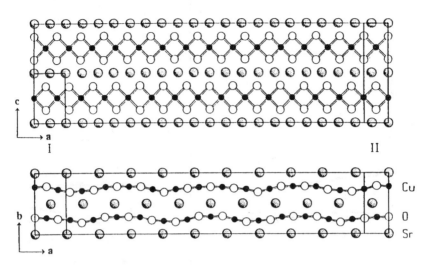

Fig. 11. Structure of $Sr_{0.73}CuO_2$ compound containing one–dimensional CuO_2 chains separated by Sr atoms

We have continued the investigation of the magnetic properties of $Ca_{0.83}CuO_2$ and $Sr_{0.73}CuO_2$. Specific heat and elastic neutron scattering of polycrystalline material and magnetic torque on single crystals have been measured. These measurements give clear evidence that the AF ordered state previously found is of long–range 3D character [24]. This finding is surprising considering the fact that the spin chains are highly doped with holes. Note that in the 2D Cu–O layers present in high-T_c cuprates, no Néel order for doping levels exceeding 0.04(2) holes/Cu nor a static stripe phase yielding long–range magnetic order was found for doping larger than ~ 0.125 holes/Cu [25].

The temperature dependence of the specific heat C/T of polycrystalline $Ca_{0.83}CuO_2$ and $Sr_{0.73}CuO_2$ is shown in Fig. 13. A peak in the specific heat is observed for $Ca_{0.83}CuO_2$ as well as for $Sr_{0.73}CuO_2$ at $T_{AF} \simeq 10\,\mathrm{K}$. This indicates a static magnetic transition. To decide whether the magnetically ordered state at $T < T_{AF}$ is of long–range character, we performed elastic neutron scattering at the DMC spectrometer at PSI Villigen. In both compounds a Bragg peak attributed to a twofold increase in periodicity along the CuO_2 chain axis is found. The temperature dependence of the integrated intensity of the $(0, 1/2, 1)$ peak of $Ca_{0.83}CuO_2$ measured at DIB spectrometer at the high flux reactor of the Institute Laue–Langevin, Grenoble, France, is shown in Fig. 14(a). A clear increase of the neutron intensity is observed at $T < 12\,\mathrm{K}$.

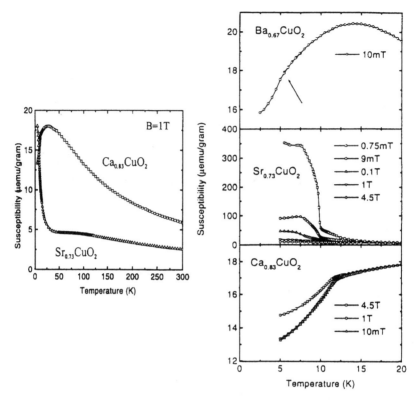

Fig. 12. Susceptibility of $A_{1-x}CuO_2$ for $A = Sr$, Ca and Ba. $Sr_{0.73}CuO_2$ shows weak–ferromagnetic transition at $T = 10\,K$, $Ca_{0.83}CuO_2$ antiferromagnetic at $T = 12\,K$ and $Ba_{0.67}CuO_2$ antiferromagnetic at $5\,K$

Angle–dependent torque measurements were performed on a twinned single crystal of $Sr_{0.73}CuO_2$ (volume $V \approx 120 \times 90 \times 8\,\mu m^3$) with a highly sensitive torquemeter (resolution $\sim 10^{-13}$ Nm). Measurements on an untwinned single crystal of $Ca_{0.83}CuO_2$ have been reported previously [22]. We use the torquemeter as a tool for our investigation because the sensitivity of a SQUID magnetometer would not be sufficient. For comparison, a SQUID magnetometer has a typical sensitivity of $\Delta m \simeq 10^{-7}$emu, whereas we achieve $\Delta m \sim 10^{-10}$emu in $B = 1$ T with the torquemeter. For $Sr_{0.73}CuO_2$ a series of torque curves for rotation of the magnetic field is presented in Fig. 15(a). The temperature dependence of the torque amplitude is shown in Fig. 15(b). The torque amplitude decreases towards zero upon approaching $T_{AF} = 10.0(2)$ K. The temperature dependence of the torque amplitude of $Ca_{0.83}CuO_2$ is shown in Fig. 14(b).

These experimental results give clear evidence of long–range 3D magnetic order. The observed gain in neutron intensity at half–integer indices is con-

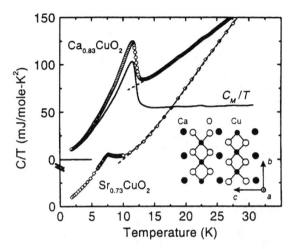

Fig. 13. Specific heat $C(T)$ of $Ca_{0.83}CuO_2$ and $Sr_{0.73}CuO_2$. The curves for $Ca_{0.83}CuO_2$ are offset for clarity. The magnetic contribution C_M/T is shown for $Ca_{0.83}CuO_2$. The inset shows part of the unit cell of $Ca_{0.83}CuO_2$

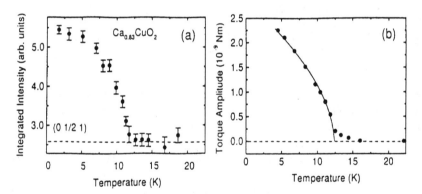

Fig. 14. (a) Temperature dependence of the $(0, 1/2, 1)$ neutron peak intensity of $Ca_{0.83}CuO_2$ measured on DIB. (b) Temperature dependence of the torque amplitude of $Ca_{0.83}CuO_2$ ($\mu_0 H = 1.5$ T)

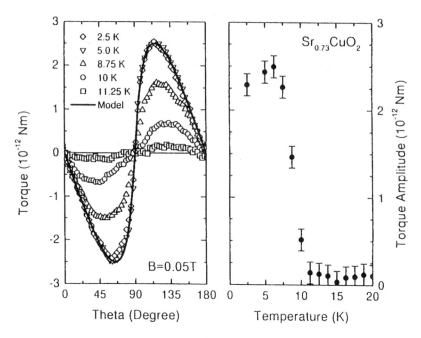

Fig. 15. (a) Angle–dependent torque ($\tau(\theta)$) of $Sr_{0.73}CuO_2$ at different T upon field rotation ($\mu_0 H = 0.05$ T). **(b)** Temperature dependence of the torque amplitude

sistent with AF ordering. A small structural reorientation of the atoms in the Cu–O subcell could, however, also account for additional neutron intensity at these positions. To differentiate between these two possibilities, note that the torque signal is sensitive only to the magnetic moment. A structural reorientation would not contribute to a torque signal. With this argument, a structural transition at $T \approx 10$ K resulting in an increased neutron intensity can be excluded. The finding of additional neutron intensity at half–integer indices together with the appearance of torque provides conclusive evidence of long–range AF order in $Ca_{0.83}CuO_2$ and $Sr_{0.73}CuO_2$.

Whether this ordered state coexists with a dimerized singlet ground state as in doped $CuGeO_3$ is not yet clear. Preliminary inelastic neutron scattering gives some evidence that this might be the case [26].

11   Conclusions

Rhenium substitution for Hg improves the irreversibility line position for a $Hg_{0.77}Re_{0.23}Ba_2Ca_2Cu_3O_{8+\delta}$ crystal only at low temperatures (below 80 K). The irreversibility field of almost optimally doped, unsubstituted Hg-1223 crystal ($T_c = 130$ K) is about two–three times larger than the one of underdoped crystal ($T_c = 120$ K). Neutron irradiation enhances the irreversibility field of Hg-1223, while also leads to a decrease of the effective mass anisotropy.

The anisotropy γ of Hg-1201 single crystals decreases monotonically from 75 to 20 with the increase of oxygen content. The penetration depth λ_{ab} has a minimum for the optimally doped crystals.

Specific heat, elastic neutron scattering of polycrystalline material of one–dimensional infinite–chain cuprates $Ca_{0.83}CuO_2$ and $Sr_{0.73}CuO_2$ and magnetic torque on single crystals give clear evidence that the AF ordered state below 10 K is of long–range 3D character.

References

1. Schwer H., Kopnin E.M., Jun J., Karpiński J. (1997)
   J Solid State Chem 134:427–430
2. Kopnin E.M., Schwer H., Jun J., Meijer G.I., Moliński R., Conder K., Karpiński J. (1997) Physica C 282–287:483–484
3. Karpiński J., Schwer H., Kopnin E., Moliński R., Meijer G.I., Conder K. (1997) Physica C 282–287:77–80
4. Karpiński J., Schwer H., Moliński R., Meijer G.I., Conder K., Kopnin E., Löhle J., Rossel C., Zecli D., Hofer J., Wiśniewski A., Puźniak R. (1997) In: Narlikar A. (Ed.) Studies of High Temperature Superconductors vol. 24, Nova Sci. Publ., Commack, NY, 165
5. Karpiński J., Schwer H., Conder K., Löhle J., Moliński R., Morawski A., Rossel C., Zech D., Hofer J. (1996) In: Klamut J., et al. (Eds.) Recent Developments in High Temperature Superconductivity, Lecture Notes in Physics vol 475, Springer, Berlin Heidelberg, 83–103
6. Karpiński J., Schwer H., Meijer G.I., Conder K., Kopnin E., Rossel C. (1997) Physica C 274:99–106
7. Chmaissem O., Jorgensen J.D., Yamaura K., Hiroi Z., Takano M., Shimoyama J., Kishio M. (1996) Phys Rev B 53:14 667
8. Puźniak R., Karpiński J., Wiśniewski A., Szymczak R., Angst M., Schwer H., Moliński R., Kopnin E. (1998) Physica C 309:161-169
9. Wiśniewski A., Puźniak R., Karpiński J., Hofer J., Szymczak R., Baran M., Sauerzopf F.M., Moliński R., Thompson J.R. (1999) Phys Rev B to be published
10. Hofer J., Karpiński J., Willemin M., Meijer G.I., Kopnin E., Moliński R., Schwer H., Rossel C., Keller H. (1998) Physica C 297:103-110
11. Schneider T., H.Keller H. (1993) Int J Mod Phys 8:487
12. Angst M. (1998) Diploma Thesis, ETH Zürich
13. Bean C.P., Livingston J.D. (1964) Phys Rev Lett 12:14
14. Burlachkov L., Geshkenbein V.B., Koshelev A.E., Larkin A.I., Vinokur V.M. (1994) Phys Rev B 50:16 770
15. Kogan V.G., Fang M.M., Mitra S. (1988) Phys Rev B 38:11 958
16. Glazman L.I., Koshelev E.E. (1992) Phys Rev B 43:2835
17. Bulaevskii L.N., Ledvij M., Kogan V.G. (1992) Phys Rev Lett 68:3773
18. Houghton A., Pelcovits R.A., Sudbo A. (1989) Phys Rev B 40:6763
19. Schwer H., Moliński R., Kopnin E., Meijer G.I., Karpiński J. (1999) J Solid State Chem 143:277–284

20. Schwer H., Moliński R., Kopnin E., Angst M., Karpiński J. (1999)
    Physica C 311:49–57
21. Oseroff S.B., Cheong S.-W., Aktas B., Hundley M.F., Fisk Z., Rupp L.W.,Jr.
    (1995) Phys Rev Lett 74:1450
22. Meijer G.I., Rossel C., Kopnin E.M., Willemin M., Karpiński J., Schwer H.,
    Conder K., Wachter P., (1998) Europhys Lett 42:339
23. Shengelaya A., Meijer G.I., Karpiński J., Zhao G.-M., Schwer H.,
    Kopnin E.M., Rossel C., Keller H., (1998) Phys Rev Lett 80:3626
24. Meijer G.I., Rossel C., Henggeler W., Keller H., Fautli F., Karpiński J.,
    Schwer H., Kopnin E.M., Wachter P., Black R.C., Diederichs J. (1998)
    Phys Rev B 58:14 452
25. Tranquada J.M., Sternlieb B.J., Axe J.D., Nakamura Y., Uchida S. (1995)
    Nature 375:561
26. Meijer G.I. et al. (1998) unpublished data

Angular Dependence of Flux Pinning in High-T_c Superconductor Thin Films and Superconductor–Ferromagnetic Heterostructures

S. Koleśnik, T. Skośkiewicz, P. Przysłupski, and M. Z. Cieplak

Institute of Physics, Polish Academy of Sciences,
Al. Lotników 32/46, PL-02-668 Warsaw, Poland

Abstract. We report on the results of measurements of the angular dependence of magnetization of high-T_c superconducting $YBa_2Cu_3O_{7-\delta}$ and $La_{1.85}Sr_{0.15}CuO_4$ thin films and superconducting/ferromagnetic $YBa_2Cu_3O_{7-\delta}/Nd_{0.67}Sr_{0.33}MnO_3$ superlattices. We compare our present results to our results, previously obtained for $YBa_2Cu_3O_{7-\delta}$ and $Bi_2Sr_2CaCu_2O_8$ single crystals. We discuss the influence of the presence of $Nd_{0.67}Sr_{0.33}MnO_3$ sublayers in $YBa_2Cu_3O_{7-\delta}/Nd_{0.67}Sr_{0.33}MnO_3$ superlattices on the critical temperature, flux pinning, and the critical field for the first vortex penetration.

1 Introduction

High-T_c superconductor (HTSC) thin films are interesting from the point of view of many possible applications, including superconducting quantum interference devices (SQUID's) and superconductive electronic circuits, which base on rapid single flux quantum (RSFQ) logic, as well as for microwave applications and many others.

Recently developed superlattices [1], consisting of layers of superconducting $YBa_2Cu_3O_{7-\delta}$ (Y–Ba–Cu–O) and ferromagnetic $Nd_{0.67}Sr_{0.33}MnO_3$ (Nd–Sr–Mn–O), which exhibit a giant magnetoresistance effect, [2] may also be attractive for fabrication of electronic chips, incorporating magnetic field sensors. Basic parameters of the materials, building such heterostructures, may lead to the coexistence of superconductivity and ferromagnetism in those particular structures [3]. The role of ferromagnetic layers in superconducting/ferromagnetic superlattices is not yet satisfactorily known. Ferromagnetic particles on the surface of a superconducting thin film can act as additional pinning centers [4]. On the other hand, injection of spin–polarized carriers from a current–carrying ferromagnetic layer, may decrease the critical current in an adjacent superconducting layer [5]. An increase of the thickness of the ferromagnetic layer leads to a sudden drop in T_c of Nb/Fe multilayers, due to decoupling of superconducting Nb layers [6].

The isothermal remanent magnetic moment measurement method was applied to determination of the lower critical field H_{c1} of niobium–doped

$SrTiO_3$ as early as in 1966 [7]. After the discovery of HTSC's, this method was used by many other groups [8–10]. By use of this method, we have determined H_{c1} of Pb–Sr–(Y,Ca)–Cu–O, [11] Bi–Sr–Ca–Cu–O, [12] and Y–Ba–Cu–O [13] single crystals. Angular dependence of the isothermal remanent magnetic moment in low magnetic fields showed us that the vortices remain locked–in parallel to the ab plane of a Bi–Sr–Ca–Cu–O crystal, for a wide range of the angle between the magnetic field direction and the c axis of the crystal [12].

Field dependence of the isothermal remanent magnetic moment just above the first flux penetration, analyzed within the framework of the extended Bean model [9], can deliver information about the flux distribution in Y–Ba–Cu–O crystals [9,13].

In this paper, we determine the superconducting parameters of a Y–Ba–Cu–O/Nd–Sr–Mn–O superlattice. We will show that the results of the magnetization measurements lead to an apparently short penetration depth in HTSC thin films and superconducting/ferromagnetic superlattices.

2   Experimental Details

Y–Ba–Cu–O thin films and Y–Ba–Cu–O/Nd–Sr–Mn–O superlattices have been grown by high pressure dc sputtering on $SrTiO_3$ and $LaAlO_3$ substrates, respectively. $La_{1.85}Sr_{0.15}CuO_4$ (La–Sr–Cu–O) thin films have been produced by pulsed laser deposition on $LaSrAlO_4$ substrates. Basic parameters of the studied samples are presented in Table I.

Table 1. Parameters of the studied thin film samples. $(1-N_c)^{-1}$ is the enhancement factor of the internal field, due to the demagnetizing factor N_c

Sample code	Compound	Thickness (nm)	T_c (K)	$(1-N_c)^{-1}$
NdY45	$[Nd_{0.67}Sr_{0.33}MnO_3(9.6\,nm)$ $/YBa_2Cu_3O_7(9.6\,nm)]_{12}$	220	43	6000
Y46	$YBa_2Cu_3O_7$	200	91	10500
D_3	$La_{1.85}Sr_{0.15}CuO_4$	450	27	5000

We performed magnetization measurements in a SQUID magnetometer for arbitrarily chosen angles between the magnetic field direction and the surface of the sample. A metallic–glass foil shield around the superconducting magnet in our magnetometer limits the residual field in the sample chamber to the value of $1\,\mu T$. A rotary sample holder, designed for studies of HTSC thin films, gives the angular resolution of approximately $1°$.

The isothermal remanent magnetic moment measurement procedure is illustrated in Fig. 1. The isothermal remanent magnetic moment m_{irm} is measured at a constant temperature in the magnetic field H_0 (herewith always equal to zero). Between the subsequent measurement steps, the sample is exposed to a magnetic field H_i and reduced to H_0. At low magnetic fields, m_{irm} remains constant and is equal to a small magnetic background, resulting from the trapped flux in a nonzero residual magnetic field. The value of the magnetic field, for which m_{irm} starts to deviate from the initial value, is defined as the critical field for the first vortex penetration H_p.

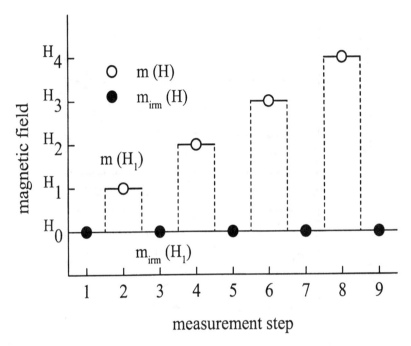

Fig. 1. Measurement procedure, described in the text. Closed circles represent the measurements of the isothermal remanent magnetic moments m_{irm}, performed after applying a magnetic field H_i and reducing the magnetic field to H_0

From the analysis of the isothermal remanent magnetization, we determined the angular dependence of the critical field for the first vortex penetration and the critical current density.

3   Results and Discussion

Temperature dependence of the "zero–field–cooled" (ZFC), "field–cooled" (FC) and remanent (REM) magnetic moments is presented in Fig. 2, for several angles ϕ between the magnetic field direction and the normal to the sample surface. At low angles, significantly more negative FC magnetic moment values can be observed for sample NdY45, which is a result of weaker pinning in this sample. At high angles, when the magnetic field is almost parallel to the thin film surface, we observe a "paramagnetic Meissner effect" and significant enhancement of the REM moment.

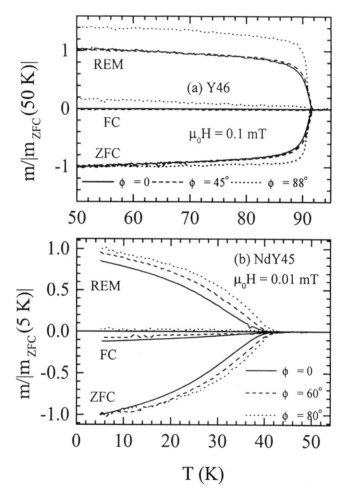

Fig. 2. Temperature dependence of the scaled magnetic moment for Y–Ba–Cu–O thin film Y46 (**a**), and Y–Ba–Cu–O/Nd–Sr–Mn–O superlattice NdY45 (**b**), for several angles between the magnetic field direction and the normal to the sample

The isothermal remanent magnetic moment m_{irm} as a function of the magnetic field is presented in Fig. 3. One may observe an initial rise of m_{irm}

at very low magnetic fields, then a rollover and saturation of the magnetic moment.

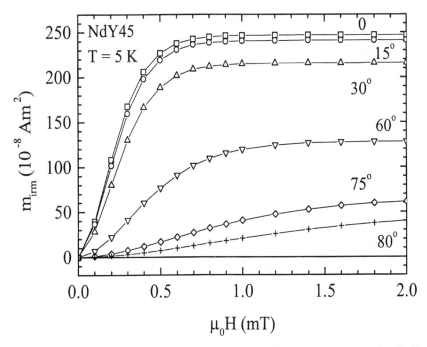

Fig. 3. Magnetic field dependence of the isothermal magnetic moment for Y–Ba–Cu–O/Nd–Sr–Mn–O superlattice NdY45, for several angles between the magnetic field direction and the normal to the sample

Angular dependence of m_{sat} for sample NdY45 is presented in Fig. 4. Except for a narrow angular region around $\phi = 0$, where a dip occurs at higher temperatures (which can be attributed to the channeling of vortices along the twin planes [14]) m_{sat} follows a cosine dependence

$$m_{sat}(\phi) = m_{sat}(0) \cos \phi. \tag{1}$$

This observation leads to a conclusion that, for any field orientation, (except of $\boldsymbol{H}$ exactly parallel to the ab plane), superconducting currents flow in the ab plane and the magnetic moment is parallel to the c axis.

From the saturation magnetic moment m_{sat} for $\boldsymbol{H} \parallel c$, we calculate the in-plane critical current density $J_c(\phi = 0)$. Taking into account the fact that m_{sat} is equal to one half of the hysteresis width at zero magnetic field, we apply a simplified formula [15]

$$J_c(0) = 3m_{sat}(0)/Vr , \tag{2}$$

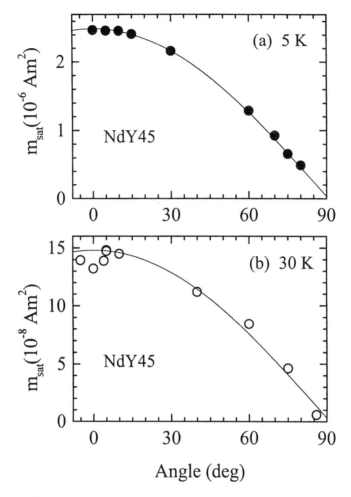

Fig. 4. Angular dependence of the saturation magnetic moment for Y–Ba–Cu–O/Nd–Sr–Mn–O superlattice NdY45

where V – the sample volume, r – the mean sample radius. The extrapolated zero–temperature values of J_c are equal to 0.8, 140 and 20×10^9 A/m^2 for samples NdY45, Y46, and D_3, respectively.

Temperature dependence of the critical current density for our samples is shown in Fig. 5. In this figure, we compare our results to the results obtained by Darhmaoui and Jung [16] for oxygen–deficient Y–Ba–Cu–O thin films. Darhmaoui and Jung observed a crossover from Ambegaokar–Baratoff–type temperature dependence of J_c to Ginzburg–Landau–like dependence with increasing oxygen deficiency. The temperature dependence of J_c for superlattice NdY45 points at a similarity in pinning properties between Y–Ba–Cu–

O/Nd–Sr–Mn–O superlattices and oxygen–deficient Y–Ba–Cu–O thin films
with same T_c.

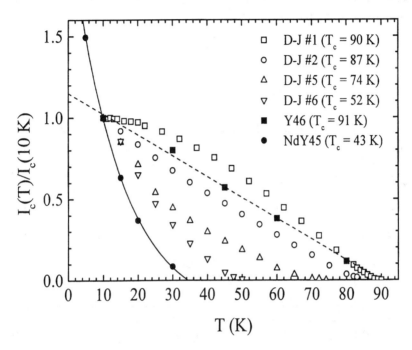

Fig. 5. Temperature dependence of the scaled critical current for Y–Ba–Cu–O thin
film Y46 and Y–Ba–Cu–O/Nd–Sr–Mn–O superlattice NdY45, compared to the results for Y–Ba–Cu–O thin films from Ref. [16]

Next, we analyze the low–field region of the magnetic field dependence
of m_{irm}. McElfresh *et al.* [9] have proposed an extension of the Bean model
to describe $m_{\text{irm}}(H)$. Their model usually fails to describe quantitatively the
entire magnetic field region; however, it gives good results in the vicinity of
the critical field for the first flux penetration H_{p} and it is a very useful tool
for determination of H_{p}. The model bases on the assumption that $J_c \propto B^{-n}$.
Then, below the full flux penetration field

$$m_{\text{irm}} \propto \left[\left(\frac{1}{2}H + \frac{1}{2}H_{\text{p}} \right)^{n+2/n+1} - H_{\text{p}}^{n+2} \right]. \qquad (3)$$

In Figure 6 we present a typical magnetic field dependence of m_{irm} for
sample NdY45. We fitted the formula from (3) to our experimental data, using
two different exponents: $n = 0$ and $n = 0.5$. Equation (3) with the exponent
$n = 0.5$ can describe $m_{\text{irm}}(H)$ for Y–Ba–Cu–O single crystals [9,13] and La–
Sr–Cu–O thin films in the present study. For thinner Y–Ba–Cu–O thin films

and Y–Ba–Cu–O/Nd–Sr–Mn–O superlattices, a better fit can be obtained by inserting the exponent $n = 0$ into (3). $n = 0$ means that J_c is independent of B as in the conventional Bean model [17]. We suggest that there is a possible crossover from the conventional Bean behavior to the extended Bean model with increasing sample thickness.

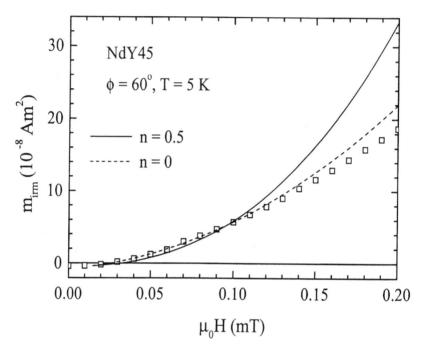

Fig. 6. Comparison of fits of the magnetic field dependence of m_{irm} to the experimental data. Two different exponents n were used

Magnetic field dependence of m_{irm} for sample NdY45 for several angles between the magnetic field direction and the normal to the surface is shown in Fig. 7. From the fits of (3) with $n = 0$ to the data, we obtain the values of H_p. Angular dependence of H_p for samples Y46 and NdY45 are presented in Fig. 8(a). For comparison, the values determined at similar reduced temperatures are shown. H_p for superlattice NdY45 is by about one order of magnitude lower than for Y–Ba–Cu–O thin film Y46. The angular dependence of H_p can be approximated by the formula, given by the 3D London theory [18]

$$H_p = H_{c1}^{\parallel}/((1 - N_c)^{-2}\gamma^{-2}\cos^2\phi + (1 - N_{ab})^{-2}\sin^2\phi)^{1/2}, \qquad (4)$$

where γ is the anisotropy ratio, N_c, N_{ab} – demagnetizing factors. ($N_{ab} = 0$ for thin films.) We calculate the demagnetizing factor N_c from the initial

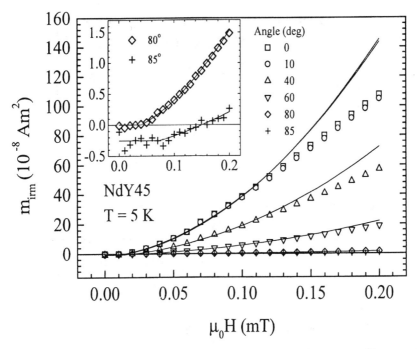

Fig. 7. Magnetic field dependence of m_{irm} at low magnetic fields. Lines show results of the fits of the conventional Bean model to the data

slope of the magnetic moment for $\phi = 0$ $dm/dH|_{H\|c}$ (below H_p) and the sample dimensions. The calculated values are listed in Table I. The measured magnetic moments are by $30 - 50\%$ lower than the values, expected from an approximation of the shape of the sample by an ellipsoid [19]. We interpret this result by considering that, in the perpendicular magnetic field, magnetic field lines are strongly curved at the surface of a thin film [20]. Therefore, below H_p, a thin surface layer (of the thickness λ_{eff}) of the thin film is already penetrated by the magnetic flux, from both sides.

By introducing the calculated demagnetizing factors N_c, we obtain the angular dependence of $H_{c1}(\phi) = H_p(\phi)(1 - N_c)^{-1} \cos\phi$, which is plotted in Fig. 8(b). H_{c1} for superlattice NdY45 is significantly lower than for Y–Ba–Cu–O thin film Y46. This is another similarity between Y–Ba–Cu–O/Nd–Sr–Mn–O superlattices and oxygen–deficient Y–Ba–Cu–O, where the penetration depth increases with increasing oxygen deficiency [21] and, hence, H_{c1} becomes lower.

The obtained values of H_{c1} for Y–Ba–Cu–O thin film are much higher than the values, previously reported for Y–Ba–Cu–O crystals. Since the lower critical field is inversely proportional to the square of the penetration depth, such a high H_{c1} would lead to a very short penetration depth λ_{eff}. To clarify

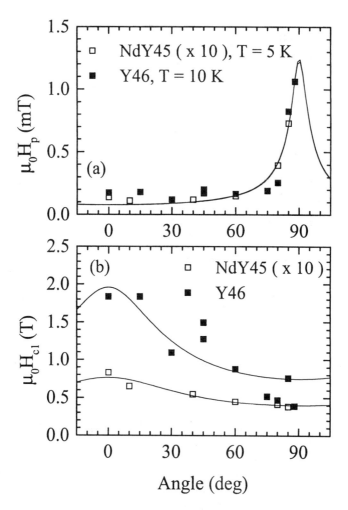

Fig. 8. Angular dependence of the critical field for the first vortex penetration H_p and the lower critical field H_{c1} for Y–Ba–Cu–O thin film Y46 and Y–Ba–Cu–O/Nd–Sr–Mn–O superlattice NdY45

this possibility, we analyze the temperature dependence of the magnetic moment, measured in low magnetic fields, which is presented in Fig. 9. Following Ref. [22], we fitted (5) to the experimental data

$$\frac{m_{\mathrm{ZFC}}(T)}{m(0)} = 1 - \frac{2\lambda(T)}{d} \tanh\left(\frac{d}{2\lambda(T)}\right) \tag{5}$$

We used the two–fluid formula

$$\frac{\lambda(T)}{\lambda(0)} = \left(1 - \left(\frac{T}{T_c}\right)^4\right)^{-1/2}. \tag{6}$$

From these measurements, we obtained a penetration depth λ_{eff} small compared to the sample thickness d. The effective penetration depth values, obtained from different experiments, are listed in Table II. Although different experiments give different values of λ_{eff}, all the values are much smaller than λ_{ab} for bulk samples.

Table 2. Penetration depth determined from different experiments

Sample code	$dm/dH|_{H\|c}$ λ_{eff} (nm)	H_{c1} λ_{eff} (nm)	$m_{ZFC}(T)$ λ_{eff} (nm)	Bulk samples λ_{ab} (nm)
NdY45	131	107	–	–
Y46	55	21	5	140
D_3	128	63	23	400

4   Summary

In summary, we have investigated magnetic properties of Y–Ba–Cu–O and La–Sr–Cu–O thin films and Y–Ba–Cu–O/Nd–Sr–Mn–O superlattices, by the isothermal remanent magnetic moment method, for various angles between the magnetic field direction and the normal to the sample surface.

The angular dependence of H_{c1} and the critical current density has been determined.

Y–Ba–Cu–O/Nd–Sr–Mn–O superlattices present many similarities to oxygen–deficient Y–Ba–Cu–O samples. In comparison to fully oxygenated Y–Ba–Cu–O thin films, T_c is lower in these superlattices, pinning is much weaker, λ is longer and the temperature dependence of J_c reflects that of deoxygenated Y–Ba–Cu–O thin films.

The following scenario of the first flux penetration emerges from the results of the present study. Due to a strong demagnetization effect in HTSC thin films, magnetic field lines are strongly curved near the sample surface. In the magnetic field lower than H_p, a thin layer of the thin film sample is already penetrated by the flux lines. The thickness of this layer is smaller than the penetration depth of a bulk superconductor and manifests itself as an apparently short effective penetration depth, giving rise to large values of H_{c1} in thin superconducting samples.

Acknowledgements

Financial support from the Polish Committee for Scientific Research (KBN) under grant No. 2 P03B 10714 is acknowledged.

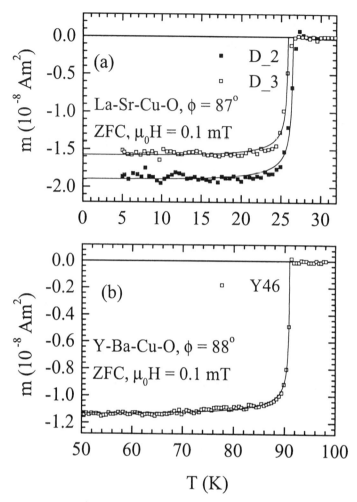

Fig. 9. Temperature dependence of the magnetic moment for Y–Ba–Cu–O thin film Y46 and La–Sr–Cu–O thin films, for the magnetic field oriented almost parallel to the thin film surface. Lines show the fits of the two–fluid temperature dependence of the penetration depth to the data

References

1. Przysłupski P., Koleśnik S., Skośkiewicz T., Dynowska E., Sawicki M. (1997) IEEE Appl Supercond 7:2192
2. Xiong G.C., Li Q., Ju H.L., Mao S.N., Senapati L., Xi X.X., Greene R.L., Venkatesan T. (1995) Appl Phys Lett 66:1427
3. Sa de Melo C.A.R. (1997) Phys Rev Lett 79:1933
4. Martin J.I., Velez M., Nogues J., Schuller I.K. (1997) Phys Rev Lett 79:1929
5. Vas'ko V.A., Larkin V.A., Kraus P.A., Nikolaev K.R., Grupp D.E., Nordman C.A., Goldman A.M. (1997) Phys Rev Lett 79:1929
6. Verbanck G., Potter C.D., Metlushko V., Schad R., Moshchalkov V.V., Bruynseraede Y. (1998) Phys Rev B 57:6029; Mühge Th., Theis-Bröhl K., Westerholt K., Zabel H., Garif'yanov N.N., Goryunov Yu.V., Garifullin I.A., Khaliullin G.G. (1998) Phys Rev B 57:5071
7. Ambler E., Colwell J.H., Hosler W.R., Schokley J.F. (1966) Phys Rev 148:280
8. Moshchalkov V.V., Zhukov A.A., Petrov D.K., Voronkova V.I., Yanovskii V.K. (1990) Physica C 166:185
9. McElfresh M.W., Yeshurun Y., Malozemoff A.P., Holtzberg F. (1990) Physica A 168:308
10. Buntar V., Sauerzopf F.M., Weber H.W. (1996) Phys Rev B 54:R9651; Böhmer C., Brandstätter G., Weber H.W. (1997) Supercond Sci Technol 10:A1
11. Korczak Z., Korczak W., Koleśnik S., Skośkiewicz T., Igalson J. (1990) Supercond Sci Technol 3:370
12. Koleśnik S., Skośkiewicz T., Igalson J., Tarnawski Z. (1996) Phys Rev B 54:13 319
13. Koleśnik S., Weber H.W., Skośkiewicz T., Sawicki M., Sadowski W. (1997) Physica C 282–287:1961
14. Oussena M., de Groot P.A.J., Porter S.J., Gagnon R., Taillefer L. (1995) Phys Rev B 51:1389
15. Wiesinger H.P., Sauerzopf F.M., Weber H.W. (1992) Physica C 203:121
16. Darhmaoui H., Jung J. (1996) Phys Rev B 53:14 621
17. Bean C.P. (1962) Phys Rev Lett 8:250; Bean C.P. (1964) Rev Mod Phys 36:31
18. Balatskii A.V., Burlachkov L.I., Gor'kov L.P. (1986) Sov Phys.–JETP 63:866
19. Osborn J.A. (1945) Phys Rev 67:351
20. Brandt E.H. (1993) Phys Rev B 48:6699
21. Janossy B., Prost D., Pekker S., Fruchter L. (1991) Physica C 181:51
22. Krusin-Elbaum L., Greene R.L., Holtzberg F., Malozemoff A.P., Yeshurun Y. (1989) Phys Rev Lett 62:217

Basic Superconducting–State Parameters and Their Influence on the Flux Pinning in Homologous Series $HgBa_2Ca_{n-1}Cu_nO_{2n+2+x}$

R. Puźniak

Institute of Physics, Polish Academy of Sciences, Al. Lotników 32/46, PL–02-668 Warsaw, Poland

Abstract. The basic parameters describing the superconducting state, such as the penetration depth and the coherence length of the members of $HgBa_2Ca_{n-1}Cu_nO_{2n+2+x}$ (n = 1, 2, and 3) are presented. The data was analyzed in order to determine superconducting carrier concentration and effective mass. The low position of the irreversibility line in $H - T$ phase diagram of $HgBa_2Ca_{n-1}Cu_nO_{2n+2+x}$ is related to relatively small values of the thermodynamic critical field in comparison with that one of $YBa_2Cu_3O_{7-x}$. The necessary conditions to be fulfilled to improve the irreversibility line position for Hg–based compounds are discussed.

1 Introduction

The irreversibility line (IL) for Hg–based compounds is located at relatively low position in the $H - T$ phase diagram [1–9]. Furthermore, it is not improved significantly at high temperatures even in the materials with artificially introduced pinning centers [10,11]. This means that the intrinsic parameters of Hg–based compounds, which control flux pinning, are not as good as those for some other high-T_c superconductors, e.g. $YBa_2Cu_3O_{7-x}$ (Y-123). On the other hand, the highest transition temperature among the high-T_c superconductors, equal to 135 K under normal pressure, was achieved for the (optimally doped) mercury–based copper–oxide compound $HgBa_2Ca_{n-1}Cu_nO_{2n+2+x}$ with n = 3 (Hg-1223) [12]. At a working temperature of 77 K, it is possible to operate with the superconductor for T/T_c ratio as low as 0.57. The value of 0.57 still is very high in comparison with one of 0.18 for classical Nb–based superconductors with T_c of 23 K, operating at liquid helium. Nevertheless, far away from the transition temperature, the fluctuation effects are significantly diminished, and the working conditions for Hg-1223 become relatively stable. Therefore, the high T_c's of Hg–based superconductors provide a strong motivation for trying to improve the flux pinning properties of these compounds.

In the present paper, the basic parameters describing the superconducting state, such as the penetration depth and the coherence length for optimally doped $HgBa_2Ca_{n-1}Cu_nO_{2n+2+x}$ superconductors with n = 1, 2, and 3, are given. The low position of the irreversibility line (IL) in the $H - T$ phase diagram, limiting possible applications for the Hg–based superconductors, is explained by relatively small values of the thermodynamic critical field. The influence of chemical substitutions on the irreversibility line position is discussed. The necessary conditions to be fulfilled to improve the IL position for Hg–based compounds are suggested. The magnetic susceptibility data obtained for the overdoped Hg-1223 under hydrostatic pressure suggest that the copper–oxide planes for the materials with $n \geq 3$ are not doped equivalently, i.e. not all the planes in the material with the highest T_c are doped optimally. This effect may be caused by various chemical environments for different Cu–O planes in the materials with $n \geq 3$.

2 Thermodynamic Parameters

The superconducting–state thermodynamic parameters of homologous series $HgBa_2Ca_{n-1}Cu_nO_{2n+2+x}$ were derived from measurements of the reversible dc magnetization $M(H, T)$ in magnetic fields H, where $H_{c1} \ll H \ll H_{c2}$, applied both parallel ($\boldsymbol{H} \parallel c$) and perpendicular ($\boldsymbol{H} \perp c$) to the c–axis of the grain–oriented samples (perpendicular and parallel to the CuO_2 planes for the first and second field configuration, respectively). For high-T_c cuprates, there exists a broad field domain where $H_{c1} \ll H \ll H_{c2}$. Within such a range of field the reversible magnetization M is known to be linearly proportional to $\ln H$ (Refs. 13 and 14) such that:

$$M(H) = -\frac{\Phi_0}{32\pi^2 \lambda_{ab}^2} \ln \frac{\eta H_{c2\parallel c}}{eH} \tag{1}$$

for H parallel to the c–axis, and

$$M(H) = -\frac{\Phi_0}{32\pi^2 \lambda_{ab}\lambda_c} \ln \frac{\eta H_{c2\perp c}}{eH} \tag{2}$$

for $\boldsymbol{H} \perp c$. Here Φ_0 is the flux quantum, λ_{ab} – the penetration depth in the ab plane, λ_c – the penetration depth along the c–axis, H_{c2} – the upper critical field in the direction of applied field, and η – a constant of the order of unity. The dependence of M on $\ln H$ is derived directly from the London model [15], which assumes that the normal vortex cores (of radius ξ) do not overlap. In order to diminish the influence of thermal fluctuations on the $M(H)$ dependence, the magnetization measurements were performed at relatively low temperatures. For the samples of Hg-1201, Hg-1212, and Hg-1223 with T_c of 96, 127, and 135 K maximum applied temperatures were equal to 88, 110 and 110 K, respectively. In such a case, the coherence length is relatively small. The interaction between cores is negligible and the conditions of the applicability of (1) and (2) are fulfilled. Therefore, the equations were applied as the basis for the analysis of the experimental data.

The magnetic measurements of grain–aligned Hg-12($n-1$)n ($n = 1$, 2 and 3) samples were performed using a commercial 5.5 T SQUID magnetometer

(Quantum Design). The magnetization M was measured at fixed temperatures as a function of magnetic field H for both field configurations, i.e.
$H \parallel c$-axis and $H \perp c$-axis [16]. From the experimental $M(H)$ data, the
penetration depth and the upper critical field were determined at different
temperatures. For the measurements performed on polycrystalline materials,
the imperfect alignment of the grains has an influence on the $M - H$ relation. Because of the anisotropy of the material, i.e. much larger λ_c than
λ_{ab}, the uncertainty in the determination of the in–plane penetration depth
and the value of $H_{c2\parallel c}$ is much smaller than the uncertainty in the determination of λ_c and $H_{c2\perp c}$, respectively. In most cases, the experimental data
for $H \parallel c$-axis is reliable within the accuracy of the order of 10%, even for
the granular materials with the grains not perfectly aligned. In the present
paper, the analysis and discussions are concentrated on the in–plane components of the anisotropic superconducting–state parameters only, i.e. on the
in–plane components of the penetration depth, coherence length, and related
parameters.

The experimental data for $\lambda_{ab}(T)$ and $H_{c2\parallel c}(T)$ of Hg-1201 is presented in Fig. 1. The data for $\lambda(T)$ is described by empirical relation $\lambda(T) =$

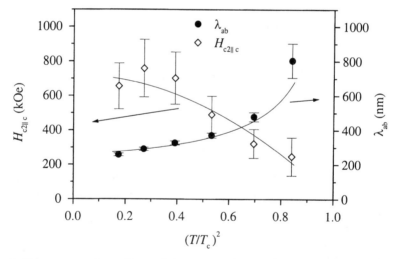

Fig. 1. Temperature dependence of the penetration depth λ_{ab} and the upper critical
field $H_{c2\parallel c}$ for Hg-1201. The experimental data for $\lambda(T)$ and $H_{c2}(T)$ is described
by empirical relations $\lambda(T) = \lambda(0)/[1 - (T/T_c)^{2.76}]^{0.6}$ and $H_{c2}(T) = H_{c2}(0)[1 - (T/T_c)^4]$, respectively

$\lambda(0)/[1 - (T/T_c)^A]^B$, where $\lambda(0)$ is a fitting parameter, and A and B are
constants equal to 2.76 and 0.6, respectively. This formula approximates well
the temperature dependence of the penetration depth, over a wide range of
temperature in the clean BCS limit. The experimental data for $H_{c2}(T)$ is de-

scribed by relation $H_{c2}(T) = H_{c2}(0)[1 - (T/T_c)^4]$. Similar results have been obtained for Hg-1212 and Hg-1223. The superconducting state parameters $\lambda_{ab}(0)$ and $H_{c2\|c}(0)$ were estimated by extrapolation of the $\lambda_{ab}(T)$ and the $H_{c2\|c}(T)$ to zero temperature. Obtained values are presented in Table 1. The ab–plane components of the coherence length ξ_{ab} were estimated applying the relation $\xi_{ab} = (\Phi_0/2\pi H_{c2\|c})^{1/2}$ for the given values of $H_{c2\|c}$. The values of $H_{c1\|c}$ were calculated using the following formula [17]:

$$H_{c2\|c} = \frac{\Phi_0}{4\pi\lambda_{ab}^2} \left(\ln\left(\frac{\lambda_{ab}}{\xi_{ab}}\right) + 0.5 \right) . \tag{3}$$

The thermodynamic critical field was estimated from the relation:

$$H_c(0) = \left(\frac{H_{c1\|c}(0)H_{c2\|c}(0)}{\ln \kappa_{\|c}} \right)^{1/2} , \tag{4}$$

where Ginzburg–Landau parameter κ for $\boldsymbol{H}$ parallel to the c–axis is defined by the relation $\kappa_{\|c} = \lambda_{ab}/\xi_{ab}$.

The relation between superconducting carrier density, n_s , effective mass, m_s^*, and measurable penetration depth, λ, is given within the clean limit approximation by formula:

$$\lambda^{-2} = \frac{4\pi n_s e^2}{m_s^* c^2} , \tag{5}$$

with m_s^* substituted by m_{ab}^* and m_c^* for the ab–plane and the c–axis components of the penetration depth, λ_{ab} and λ_c , respectively. To determine the values of n_s and m_{ab}^* it is necessary to derive at least one more equation relating measurable variables with carrier concentration and effective mass. Since the coherence length, ξ, is proportional to the velocity at the Fermi surface, v_F, and v_F is related to n and m^*, additional equation relating measurable ξ_{ab} with superconducting carrier density and effective mass at the temperature of $0\,K$ can be derived as follows [18]:

$$\xi_{ab} = \frac{a\hbar^2(3\pi^2 n_s)^{1/3}}{(m_{ab}^{*\,5}m_c^*)^{1/6}k_B T_c} \tag{6}$$

with $a = 0.18$ in BCS theory.

The values of thermodynamic parameters for Hg-12$(n-1)n$ family with $n = 1, 2, 3$ [16,18,19] presented in Table 1 are compared with those for optimally doped YBa$_2$Cu$_3$O$_{7-x}$ [20–22]. The values obtained by us are rather well correlated with the values published recently by other authors for the in–plane components of the parameters describing superconducting–state of the optimally doped Hg-12$(n-1)n$ [23–27]. The most reliable data concerning the anisotropy of Hg-12$(n-1)n$ were found in torque measurements performed on single crystals by Hofer $et\ al.$ [28] and by Rossel $et\ al.$ [29]. Their results

Table 1. Basic thermodynamic parameters characterizing superconducting state of Hg–based superconductors and Y-123. v_F is the velocity at the Fermi surface, E_F – the Fermi energy, T_F – the Fermi temperature, and m_e – the electron mass

Parameter	Hg-1201[*]	Hg-1212[*]	Hg-1223[*]	Y-123[*]
T_c (K)	96^a	127^b	135^b	92.2^d
$\lambda_{ab}(0)$ (Å)	2600^a	2050^b	1540^b	890^e
$\mu_0 H_{c2\|\|c}(0)$ (T)	72^a	113^b	88^b	122^d
$\xi_{ab}(0)$ (Å)	21.1	16.6	19.3	16.4^d
$H_{c1\|\|c}(0)$ (Oe)	129	208	339	900^f
$H_c(0)$ (Oe)	4400	7000	8300	$16\,000^g$
$\kappa_{\|\|c}(0)$	123	123	80	$74^{e,h}$
$n_s(0)$ (cm^{-3})	$7.8 \times 10^{20\ a}$	$9.8 \times 10^{20\ c}$	$1.6 \times 10^{21\ c}$	$1.6 \times 10^{22\ i}$
$m_{ab}^*(0)/m_e$	1.9^a	1.5^c	1.4^c	4.5^i
$v_{F\ ab}$ (cm/s)	$1.5 \times 10^{7\ c}$	$1.5 \times 10^{7\ c}$	$1.9 \times 10^{7\ c}$	1.1×10^7
E_F (eV)	0.11	0.10	0.14	0.14
T_F (K)	1300	1100	1600	1600
T_F/T_c	14	9	12	18

[*] All materials optimally doped; [a] Ref. 18; [b] Ref. 16; [c] Ref. 19; [d] Ref. 20; [e] Ref. 21; [f] Ref. 22; [g] Calculated using presented values of $H_{c1\|\|c}(0)$, $H_{c2\|\|c}(0)$, and $\kappa_{\|\|c}(0)$; [h] The value of $\kappa_{\|\|c}(0) = \lambda_{ab}(0)/\xi_{ab}(0)$ calculated using the values of $\lambda_{ab}(0)$ and $\xi_{ab}(0)$ from Refs. 21 and 20, respectively, is equal to 54; [i] Calculated using $\gamma = \lambda_{ab}/\lambda_c$ value from Ref. 21.

indicate that the anisotropy value $\gamma = (m_c/m_{ab})^{1/2}$ increases with increasing number of copper–oxide planes and varies from 27 for Hg-1201 to 44 for Hg-1223 [28,29].

A comparison of superconducting parameters for $HgBa_2Ca_{n-1}Cu_nO_{2n+2+x}$ with those for Y-123 yields the following conclusions: The value of the ab–plane components of the penetration depth for Hg-12$(n-1)n$ phases (n = 1, 2, 3) are generally higher than that for Y-123, whereas the ab–plane component of the coherence length for Hg–based superconductors and that for Y-123 are similar. The above relations are very well correlated with the estimated values of superconducting carrier concentration being one order of magnitude smaller for Hg–based family than for Y-123. The differences in the superconducting carrier effective mass are much smaller. The Ginzburg–Landau parameters $\kappa = \lambda/\xi$, for Hg–based superconductors are higher than that for Y-123. Finally, the isotropic thermodynamic critical fields, H_c, for Hg-12$(n-1)n$ are lower than that for Y-123.

The differences in the basic thermodynamic superconducting–state parameters between Hg-12$(n-1)n$ and Y-123 have a significant influence on the properties that determine the possible applications of the superconducting material. The lower position for the irreversibility line of Hg–based superconductors, in comparison with that of Y-123, can be related to the differences in the values of thermodynamic critical field between Hg-12$(n-1)n$ and Y-123. In terms of flux creep description, the irreversibility field is proportional to the activation energy, U, which is proportional to the product of H_{c2} and ξ [30,31]. This is significantly smaller for the Hg-12$(n-1)n$ compounds than for Y-123, because of smaller values of H_c. However, higher anisotropy for the Hg–based compounds, in comparison with that for Y-123, also leads to worse pinning properties of Hg-12$(n-1)n$ superconductors.

3   Irreversibility Line, Critical Current and the Possibility of Their Modification

At high temperatures and in low fields, the vortex structure for the magnetic field $H \parallel c$–axis can be well described by three–dimensional (3D) flux lines behavior. In higher fields and at lower temperatures, a crossover to two–dimensional (2D) "pancake" vortex structure is observed for layered materials. Using the description of Houghton et $al.$ [32], based upon a thermodynamic melting of the flux lattice (in the limit of negligible pinning), the irreversibility line in high temperature range can be described by:

$$H_{\mathrm{irr}}(T) = H_0(1 - T/T_{\mathrm{c}})^n, \tag{7}$$

where $H_0 \propto 1/(\lambda_{ab}^4(0)\gamma^2)$. Here $\gamma = (m_c^*/m_{ab}^*)^{1/2}$ is the effective mass anisotropy. From the above, it is obvious that an improvement of the irreversibility line position can be achieved by decrease of the penetration depth or by decrease of the effective mass anisotropy. A decrease of λ can be achieved by an increase in n_s and/or a decrease in m^*. The changes in the effective mass anisotropy are related to the changes in the coupling between superconducting layers of the material. Since this coupling is Josephson–like, its strength is exponentially reduced with increasing interlayer distance [33,34]. The interlayer distance generally increases from $YBa_2Cu_3O_{7-x}$ and $CuBa_2Ca_{n-1}Cu_nO_{2n+2+x}$ (Cu-12$(n-1)n$), across $La_{2-x}Sr_xCuO_4$ (La-214) and $HgBa_2Ca_{n-1}Cu_nO_{2n+2+x}$, to $Bi_2Sr_2Ca_{n-1}Cu_nO_{2n+4+x}$ (Bi–22$(n-1)n$), and so the irreversibility line position decreases. The interlayer coupling in the material can be improved by shortening the "blocking layer" distance (the distance between groups of n CuO_2 conducting planes) and by the increase of electrical conductivity in the blocking layer, which can eventually lead to proximity–induced weak superconductivity in the blocking layer [35]. The above factors controlling the position of the irreversibility line are related to the intrinsic parameters of the material. There is another, extrinsic possibility of an improvement of the irreversibility line position – the introduction of more effective pinning centers into material. Such centers are usually very effective at relatively low temperatures, where the thermal activation energy is small in comparison with the pinning energy. However, the effectiveness of the pinning centers may decrease rather fast with increasing temperature, and pinning may be ineffective close to T_c. Finally, a shift of the irreversibility line to higher fields and temperatures in the $H - T$ phase diagram may originate from the existence of a surface barrier making the movement of vortices more difficult.

Recently, significant improvement in the flux pinning of the mercury based ceramic samples with Hg partially replaced by Re or Cr and with Ba replaced by Sr was reported in the literature several times [36–42]. Shimoyama *et al.* [36–40] have argued that: (1) the substitution of Sr for Ba significantly shortens the blocking layer by about $0.8 - 0.9$ Å [39]; and (2) the chemical substitution at the Hg site may make the blocking layer more metallic [35,40]. Chmaissem *et al.* explored the possibility that extended defects in a chemically substituted $HgSr_2CuO_{4+\delta}$ compound could act additionally as pinning centers [41]. Studies of Fabrega *et al.* [42] performed on grain–aligned $Hg_{1-x}Re_xBa_2Ca_2Cu_3O_{8+x}$ samples showed that Re substitution enhances bulk pinning at high temperatures as well as at low temperatures.

In order to study the influence of different factors on the irreversibility line position we determined the $H_{irr}(T)$ in a wide temperature range for Re substituted $HgBa_2Ca_2Cu_3O_{8+x}$ single crystals with T_c of 130 K [10]. The results were compared with the data obtained for $HgBa_2Ca_2Cu_3O_{8+x}$ single crystals with the same transition temperature. The characteristic features of Re–substituted crystals, in comparison with single crystals of the parent compounds, are (1) planar defects (faults) caused by Re substitution, acting as pinning centers, (2) a small shortening of the c–axis lattice constant in comparison with that one for the material without substitution (Re substitution decreases the blocking layer distance Ba–Ba by about 0.2 Å [43]), and (3) a possibly more metallic character of the blocking layer obtained as a result of chemical substitution at the Hg site.

The temperature dependence of the irreversibility field for $(Hg,Re)Ba_2Ca_2Cu_3O_{8+x}$ (HgRe-1223) in the magnetic field applied along the c–axis of the crystal is presented in Figure 2. The values of H_{irr} plotted as a function of reduced temperature $(1 - T/T_c)$ are compared with the data obtained for almost optimally doped Hg-1223 crystal with $T_c \approx 130$ K. At high temperatures $(T > 90 K)$, the IL for HgRe-1223 is essentially the same as the IL for the parent compound. In this temperature region, presumably the pinning energy of defects caused by Re substitution is low in comparison with the thermal activation energy. The comparison of H_{irr} data obtained for Re doped HgRe-1223 and for undoped parent compound of Hg-1223 excludes any significant direct influence of Re substitution on Hg site on the position of the irreversibility line at high temperatures. Hence, no improvement of the coupling between superconducting layers was achieved by making the blocking layer more metallic as a result of Re substitution into $HgBa_2Ca_2Cu_3O_{8+x}$. Additionally, planar defects separating domains formatted by the $Re-O6$ octahedra do not improve the position of the irreversibility line of HgRe-1223 crystals at high temperatures. Only a significant shortening of the blocking layer may be effective to improve position of the H_{irr} at high temperatures in $H - T$ phase diagram.

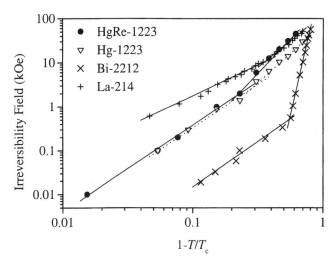

Fig. 2. The temperature dependence of the irreversibility field in the magnetic field applied along the c-axis for HgRe-1223 crystal, compared with the data obtained for Hg-1223, $Bi_2Sr_2CaCu_2O_{8+x}$ [44], and $La_{1.86}Sr_{0.14}CuO_4$ [45]

Figure 2 presents also a comparison of the irreversibility lines for HgRe-1223, $Bi_2Sr_2CaCu_2O_{8+x}$ [44], and $La_{1.86}Sr_{0.14}CuO_4$ [45]. All the ILs were determined from the magnetization measurements and correspond to the critical current density criterion $10\,A/cm^2$. The IL of HgRe-1223 is apparently located in higher magnetic fields and at higher temperatures than the IL of Bi-2212 – the compound with much higher anisotropy and greater interlayer spacing. On the other hand, the comparison of HgRe-1223 and La-214 (system with lower anisotropy than HgRe-1223 and much better intercell coupling strength) shows a difference in the position of the IL only at high temperatures. At low temperatures, the effectiveness of pinning centers introduced by Re substitution leads to similar values of H_{irr} for both compounds.

Figure 3 presents the width of hysteresis loops for HgRe-1223 determined in magnetic field of 6, 8, and 10 kOe at temperatures where the critical state inside the superconductor is well established. The width of the loop is proportional to the critical current density. The temperature dependence of the critical current density J_c is often described approximately by the formula: $J_c(T) = J_c(0)\exp(-T/T_0)$ over a certain magnetic field range [46,47]. The fitting parameter T_0 characterizes the decrease of J_c with increasing temperature. The magnitude of T_0 changes slightly over the field (see for example, behavior of $J_c(T)$ found for $CuBa_2Ca_3Cu_4O_{10+x}$ ($T_c = 117\,K$) with T_0 of about 12 K [48]). The magnitude of T_0 may be used to estimate the capability of the HTSC to be operated at high temperature, and definitely a higher value of T_0 would be preferred for application. According to the value of T_0, the materials can be classified into various groups. The $YBa_2Cu_3O_{7-x}$ and $CuBa_2Ca_3Cu_4O_{10+x}$ compounds characterized by an average $T_0 \approx 14\,K$ rank in the most steady ones, Bi-2212 system characterizes the lowest value of $T_0 \approx 4\,K$ [49]. As one can see in Fig. 3, for HgRe-1223 crystal we found $T_0 \approx 10\,K$ in the temperature range $50 - 80\,K$. This confirms that from the point of view of applications, HgRe-1223 has properties still worse than those of Y-123.

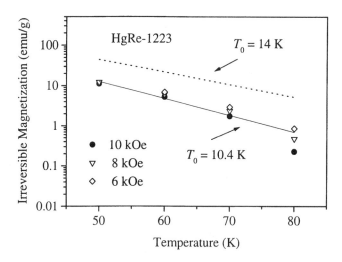

Fig. 3. The width of hysteresis loops for HgRe-1223 determined in magnetic field of 6, 8, and 10 kOe at temperatures where the critical state inside the superconductor is well established

Our studies indicate that, despite the existence of new pinning centers, the improvement of the irreversibility line position in the $H - T$ phase diagram may be achieved in the higher temperature region by shortening the blocking layer thickness in the material only. This may be achieved by substitution into the Ba sites of Sr atoms or other elements with small ionic radius.

4   Pressure Effects

For layered high-T_c superconductor, the increase in the T_c under hydrostatic pressure may indicate the charge transfer from the charge reservoir layer to the conducting layer. For mercury–based superconductors, the charge transfer may be governed by charge balance between the mixed–valence cations, namely, the mercury in the charge reservoir of Hg-12$(n-1)n$ and the copper in the conducting layers [50]. It may suggest the possibility of the achievement of higher transition temperature for the material with optimized doping.

The high–pressure synthesis and characterization of a series of Hg-1223 samples in which the doping state was successfully varied from a strongly overdoped side to the underdoped region by tuning the oxygen content only has been reported recently [51,52]. For the present magnetization studies under hydrostatic pressure, three Hg-1223 samples with clearly different doping states were selected.

Zero–field–cooled dc magnetization measurements under hydrostatic pressure up to 1 GPa have been performed with commercial vibrating sample magnetometer in magnetic field of 10 Oe. A small cylindrical sample with diameter of about 1 mm and the length up to 6 mm was placed into a bronze container where hydrostatic pressure up to 1.2 GPa was applied at room temperature (see, for example paper by Dyakonov *et al.* [53]). The sample was cooled down to 4.2 K at zero magnetic field and the $M(T)$ dependence was recorded for increasing temperature in magnetic field of 10 Oe. The increase of the T_c under hydrostatic pressure was found for all of the investigated materials, i.e. underdoped Hg-1223 with T_c of about 119 K, almost optimally doped with T_c of about 131 K, and overdoped with T_c of about 108 K. The largest shift of T_c of about 2 K at a pressure of the order of 1 GPa (10 kbar) was found for underdoped Hg-1223. The shift of T_c under hydrostatic pressure is least spectacular for the overdoped material. However, as it is shown in Figure 4, the shift is equal to about 1 K under the pressure of about 8.9 kbar (0.89 GPa). In the overdoped region for the materials with smaller number of copper–oxide layers, the decrease of T_c under hydrostatic pressure was reported [54]. Since the increase of T_c under hydrostatic pressure was observed in the overdoped range of Hg-1223, i.e. for the sample with the carrier concentration higher than that one corresponding the maximum of T_c in the series, the obtained results may indicate that the copper–oxide planes for the materials with $n \geq 3$ are not doped equivalently, i.e. not all the planes in the material with the highest T_c are doped optimally.

5   Conclusions

Mercury–based copper–oxide superconductors are interesting from a point of view of applications because of their high transition temperature values. Unfortunately, the intrinsic flux pinning properties of the material are not as good as properties of some other high-T_c superconductors. The low position of the irreversibility line in the $H - T$ phase diagram, limiting possible applications for the Hg–based superconductors, is related to relatively small values of the thermodynamic critical field. The improvement of the irreversibility line position may be achieved in the higher temperature region by shortening the blocking layer thickness in the material only. This may be achieved

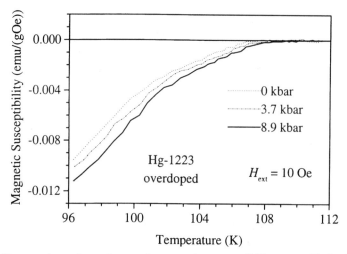

Fig. 4. Temperature dependence of magnetic susceptibility near T_c for Hg-1223 with carrier concentration above that one corresponding the maximum T_c in the series: overdoped Hg-1223, under normal pressure and the pressure of 3.7 and 8.9 kbar above it

by substitution into the Ba sites of Sr atoms or other elements with small ionic radius. The dc magnetic susceptibility data obtained for the overdoped Hg-1223 under hydrostatic pressure suggest that the copper–oxide planes for the materials with $n \geq 3$ are not doped equivalently, i.e. not all the planes in the material with the highest T_c are doped optimally. The effect may be caused by chemical differences in the vicinity of different Cu–O planes for the materials with $n \geq 3$.

Acknowledgment

This work was partially supported by the Polish Government Agency KBN under Contract No. 722P039509.

References

1. Umezawa A., Zhang W., Gurevich A., Feng Y., Hellstrom E.E., Larbalestier D.C. (1993) Nature 364:129
2. Welp U., Crabtree G.W., Wagner J.L., Hinks D.G., Radaelli P.G., Jorgensen J.D., Mitchell J.F., Dabrowski B. (1993) Appl Phys Lett 63:693
3. Gao L., Huang Z.J., Meng R.L., Lin J.G., Chen F., Beauvais L., Sun Y.Y., Xue Y.Y., Chu C.W. (1993) Physica C 213:261
4. Isawa K., Higuchi T., Machi T., Tokiwa-Yamamoto A., Adachi S., Murakami M., Yamauchi H. (1994) Appl Phys Lett 64:1301
5. Laborde O., Souletie B., Tholence J.L., Capponi J.J. (1994) Solid State Commun 90:443

6. Schilling A., Jeandupex O., Buchi S., Ott H.R., Rossel C., (1994)
   Physica C 235–240:229
7. Estrela P., Abilio C., Godinho M., Tholence J.L., Capponi J.J. (1994)
   Physica C 235–240 2731
8. Maignan A., Putilin S.N., Hardy V., Simon C., Raveau B. (1996)
   Physica C 266:173
9. Wisniewski A., Szymczak R., Baran M., Puzniak R., Karpinski J.,
   Molinski R., Schwer H., Conder K., Meijer G.I. (1996)
   Czech J Phys 46(S3):1649
10. Puzniak R., Karpinski J., Wisniewski A., Szymczak R., Angst M., Schwer H.,
    Molinski R., Kopnin E.M. (1998) Physica C 309:161
11. Wisniewski A., Puzniak R., Karpinski J., Hofer J., Szymczak R., Baran M.,
    Sauerzopf F.M., Molinski R., Kopnin E.M., Thompson J.R. (1999)
    Phys Rev B, submitted
12. Schilling A., Cantoni M., Guo J.D., Ott H.R., (1993) Nature 363:56;
    Putilin S.N., Antipov E.V., Chmaissem O., Marezio M. (1993) Nature 362:226
13. Kogan V.G., Fang M.M., Mitra Sreeparna (1988) Phys Rev B 38:11 958
14. Kogan V.G., Ledvij M., Simonov A.Yu., Cho J.H., Johnston D.C. (1993)
    Phys Rev Lett 70:1870
15. de Gennes P.G. (1966) Superconductivity of Metals and Alloys, Benjamin,
    New York
16. Puzniak R., Usami R., Isawa K., Yamauchi H. (1995) Phys Rev B 52:3756
17. Burns G. (1992) High–Temperature Superconductivity, Academic Press,
    Boston
18. Puzniak R., Usami R., Yamauchi H. (1996) Phys Rev B 53:86
19. Puzniak R. (1997) Physica C 282–287:1459
20. Welp U., Kwok W.K., Crabtree G.W., Vandervoort K.G., Liu J.Z. (1989)
    Phys Rev Lett 62:1908
21. Salamon M.B. (1989) In: Ginsberg D.M. (Ed.) Physical Properties of High
    Temperature Superconductors I, World Sci., Singapore, 39
22. Yeshurun Y., Malozemoff A.P., Holzberg F., Dinger T.R. (1988)
    Phys Rev B 38:11 828
23. Le Bras G., Fruchter L., Vulcanescu V., Viallet V., Bertinotti A., Forget A.,
    Hammann J., Marucco J.F., Colson D. (1996) Physica C 271:205
24. Thompson J.R. (1998) In: Narlikar A. (Ed.) Studies of High Temperature
    Superconductors, Nova Sci. Publ. , Commack, NY, 26–27
25. Kim Mun-Seog, Lee Sung-Ik, Yu Seong-Cho, Hur N.H. (1996)
    Phys Rev B 53:9460
26. Zech D., Hofer J., Keller H., Rossel C., Bauer P., Karpinski J., (1996)
    Phys Rev B 53:R6026
27. Vulcanescu V., Fruchter L., Bertinotti A., Colson D., Le Bras G.,
    Marucco J.F. (1996) Physica C 259:131
28. Hofer J., Karpinski J., Willemin M., Meijer G.I., Kopnin E.M., Molinski R.,
    Schwer H., Rossel C., Keller H., (1998) Physica C 297:103
29. Rossel C., Bauer P., Zech D., Hofer J., Willemin M., Keller H. (1996)
    J Appl Phys 79:816
30. Yeshurun Y., Malozemoff A.P. (1988) Phys Rev Lett 60:2202
31. Thinkham M. (1988) Phys Rev Lett 61:1658
32. Houghton A., Pelcovits R.A., Sudbø A. (1989) Phys Rev B 40:6763

33. Kim D.H., Gray K.E., Kampwirth R.T., Smith J.C., Richeson D.S., Marks T.J., Kang J.H., Talvacchio J., Eddy M. (1991) Physica B 177:431
34. Clem J.R. (1991) Phys Rev B 43:7837
35. Tallon J.L., Berhard C., Niedermayer Ch., Shimoyama J., Hahakura S., Yamaura K., Hiroi Z., Takano M., Kishio K. (1996) J Low Temp Phys 105:1379
36. Shimoyama J., Hahakura S., Kobayashi R., Kitazawa K., Kishio K. (1994) Physica C 235–240:2795
37. Shimoyama J., Kishio K., Hahakura S., Kitazawa K., Yamaura K., Hiroi Z., Takano M. (1995) In: Yamafuji K., Morishita T. (Eds.) Advances in Superconductivity VII, Springer, Tokyo, 287
38. Shimoyama J., Hahakura S., Kitazawa K., Yamafuji K., Kishio K. (1994) Physica C 224:1
39. Chmaissem O., Jorgensen J.D., Yamaura K., Hiroi Z., Takano M., Shimoyama J., Kishio K. (1996) Phys Rev B 53 14 667
40. Yamaura K., Shimoyama J., Hahakura S., Hiroi Z., Takano M., Kishio K. (1995) Physica C 246:351
41. Chmaissem O., Argyriou D.N., Hinks D.G., Jorgensen J.D., Storey B.G., Zhang H., Marks L.D., Wang Y.Y., Dravid V.P., Dabrowski B. (1995) Phys Rev B 52:15 636
42. Fabrega L., Martinez B., Fontcuberta J., Sin A., Pinol S., Obrados X. (1998) Physica C 296:29
43. Schwer H., Molinski R., Kopnin E.M., Meijer G.I., Karpinski J. (1999) J Solid State Chem, submitted
44. Ricketts J., Puzniak R., Liu C.-J., Gu G.D., Koshizuka N., Yamauchi H. (1994) Appl Phys Lett 65:3284
45. Schilling A., Jin R., Guo J.D., Ott H.R., Tanaka I., Kojima H. (1994) Physica B 194–196 1555
46. Senoussi S., OussĆna M., Collin G., Campbell I.A. (1988) Phys Rev B 37:9792
47. Christen D.K., Thompson J.R. (1993) Nature 364:98
48. Jin C.-Q., Adachi S., Wu X.-J., Yamauchi H., Tanaka S. (1995) In: Yamafuji K., Morishita T. (Eds.) Advances in Superconductivity VII, Springer, Tokyo, 249
49. Jin C.-Q., et al. (1999) to be published
50. Shil'shtein S.S. (1998) Phys Sol State 40:1793 [translated from Russian: Fiz Tverd Tela]
51. Fujinami K., Ito T., Suematsu H., Matsuura K., Karppinen M., Yamauchi H. (1997) Phys Rev B 56:14 790
52. Fujinami K., Suematsu H., Karppinen M., Yamauchi H. (1998) Physica C 307:202
53. Dyakonov V.P., Markovich V.I., Puzniak R., Szymczak H., Doroshenko N.A., Yuzhelevskii Ya.I. (1994) Physica C 225:51
54. Qiu X.D., Xiong Q., Gao L., Cao Y., Xue Y.Y., Chu C.W. (1997) Physica C 282–287:885

Oxygen Isotope Effects in the Manganates and Cuprates Studied by Electron Paramagnetic Resonance

A. Shengelaya[1], Guo–meng Zhao[1], K. Conder[2], H. Keller[1], and K.A. Müller[1]

[1] Physik–Institut der Universität Zürich, CH-8057 Zürich, Switzerland
[2] Laboratorium für Festkörperphysik, ETH Zürich, CH-8093 Zürich, Switzerland

Abstract. We present the results of the oxygen isotope effect study in colossal magnetoresistive manganate $La_{1-x}Ca_xMnO_{3+y}$ using electron paramagnetic resonance (EPR). We observed strong isotope effects on EPR intensity and linewidth which can be explained by a model where a bottlenecked spin relaxation takes place from the exchange–coupled constituent Mn^{4+} ions via the Mn^{3+} Jahn–Teller ions to the lattice. For $x = 0.2$ the ferromagnetic exchange energy J exhibits a $^{16}O/^{18}O$ oxygen isotope effect of $\sim -10\%$. The observed isotope effects suggest the presence of Jahn–Teller polarons in these materials.

We also report the results of the similar study in cuprate superconductors $La_{2-x}Sr_xCuO_4$. Experiments showed large oxygen isotope effect on EPR linewidth. It was found that isotope effect is strong in samples with small Sr doping and decreases with Sr concentration increase. These results provide the first microscopic evidence for the polaronic charge carriers in the cuprate superconductors.

1   Introduction

In most materials, magnetic and electronic phenomena at room temperature and below are essentially unaffected by lattice vibrations because the electronic and lattice subsystems are decoupled according to the Born–Oppenheimer adiabatic approximation. The atoms can usually be considered as infinitely heavy and static in theoretical descriptions of electronic phenomena. However, this approximation would break down in compounds where there is a strong Jahn–Teller (JT) effect. Höck et al. [1] studied theoretically a one–dimensional conductor with JT ions. They showed that small JT polarons can be formed when the JT stabilization energy is comparable with the bare conduction bandwidth. Polarons are not "bare" charge carriers, but carriers dressed by local lattice distortions. In other words, the electronic and lattice subsystems are no longer decoupled, so one would expect that lattice vibrations should affect electronic quantities.

To be more specific, the polaronic nature of the charge carriers can be demonstrated by the oxygen isotope effect on the effective bandwidth W_{eff} of polarons, which in turn depends on the isotope mass M [2]:

$$W_{eff} \propto W_{exp}(-\gamma E_b/\hbar\omega) , \qquad (1)$$

where W is the bare conduction bandwidth, E_b is the binding energy of polarons, ω is the characteristic frequency of the optical phonons depending on the isotope mass M ($\omega \propto M^{-1/2}$). The dimensionless parameter γ is a function of E_b/W with $0 < \gamma \leq 1$.

2   Magnetoresistive Manganates

Recently, the ferromagnetic systems $La_{1-x}Me_xMnO_{3+y}$ (where Me = Ca, Sr,Ba) have become the focus of scientific and technological interest because of the colossal magnetoresistance (CMR) effects found in these materials [3].

Doped manganese perovskites are mixed–valent systems containing Mn^{3+} and Mn^{4+} ions. The magnetic and electronic properties in these compounds have traditionally been examined with the double exchange (DE) model, which considers the transfer of an electron between neighboring Mn^{3+} and Mn^{4+} ions through the Mn–O–Mn path [4]. The electron transfer depends on the relative alignment of the electron spin and localized Mn^{4+} spin. When the two spins are aligned, the carrier avoids the strong on–site Hund exchange energy and hops easily. Thus the DE model provides an explanation for a strong coupling between the charge carriers and the localized manganese moments. However, recent theoretical considerations indicated that DE alone does not explain the CMR, and that polaronic effects due to a very strong electron–phonon coupling should be included [5,6]. The strong electron–phonon coupling is expected because the electronic ground state of the Mn^{3+} ions is degenerate and this degeneracy is removed by a spontaneous distortion of the surrounding lattice, known as the Jahn–Teller (JT) effect [7]. A recent demonstration of a giant oxygen isotope shift of $> 20\,K$ on the ferromagnetic transition temperature T_C by Zhao et al. [8] provided direct experimental evidence of the strong coupling of the charge carriers to JT lattice distortions and of JT polaron formation in $La_{1-x}Ca_xMnO_{3+y}$.

We measured the temperature dependence ($T_c < T < 3T_c$) of the EPR for ceramic powder samples of $La_{1-x}Ca_xMnO_{3+y}$ with $x = 0.1, 0.2$ which were substituted by different oxygen isotopes (^{16}O and ^{18}O). The samples used here are the same as those studied by Zhao et al. [8]. The EPR measurements were performed at $9.4\,GHz$ using BRUKER ER-200D spectrometer. A strong symmetric EPR signal with a lineshape very close to Lorentzian and g–value of 2.0 was observed over the whole range of temperature investigated. Two typical EPR signals are shown in Fig. 1.

The temperature dependence of the linewidth for the $x = 0.2$ sample with different oxygen isotopes is shown in Fig. 2. With decreasing temperature the linewidth decreases, passes through the minimum at some temperature T_{min} and increases on further cooling to T_c. It is interesting that T_{min} in the ^{18}O sample is shifted to lower temperatures in comparison with the ^{16}O sample and that there are significant differences in linewidths below T_{min}. The integral intensity I of the EPR signal decreases with temperature much

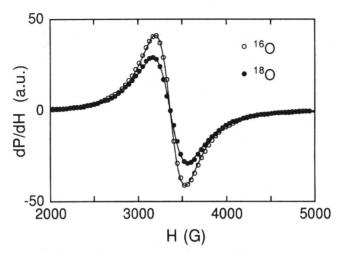

Fig. 1. EPR signal of ^{16}O and ^{18}O samples of La$_{0.8}$Ca$_{0.2}$MnO$_{3+y}$ measured at $T = 300$ K under identical experimental conditions. The fits with Lorentzian line shape are indicated by solid lines

faster than would be expected according to the Curie law. To show this, we plot in Fig. 3 the product $I \times T$ versus temperature. From Fig. 3 one can see that the intensity of the EPR signal in the ^{16}O sample is higher than in the ^{18}O sample.

In order to understand the striking differences of resonance linewidths and intensities in different oxygen isotope samples, it is necessary to clarify first the origin of the EPR signal in manganese perovskites. We propose that the EPR signal observed in La$_{1-x}$Ca$_x$MnO$_{3+y}$ is primarily due to Mn^{4+} (3d^3 with $S = 3/2$) ions. In an octahedral anion crystal electric field this ion has a ground state, corresponding to an orbital singlet A_2. Consequently the spin–lattice relaxation is weak and this makes EPR of Mn^{4+} easy to observe even at high temperatures [9]. The Mn^{3+} (3d^4 with $S = 2$) is unlikely to have an observable EPR signal as it exhibits a large zero–field splitting and strong spin–lattice relaxation (the ground state of Mn^{3+} ion is the orbital doublet) [10].

However, it is clear that the observed signal cannot be attributed to isolated Mn^{4+} ions. To construct a model of paramagnetic centers responsible for these EPR signals, it is important to point out that doped manganese perovskites are mixed valence compounds with Mn^{4+} and Mn^{3+} ions and strong ferromagnetic DE interaction between them. Thus, we should consider the EPR response of the system to contain three distinct components: Mn^{4+} ions: s, Mn^{3+} ions: σ, and the lattice: L. Figure 4 shows a standard schematic picture for such a system, with arrows indicating possible relaxation paths between components. The theory to describe such a system was developed in connection with the EPR of localized magnetic moments in

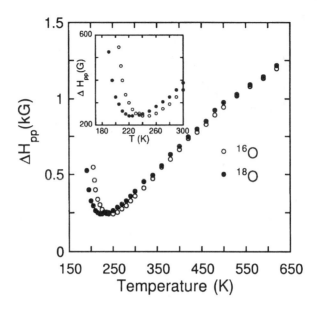

Fig. 2. Temperature dependence of the peak–to–peak EPR linewidth ΔH_{pp} for ^{16}O and ^{18}O samples of $La_{0.8}Ca_{0.2}MnO_{3+y}$. The inset shows the low temperature region on an enlarged scale

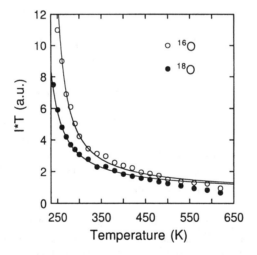

Fig. 3. Temperature dependence of the integral intensity of EPR signal times temperature $(I \times T)$ for ^{16}O and ^{18}O samples of $La_{0.8}Ca_{0.2}MnO_{3+y}$. The solid lines represent the best fit to (1) in the temperature range $250\,K \leq T \leq 500\,K$

metals [11]. Assuming that the relaxation rate $R_{\sigma L}$ of the Mn^{3+} spins to the lattice is much smaller than the exchange–induced cross relaxation rate $R_{\sigma s}$ (and back $R_{s\sigma}$) between Mn^{3+} and Mn^{4+}, and if the direct relaxation of Mn^{4+} ions to the lattice R_{sL} is negligible ($R_{\sigma L}$, $R_{sL} \ll R_{\sigma s}$, $R_{s\sigma}$), then a so–called "bottleneck" effect will take place in the transfer of energy between the spin subsystems [11]. In this limit magnetic energy which is transferred from the Mn^{4+} to the Mn^{3+} spin system is quite likely to be returned back rather than passed on to the lattice. Consequently, the relaxation of the system is dominated by the bottleneck due to the slow Mn^{3+}–lattice relaxation process. The concept of a "bottleneck" allowed us to explain many peculiar EPR features in $La_{1-x}Ca_xMnO_{3+y}$ [12].

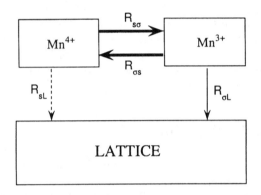

Fig. 4. Block diagram showing the energy flow paths for the Mn^{4+} and Mn^{3+} spin subsystems and the lattice. The relaxation rate R_{ab} represents relaxation from subsystem a to subsystem b. The thickness of the arrows is a measure of the magnitude of the particular relaxation rate R_{ab}

Using this model we fitted the temperature dependence of the EPR intensity. The result of the fit using only a single free parameter, namely, the exchange integral J, is shown in Fig. 3. The agreement with the experimental data is fairly good below 500 K. The fit yields $J = 78(1)$ K for ^{16}O and $J = 71(1)$ K for ^{18}O. As can be seen from Fig. 3, above 500 K the intensity of the EPR signal drops faster than is predicted by (1). This deviation can be associated with a gradual transition from the bottleneck to the isothermal regime.

In terms of the bottleneck model it is easy to explain the striking changes of the EPR linewidth and the intensity with oxygen isotope substitution if we assume that the exchange integral J is larger in ^{16}O samples than in ^{18}O samples. Indeed, according to this model, for ferromagnetic exchange ($J > 0$) a larger J corresponds to a higher EPR signal intensity [12]. It is worthy to note that a fit to the data in Fig. 3 gives a value of J for the ^{16}O sample which is about 10% larger than for the ^{18}O sample. This difference in J

should be compared with the 10% difference of T_c observed in these samples ($T_c \approx 207\,K$ and $186\,K$ for the ^{16}O and ^{18}O samples, respectively) [8]. The shift of T_{min} to lower temperatures in the ^{18}O sample (see Fig. 2) can also be easily understood. This sample has a lower T_c due to the smaller J, thus critical broadening starts at lower temperatures in comparison with the ^{16}O sample. Here we should note that qualitatively the same results were obtained for the $x = 0.1$ sample. The EPR linewidth and intensity also depend on the oxygen isotope mass, but the changes observed were slightly smaller then in the $x = 0.2$ sample, in accordance with the smaller oxygen isotope effect on T_c found for $x = 0.1$ [8].

The analysis of our results leads to the conclusion that the ferromagnetic exchange integral between Mn^{4+} and Mn^{3+} ions depends on the oxygen mass and $J(^{16}O) > J(^{18}O)$. This important fact can be understood in terms of JT polaron formation in $La_{1-x}Ca_xMnO_{3+y}$. In the DE model the exchange integral J is proportional to the effective bandwidth W_{eff} of JT polarons, which in turn depends on the isotope mass. According to (1), W_{eff}, and in turn J, decreases with enhanced oxygen isotope mass, in agreement with the present experimental observations.

3 Superconducting Cuprates

The microscopic pairing mechanism for high-T_c superconductivity (HTSC) is one of the most controversial issues in condensed matter physics. Eleven years after the discovery of the HTSC by Bednorz and Müller [13], there have been no microscopic theories that can describe the physics of HTSC completely and unambiguously. Because of high values of T_c many non–phonon mediated mechanisms have been proposed. On the other hand, there is increasing experimental evidence that a strong electron–phonon coupling is presented in cuprates [14], pointing towards an important role of phonons in the pairing mechanism.

In previous chapter we presented experimental evidence of polaronic charge carriers in the magnetoresistive manganates which share many common features with cuprate superconductors: both systems, for example, show a strong JT effect. The formation of polaronic charge carriers in manganates arises from a JT effect. This may imply that such polaronic charge carriers should also exist in cuprates, because Cu^{2+} is even stronger JT ion than Mn^{3+}. Several independent experimental results have indeed pointed towards this possibility [14–16].

To look for possible polaronic effects in cuprate superconductors, we performed EPR measurements in the $La_{2-x}Sr_xCuO_4$ (LSCO) superconductor with different oxygen isotopes. In order to observe the EPR signal, this compound was doped with a few percent of Mn ions which replace Cu ions in copper–oxygen layer and serve as an EPR probe. Recently, Kochelaev *et al.* [17] have studied EPR of Mn^{2+} in LSCO and showed that the Mn^{2+} relaxation in this compound is bottleneck dominated. In these case the block diagram shown in Fig. 4 for manganates should also be applicable to cuprates.

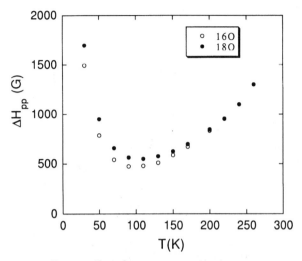

Fig. 5. EPR signal of ^{16}O and ^{18}O samples of $La_{1.94}Sr_{0.06}Cu_{0.98}Mn_{0.02}O_4$ measured at $T = 50\,K$ under identical experimental conditions. The fits with Lorentzian line shape are indicated by solid lines

We studied several samples with different concentration of Mn (1–2 at.%) and Sr (0.06–0.20). The EPR signal was observed in all examined samples. The lineshape of the signal is Lorentzian and symmetric throughout the whole temperature range. The resonance field of the spectra corresponds to $g \sim 2$, a value very close to the g–factor for the Mn^{2+} ion. Figure 5 shows typical EPR spectra for samples with different oxygen isotopes. Analysis of the spectra showed that the integral intensities of the EPR signals in two isotope samples is the same, but linewidths are different. This gives different amplitudes of signal for two isotope samples. The linewidth for ^{18}O sample is larger than for ^{16}O. The temperature dependence of the linewidth for the $x = 0.06$ sample with different oxygen isotopes is shown in Fig. 6. One can see from Fig. 6 that the isotope effect is temperature dependent. It is zero at high temperatures and increases with decreasing temperature. We studied isotope effect as a function of Sr concentration. It was found that isotope effect is very large at small Sr concentration (underdoped region) and decreases with Sr doping to zero in overdoped region. This is similar to an isotope effect on T_c in cuprate superconductors, which is also large in underdoped regime and decreases towards optimum doping [18]. At the moment there is no well established theoretical model to describe spin dynamics in doped cuprates. Therefore, it is difficult to fit the temperature dependence of the

EPR linewidth to extract some quantitative characteristics of the observed isotope effect. However, the observation of an oxygen isotope effect on EPR linewidth indicates an intimate connection between the lattice vibrations and magnetism and can be considered as a *microscopic* evidence for the polaronic charge carriers in cuprate superconductors.

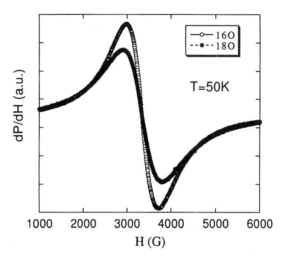

Fig. 6. Temperature dependence of the peak–to–peak EPR linewidth ΔH_{pp} for ^{16}O and ^{18}O samples of $La_{1.94}Sr_{0.06}Cu_{0.98}Mn_{0.02}O_4$

4   Conclusions

We studied the oxygen isotope effects (OIE) on EPR signal in the mixed valence perovskite $La_{1-x}Ca_xMnO_{3+y}$ for ^{16}O and ^{18}O substituted compounds. We observed large OIE on the EPR intensity and linewidth which can be explained by a model where a bottlenecked spin relaxation takes place from the exchange–coupled constituent Mn^{4+} ions *via* the Mn^{3+} Jahn–Teller ions to the lattice. For $x = 0.2$ the ferromagnetic exchange energy J exhibits a $^{16}O/^{18}O$ OIE of $\sim -10\%$. The observed OIE provide *microscopic* evidence of JT polaron formation in CMR manganates.

Experiments on Mn–doped $La_{2-x}Sr_xCuO_4$ superconductor also show the OIE on EPR signal, which suggests the presence of polaronic charge carriers in high-T_c cuprates.

References

1. Höck K.- H., Nickisch H., Thomas H. (1983) Helv phys Acta 50:237
2. Alexandrov A.S., Mott N.F. (1994) Int J Mod Phys 8:2075–2109
3. Chahara K., Ohno T., Kassai M., Kozono Y. (1993) Appl Phys Lett 63:1990
4. Zener C. (1951) Phys Rev 82:403;
   De Gennes P.G. (1960) Phys Rev 118:141
5. Millis A.J., Littlewood P.B., Shairman B.I. (1995) Phys Rev Lett 75:5144
6. Röder H., Zang Jun, Bishop A.R. (1996) Phys Rev Lett 76:1356
7. Jahn H.A., Teller E. (1937) Proc Roy Soc (Lond) A 161:220
8. Guo-meng Zhao, Conder K., Keller H., Müller K.A., (1996)
   Nature (Lond) 381:676
9. Müller K.A. (1959) Phys Rev Lett 2:341
10. Abragam A., Bleaney B. (1970) Electron Paramagnetic Resonance of
    Transition Ions, Clarendon Press, Oxford
11. Barnes S.E. (1981) Adv Phys 30:801
12. Shengelaya A., Zhao Guo-meng, Keller H., Müller K.A. (1996)
    Phys Rev Lett 77:5296
13. Bednorz J.G., Müller K.A. (1986) Z Phys B 64:189
14. Mihailovic D., Ruani G., Kaldis E., Müller K.A. (Eds) (1994) Proc. Int.
    Workshop on Anharmonic Properties of High-T_c Cuprates, World Sci.,
    Singapore
15. Zhao Guo-meng, Hunt M.B., Keller H., Müller K.A., (1997)
    Nature (Lond) 385:236
16. Mihailovic D., Mertelj T., Müller K.A. (1998) Phys Rev B 57:6116
17. Kochelaev B.I., Kan L., Elschner B., Elschner S., (1994) Phys Rev B 49:13 106
18. Franck J.P. (1994) In: Ginsberg D.M. (Ed) Physical Properties of High
    Temperature Superconductors IV, World Sci., Singapore

Quasiparticle Thermalization and Recombination in High–Temperature Superconductors Excited by Femtosecond Optical Pulses

Roman Sobolewski

Department of Electrical and Computer Engineering and Laboratory of Laser Energetics University of Rochester, Rochester, NY 14627–0231, USA
and
Institute of Physics, Polish Academy of Sciences, PL–02-668 Warsaw, Poland

Abstract. The discovery of high–temperature superconductors (HTS) and associated expectations of application of these materials in ultrafast electronics and optoelectronics has created an urgent need for a comprehensive experimental characterization of time–resolved dynamics of carriers in HTS and their response to pulsed, external optical perturbations. The above goal was accomplished via comprehensive transient photoexcitation measurements of light–induced nonequilibrium phenomena in high–quality, epitaxial $YBa_2Cu_3O_{7-x}$ (YBCO) thin–film microbridge samples. The photoresponse from < 100 fs–wide laser pulses was measured in the temperature range from 20 K to 80 K, using a subpicosecond electro–optic sampling system. The physical origin of the signal was attributed to the nonequilibrium electron heating effect, in which only electron states are perturbed by laser radiation, while the film phonons remain in thermal equilibrium. From the observed single–picosecond electrical transients, measured in the resistive state, we were able to extract, using the two–temperature model, the characteristic electron thermalization and electron–phonon relaxation time constants to be 0.56 ps and 1.1 ps, respectively. The nonequilibrium kinetic–inductive response was measured in the superconducting state, fitted into both the two–temperature and Rothwarf–Taylor models, and compared to the predictions of s– and d–wave pairing mechanism models. No phonon trapping effect (typical for low–temperature superconductors) was observed in YBCO; thus, the quasiparticle lifetime was given by the quasiparticle recombination time and estimated from the Rothwarf–Taylor equations to be well below 1 ps, and approximately 1.8 ps from the two–temperature model. From the point of view of applications, the single–picosecond intrinsic response of a YBCO superconductor demonstrates that hot–electron HTS photodetectors should exhibit intrinsic bit rates exceeding 300 Gbit/s, making them one of the fastest optoelectronic switches, well suited for digital and communication applications.

1 Introduction

Studies of nonequilibrium phenomena in superconductors have been a subject of intense investigations for the last 20 years [1,2]. Early transient optical experiments, performed on metallic superconductors using nanosecond and picosecond pulses, were aimed towards the understanding of the dynamics of the photon–induced, nonequilibrium superconducting–to–normal transition [3,4], while the development of the submillimeter (GHz) modulation spectroscopy resulted in studies of the electron gas relaxation times [5] and in the indirect observation of electron–phonon relaxation times at low temperatures [6]. The discovery of HTS materials has prompted a series of femtosecond photoexcitation experiments, with both pump and probe frequencies in the optical or near–infrared ranges [7–13]. These experiments were aimed towards the understanding of the hot–electron distribution at the Fermi surface [14–16], as well as the measurement of transient energy relaxation times. They have also demonstrated for the first time the existence of an ultrafast (subpicosecond) optical response in HTS materials [7–11]. The results from HTS pump/probe experiments remain controversial, since it is necessary to have a detailed understanding of the energy band structure in order to convert the reflectivity data into a time–dependent electron temperature, but we can state that the optical response of $YBa_2Cu_3O_{7-x}$ (YBCO) at $T > T_c$ under all levels of excitation, and at $T < T_c$ under high–level excitations, consists of a fast, laser–pulse–limited rise (nonequilibrium electron heating), followed by an ultrafast electron–phonon scattering, and a much slower bolometric background. The nonequilibrium photoresponse transient without a thermal background can be observed in YBCO only at $T < T_c$ under very low–fluency conditions [8,9]. The femtosecond response of the oxygen–poor and oxygen–depleted ($T_c < 50\,K$) YBCO has been shown to be qualitatively different from that of the oxygen–rich material [11–13]. This latter result was directly associated with the presence of the charge transfer gap [12], as well as with the pseudogap–like features in the time–resolved excitation spectrum [17].

Direct measurements of the transient photoimpedance response in HTS, using a pulsed laser source and a high–speed oscilloscope were also performed [18–30] and resulted in determination of electron–phonon inelastic scattering time near the transition temperature. They showed non–trivial, linear temperature dependence of the quasiparticle (QP) recombination time at low temperatures [19,20]. The relatively high–speed response of photoexcited YBCO near T_c [21] was initially surprising and many authors attributed nanosecond transients to nonequilibrium phenomena [22,23] while only a few claimed that all observations could be explained by equilibrium (bolometric) heating of the whole microbridge [24,25]. Fast photoresponse signals were also observed in the superconducting state far below T_c, and were interpreted assuming nonequilibrium conditions [26–27], although an equilibrium mechanism was also suggested [28]. Finally, in some experiments, a nonequilibrium signal was observe to be superimposed on a large and broad bolometric background [29,30]. A common disadvantage of direct, voltage–recording methods is a limited bandwidth of almost all–experimental arrangements used so far. A sophisticated circuit analysis has to be used in order to retrieve intrinsic relaxation times from experimental data. Nevertheless, final results are always obscured by relatively large errors, and the difficulty to properly distinguish between the nonequilibrium and bolometric responses. In the above context, femtosecond optoelectronic techniques are the most advantageous for HTS photoresponse experiments, since they exhibit up to 1 THz bandwidth (sub-picosecond time resolution) and high sensitivity [31]. Thus, they enable the direct determination of intrinsic thermalization and relaxation times.

The aim of this work is to review our very recent optoelectronic studies on time–resolved QP dynamics in current–biased very–high–quality epitaxial YBCO thin films exposed to < 100 fs–wide laser pulses (390 nm wavelength) [32–34]. The experiments were performed using our cryogenic electro–optic (EO) sampling system, which can be regarded as an ultrafast (< 200 fs temporal resolution) and ultrasensitive (< 150 µV voltage sensitivity) sampling oscilloscope. Thus, we were able to resolve the intrinsic response times involved in the photogenerated signal. We have measured photogeneration of single–picosecond electrical transients in microbridges in both the resistive (switched) and superconducting (flux–flow) states. In the resistive state, the photoresponse could be very well understand, both qualitatively and quantitatively, in terms of the nonequilibrium electron heating model. In the superconducting state, light–induced pair breaking led to a rapid change of superfluid density which in the presence of a bias current gave rise to an oscillatory transient due to nonequilibrium kinetic inductance. The kinetic–inductive response demonstrated an extremely fast QP recombination, and, contrary to LTS, no phonon trapping effect was observed in YBCO. We believe that our findings provide important information about the physics of energy relaxation processes under nonequilibrium conditions in HTS. They also demonstrate the potential of YBCO for ultrafast detection of optical transients. From the point of view of applications, our research indicates new opportunities for applying superconductors in optoelectronics [35]. Superconducting optoelectronics offers (similarly to superconducting electronics) the lowest value of the switching time–power consumption product. Optoelectronic HTS devices are ideally suited for photodetection applications because of their very high absorption coefficient in the entire wavelength range from the ultraviolet (UV) to 10 µm, relatively high sensitivity, and, first of all, ultrafast response. Simple YBCO microbridges exhibit intrinsic bit rate exceeds 300 Gbit/s, making them not only the cheapest, but also one of the fastest optoelectronic switches [33]. They are also ideal as photodetectors and optical–to–electrical transducers. They can transform to the electrical domain the input information coded in the form of a train of ultrafast optical pulses and, subsequently, feed it into the ultrafast superconducting processor, based, e.g. on a single–flux–quantum (SFQ) logic. Simultaneously, HTS operating temperature range enables the superconducting optoelectronics to be fully integrable with the conventional (cooled) semiconductor electronics.

Our paper is organized as follows. In the next Sect. we briefly review the two–temperature and Rothwarf and Taylor models, used to describe nonequilibrium conditions in superconductors. Section 3 presents our experimental procedures and techniques with the special emphasis put on our electro–optic sampling system. Section 4 briefly presents our experimental results, while Sect. 5 compares them with both the predictions of the nonequilibrium models discussed in Sect. 2, and our understanding of the dynamics of quasiparticles in HTS, based on the d–wave pairing symmetry. Finally, conclusions are presented in Sect. 6.

2   Nonequilibrium Thermalization and Recombination of Quasiparticles

Upon absorption of a light quantum by a Cooper pair or a single electron, the highly excited electron, with the energy close to the incident photon energy is created (due to large physical size of a Cooper pair, only one electron absorbs a photon, while the second one becomes a low–energy QP). Next, this excited (very hot) electron extremely rapidly (on tens of femtoseconds time scale, according to all–optical pump/probe experiments) looses its energy *via* electron–electron (e–e) scattering and creation of secondary excited electrons. In ordinary, metallic superconductors like Pb or NbN, the above process continues until approximately $0.1\,\mathrm{eV}$ (approx. Debye temperature), when the most efficient mechanism for redistribution of energy within the electron subsystem becomes emission of Debye phonons by electrons. The mean free path of those phonons is very small, and they efficiently excite additional electrons (break additional Cooper pairs). As the average energy of the electrons in the avalanche decreases to $\approx 1\,\mathrm{meV}$ ($T \approx 10\,\mathrm{K}$), their further multiplication due to absorption of phonons is replaced by multiplication due to e–e collisions, either in the QP–QP, or QP–Cooper–pair form. At that moment of the relaxation, which corresponds to thermalization time $\tau_{\mathrm{et}} \approx 7\,\mathrm{ps}$ for NbN [36], the global electron temperature T_{e}, somewhat above the sample phonon (lattice) temperature T_{ph} is established. The above scenario was first experimentally studied by Chi *et al.* [37] in Pb tunnel junctions, and most recently used by us to explain the extremely high quantum efficiency of NbN photodetectors [38].

In YBCO, we measured τ_{et} to be below $1\,\mathrm{ps}$ [33]. Whatever avalanche scenario could be implemented in HTS, the common wisdom dictates that it should be similar to the QP multiplication processes described above, with the thermalization process occurring *via* both e–e and electron–phonon (e–ph) scattering processes. However, the questions regarding the interplay between these two energy relaxation channels, their relative strength, and the associated values of energy–dependent scattering times remain open. The theory [43] predicts that the e–e interaction should be strongly modified in a d–wave superconductor, since there is an increase of the e–e scattering time, according to $1/\tau_{\mathrm{e-e}} \propto (T/\Delta)(1/\tau_{\mathrm{n}})$ where τ_{n} is the e–e interaction time in the normal state. Thus, the domination of the e–e channel in the QP avalanche

would result in the increase of the multiplication time below the superconducting transition. Additional features inherent to cuprates, such as the non–
flat electronic band structure [40] and the presence of strongly nonharmonic
phonon modes well coupled to electrons [41], should also be taken into account. Temperature–dependent nonharmonic broadening of phonon modes is
expected to cause the dependence of multiplication efficiency on excitation
pulse energy, as well as should lead to its increase at low temperatures. Indeed,
a non–monotonic change of the magnitude of the inductive photoimpedance
response at low temperatures has been recently observed [41].

Once the thermal distribution of electrons with the effective, elevated as
compared to phonons, T_e is established, further cooling of the electron subsystem towards T_{ph} is due to QP relaxation and recombination processes. Relaxation occurs via the e–ph interaction, by emitting of thermal (long–wave)
phonons, and can be described by either a two–temperature (2–T) model
[43,44] or, equivalently, by the set of Rathworf–Taylor (R–T) equations [45].

In the 2–T model, we study the time evolution of the electron and phonon
subsystems, with T_e and T_{ph}, respectively, used as measures of the average
energy in each system. The balance between T_e and T_{ph} is governed by the
set of two coupled differential equations:

$$C_e \frac{dT_e}{dt} = \frac{\alpha P_{in}(t)}{V} - \frac{C_e}{\tau_{e-ph}}(T_e - T_{ph}),$$

$$C_{ph} \frac{dT_{ph}}{dt} = \frac{C_e}{\tau_{e-ph}}(T_e - T_{ph}) - \frac{C_{ph}}{\tau_{es}}(T_{ph} - T_s), \qquad (1)$$

where C_e and C_{ph} are the electron and phonon specific heats, a is the radiation
absorption coefficient, V is the volume of the bridge, and T_s is the sample
temperature. $P_{in}(t)$ is the incident optical power, modeled as a Gaussian–
shaped pulse. The equations also contain the characteristic times τ_{e-ph} for
electron–phonon relaxation, and τ_{es} for phonon escape to the substrate. In
deriving (1) we used the energy balance equation $\tau_{e-ph} = \tau_{ph-e}(C_e/C_{ph})$,
where τ_{ph-e} is phonon–electron scattering time. The time evolution of T_e
and T_{ph} resulting from numerically solving (1) for a superconductor exposed
to sub–ps optical excitation is presented in Fig. 1. We note that the large
difference between T_e and T_{ph} is observed only during the first few picoseconds (nonequilibrium electron heating) after the perturbation, and later
there is only the bolometric (thermal) response associated with $T_e = T_{ph}$.
Thus, any attempt to experimentally observe nonequilibrium effects in HTS
materials requires a correspondingly fast detection system. As we will show
later, the 2–T model has been very successfully used by us to explain both
qualitatively and quantitatively experimental, single–picosecond responses of
a YBCO microbridge, current biased in the resistive state.

In the superconducting, zero–resistance state, under the external perturbation weak enough that it is not able to transform the superconductor into
the normal state, photoinduced decrease in the Cooper pair density gives rise

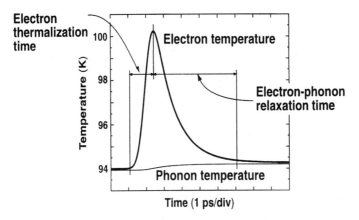

Fig. 1. Solution of the 2–T model (1) with a Gaussian–shaped excitation pulse, illustrating the nonequilibrium heating conditions when T_e exceeds T_{ph}

to a voltage transient according to kinetic inductance mechanism [26,46]

$$V_{kin} = I\frac{dL_{kin}}{dt},$$ (2)

where

$$L_{kin} = \frac{1}{\varepsilon_0 \omega_p^2}\frac{1}{f_{sc}}\frac{l}{wd},$$ (3)

and ε_0 is the vacuum permittivity, $\omega_p = 1.67 \times 10^{15}\,\text{s}^{-1}$ is the plasma frequency of YBCO, and l, w, and d are the bridge length, width, and film thickness, respectively. f_{sc} is the superfluid fraction of electrons and at $T < T_c$ can be calculated by either using the 2–T or R–T model. In the 2–T model, we need to assume that in the superconducting state, the same conditions apply to Cooper pairs as to electrons in the resistive state and the nonequilibrium electron heating can be held accountable for the f_{sc} change. Thus, we state that the superfluid fraction is given by

$$f_{sc} = 1 - \left(\frac{T_e}{T_c}\right)^2.$$ (4)

We note that T_e couples (1) and (4).

The superfluid fraction of electrons can also be independently calculated from the R–T model, noting that

$$f_{sc} = \frac{N_0 - N_{qp}}{N_0},$$ (5)

where N_0 and N_{qp} are numbers per unit volume of the total number of carriers and QPs in the superconductor, respectively. N_{qp} includes optically generated QPs and is the solution of the R–T equations:

$$\frac{dN_{qp}}{dt} = I_{qp} - \mathcal{R}N_{qp}^2 + \frac{2}{\tau_b}N_\omega ,$$

$$\frac{dN_\omega}{dt} = I_\omega + \frac{1}{2}\mathcal{R}N_{qp}^2 - \frac{1}{\tau_b}N_\omega - \frac{1}{\tau_{es}}(N_\omega - N_{\omega T}) , \qquad (6)$$

where N_ω and $N_{\omega T}$ are the numbers per unit volume of phonons with energy greater than or equal to twice the superconducting gap Δ (called 2Δ phonons) and equilibrium thermal phonons, respectively. I_{qp} is the external generation rate (optical pulse) for the QPs, I_ω represents the 2Δ phonons generation rate, $\mathcal{R}$ is the recombination rate for the QPs into Cooper pairs, and τ_b and τ_{es} are the phonon pair breaking and the phonon escape times, respectively.

The R–T equations are nonlinear and they are best suited to describe nonequilibrium conditions in a superconductor far below T_c and for moderate to strong external perturbations. The R-T model provides a different, but equivalent to 2–T description of the QP dynamics in a perturbed superconductor. In fact, one can note that (1) can be regarded as a linearized form of (6) and under low–levels of excitation both approaches coincide. The solution of (6) for a superconductor exposed to pulsed optical perturbation is shown in Fig. 2. We note again that the nonequilibrium conditions last less than 3 ps.

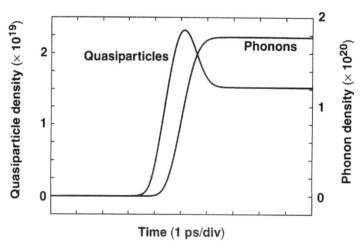

Fig. 2. Solution of the R–T model (6) with a Gaussian–shaped excitation pulse, illustrating the nonequilibrium QP generation conditions when N_{qp} exceeds N_ω

Finally, Fig. 3 illustrates the nonequilibrium kinetic–inductive response. The top panel shows f_{sc} calculated from either (4) (2–T model) or (5) (R–T model). The middle panel presents L_{kin} calculated from (3), while the bottom graph displays the resulting voltage signal as a derivative of L_{kin} multiplied by the bias current, in accordance to (3). The entire V_{kin} transient is only about 3 ps–wide, after which the superconducting, zero–voltage state is promptly restored; thus, an oscilloscope with an ~ 1 ps time resolution is required to observe kinetic–inductive transients generated by femtosecond optical pulses. Detailed comparison between predictions of the nonequilibrium kinetic inductance model and our experiments will be presented in Sect. 5.

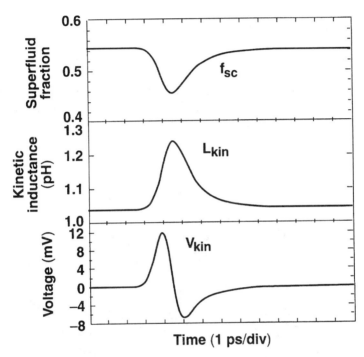

Fig. 3. Waveforms resulting from solving (4),(5) and (2),(3), illustrating the nonequilibrium kinetic inductive photoresponse mechanism

3 Experimental Procedures and Techniques

Our typical experimental structure is shown in Fig. 4 and consisted of a 5 μm–wide, 7 μm–long microbridge placed in the middle of a coplanar waveguide (CPW) center line. Test structures were patterned in high–quality, 100 nm–thick epitaxial YBCO films, grown on 0.5 mm–thick, 6×6 mm² LaAlO₃ substrates using pulsed laser deposition. Gold contact pads were deposited at both ends of CPW, using *ex situ* evaporation and lift–off. After processing, the microbridges exhibited zero–resistance critical temperature $T_{c0} \approx 90$ K, a superconducting transition width $< 1.5\,\mathrm{K}$, and a critical current density $j_c > 2 \times 10^6\,\mathrm{A/cm^2}$ at $77\,\mathrm{K}$. The CPW was $4\,\mathrm{mm}$ long, with a $30\,\mu\mathrm{m}$–wide center line and $7\,\mu\mathrm{m}$–wide gaps to the ground planes. One end of the CPW was wire–bonded directly to a semirigid, $50\,\Omega$ coaxial cable, while the other end was wire–bonded to the ground. The $4\,\mathrm{mm}$ length of the transmission line assured an $80\,\mathrm{ps}$–long reflection–free time window for our EO measurements, eliminating from the measured waveforms the artifacts caused by reflections at the CPW ends. The sample was mounted on a gold–plated alumina substrate, attached to a copper block, and placed in the He exchange gas inside a liquid–helium dewar. The dewar was temperature–stabilized to $\pm 0.2\,\mathrm{K}$ by a temperature controller and had an optical access through a pair of fused–silica windows. As shown in Fig. 4, the entire CPW structure was overlaid with an EO LiTaO$_3$ crystal to facilitate the EO measurements. The bottom face of the LiTaO$_3$ crystal had a dielectric, wavelength–selective high–reflectivity (HR) coating, which reflected the $780\,\mathrm{nm}$–wavelength sampling beam but allowed the frequency–doubled excitation beam to pass through.

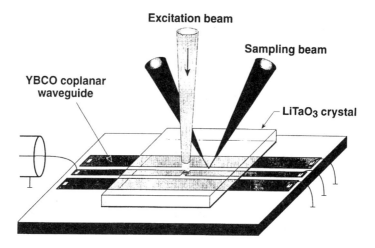

Fig. 4. Experimental sample configuration, including the excitation and probe optical beams, sample biasing, and the LiTaO$_3$ EO crystal

During the measurements, the microbridge was biased through the bias–tee using a voltage source. Current–voltage $(I - V)$ characteristics measured in a four–point configuration at different temperatures are shown in Fig. 5. One can clearly distinguish the two different voltage states – the *superconducting/flux–flow* state with zero/low voltage across the bridge and the *switched* (resistive) state where the current is almost constant while the voltage across the bridge increases rapidly. The switched state is caused by formation of a hot spot in the bridge. As the bias current through the bridge is increased, entering the flux–flow regime, the bridge becomes lossy and dissipates heat. When the generated heat exceeds the amount the surrounding cryogen can dissipate, a hot spot forms in the microbridge and its $I - V$ curve switches to a state with lower current and higher voltage. Inside the hot spot, the temperature is roughly constant, equal to T_c^2/T_b, where T_b is the bath temperature [47]. If the microbridge is long, the hot spot initially covers only part of the bridge, giving rise to a constant–current (plateau) region as the hot–spot size increases. For short microbridges, the hot spot instantly covers a very large portion (or all) of the bridge and the current plateau is quite small (see, e.g. the 80 K curve in Fig. 6). Since our microbridges are only 7 µm long and we set the bias point at the upper end of the plateau, we always assume that in the switched state, the entire bridge is in the normal state with the uniform hot–spot temperature $T_s = T_c^2/T_b$.

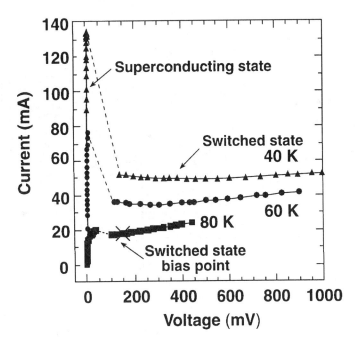

Fig. 5. $I - V$ characteristics measured in a four–point configuration for one of the microbridges used in the experiments

Femtosecond excitation and time–resolved detection experiments were performed using our cryogenic EO sampling system, shown in detail in Fig. 6. A commercial Ti:sapphire laser, pumped by an Ar–ion laser, provided ∼ 100 fs–wide optical pulses with 780 nm wavelength and 76 MHz repetition rate, at an average power of 1 W. The train of pulses from the laser was split into two paths by a 70/30 beam splitter. The first (excitation) beam (700 mW) was frequency doubled in a nonlinear β–barium borate (BBO) crystal, filtered to eliminate the remaining 780 nm light, intensity modulated by an acousto–optic modulator, and focused by a microscope objective to a 10 µm–diameter spot on the microbridge. The microscope objective was also a part of the viewing arrangement, which allowed us to observe the sample during positioning of the beams. The average optical power of the blue light, measured at a position just outside the dewar, was $\sim 2\,\mathrm{mW}$, what corresponded to a fluency of $34\,\mu\mathrm{J/cm}^2$. By calculating the amount of light absorbed in our optical beam path, we estimated that the power actually delivered to the microbridge was only 3% or 60 µW, what is the equivalent to approximately 2×10^{17} 3 eV photons per cm^3. According to our previous experiments [31], the permanent temperature increase due to laser illumination is $\sim 3\,\mathrm{K/mW}$; thus, in our case, the temperature of the light–illuminated bridge increased below $0.2\,\mathrm{K}$.

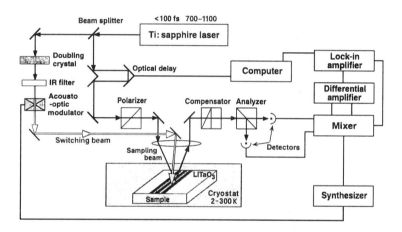

Fig. 6. Schematic of the complete cryogenic EO sampling system, including the optical beam paths and the data acquisition system

The second (sampling) beam acted as a subpicosecond sampling gate, since it sensed the birefringence induced in the LiTaO_3 crystal by the electric photoresponse transient, generated by the bridge and propagated in the CPW. The sampling beam had $\sim 2\,\mathrm{mW}$ of incident average power, traveled through a computer–controlled delay line with $< 1\,\mu\mathrm{m}$ resolution and 180 mm total travel, what corresponded to a time delay resolution of $< 6\,\mathrm{fs}$ and a maximum time window of 1200 ps, and was focused to a $\sim 10\,\mu\mathrm{m}$–diameter spot at the gap between the CPW center line and ground plane, only $\sim 20\,\mu\mathrm{m}$ away from the microbridge (see Fig. 4). The sampling beam was reflected by HR coating and directed to an analyzer/detection section. The electric field of the pulse, which was photogenerated in the bridge and propagated in the CPW underneath LiTaO$_3$, was parallel to the crystal optical axis and induced extra birefringence. In the analyzer section, the polarization change in the sampling beam due to the induced birefringence was converted into an intensity change and measured differentially. The electric–field sensitivity was maximized and linearized by choosing the polarization of the incident sampling beam to be at a 45° angle with the LiTaO$_3$ crystal axis and by adjusting the compensator that was placed in the sampling–beam path before the analyzer.

In the electronic part of the experimental setup (Fig. 6), a mixer block allowed for the use of 3 MHz frequency modulation (higher than frequencies where the $1/f$ noise of the laser dominates) and the differential measurement scheme based a lock–in amplifier. The computer recorded the time–domain evolution of the electric field at the sampling point by controlling a time delay between the excitation and sampling beams and by measuring the output of the lock–in amplifier. The time–resolved mapping of the microbridge transient photoresponse was displayed on a monitor. In the experiments presented here, averaging of up to 100 traces was used to increase the signal–to–noise ratio. By introducing a known voltage on the CPW and measuring the resulting sampling–beam intensity change, the system could be calibrated, so the computer calculated and directly displayed the measured signal magnitude in mV. From a practical point of view, our EO sampling system can be regarded as an ultrafast ($<$ 2 fs temporal resolution) and ultrasensitive ($<$ 150 µV voltage sensitivity) sampling oscilloscope. The system was augmented by a conventional oscilloscope/amplifier system with the effective bandwidth of 14 GHz, which allowed us to observe the bolometric part of the bridge response. The oscilloscope signal was also used to optimize the alignment of the excitation beam for maximum response.

4 Experimental Results

4.1 Switched State Photoresponse

Figure 7 presents a typical photoresponse transient, taken directly from the EO sampler monitor, measured when the bridge was biased in the switched state (see Fig. 5). The main figure displays the single–spike response with a low–noise baseline, indicating the absence of reflections and other artifacts caused by the measurement technique. The inset shows the same pulse at a higher time resolution. We note that it has a near–Gaussian shape and that the measured 1.3 ps–wide response is not limited by the resolution ($<$ 200 fs) of our EO sampler. The trace shown in Fig. 7 was acquired at $T_b = 50$ K, what corresponds in the hot–spot to $T_s = 158$ K. Very similar waveforms with full–width at half–maximum (FWHM) ranging from 1.1 to 1.6 ps were observed at all test points in the 20 to 80 K temperature range. This suggests that the response was not directly related to the (above T_c) hot spot temperature. The inset also shows that the main spike is followed an approximately 200 μV–level plateau on the 2 ps/div time scale) associated with the bolometric response. The bolometric signal could be observed (not shown) on a 14 GHz bandwidth oscilloscope/amplifier system and was characterized by a nanosecond fall time. The scope signal was routinely used to optimize the alignment of the excitation beam for the maximum response.

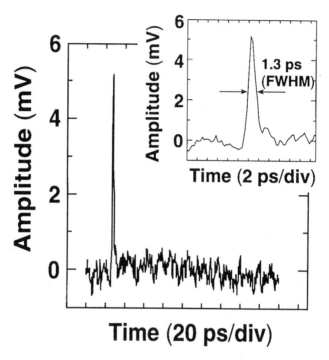

Fig. 7. Electrical single–pulse–type photoresponse transient measured with the bridge biased in the switched state at $T_b = 50\,K$. The inset shows the pulse in detail

4.2   Superconducting State Photoresponse

When the bridge was biased in the superconducting state, we recorded on our EO sampler an approximately 2 ps–wide transient of the shape shown in Fig. 8. No signal could be seen on the oscilloscope in this case. We note that the waveform is markedly different from that measured in the switched state (presented in Fig. 7) and consists of a large positive component, immediately followed by a negative part. This bipolar signal is characteristic for the kinetic inductive response [26,28,31]. For low bias currents, nearly identical transients were observed in the superconducting state in the entire tested temperature range. After the main transient, secondary oscillations could be observed in all traces collected under low–noise conditions. These trailing oscillations were not observed in earlier experiments [26–30] due to the limited bandwidth of the used detection systems, while in [31] we were afraid that they might be a result of the pulse distortion due to the sample experimental configuration. The period of these oscillations did not correspond to any sample feature size, as it would, if the reflections were the cause; thus, we believe that the oscillations are intrinsic to the YBCO photoresponse in the superconducting state and they will be discussed in more detail in Scct. 5.

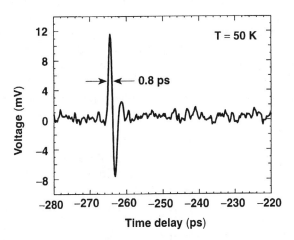

Fig. 8. Electrical oscillatory–type photoresponse transient measured with the bridge biased in the superconducting state at $T_b = 50\,\mathrm{K}$

5   Discussion of Experimental Results

5.1   Switched State Photoresponse

Figure 9 shows again a photoresponse transient (dots), measured when the bridge was biased in the switched state, but this time the experimental data is compared to simulations (solid line) based on the 2–T model (see also Fig. 1). The main figure displays the single–spike response superimposed on a bolo-metric signal, which on the picosecond scale of Fig. 9, led to a $\sim 250\,\mathrm{meV}$ ($\sim 0.3\,\mathrm{K}$) constant–level pedestal. The bath temperature was 80 K (see Fig. 5 for the bias point), leading to the hot–spot temperature $T_s = 94\,\mathrm{K}$. We note the excellent agreement between the experiment and simulations based on the numerical solution of (1). In the calculations, we used the temperat-ure dependence of the electron specific heat, $C_e = \gamma T_e$, and the values for γ and C_{ph} from [48,49]. Since the film thickness was roughly equal to the light–penetration depth and the spot size was larger than the bridge size, we assumed that the radiation is uniformly absorbed. The simulated waveform is in units of T_e, but it can be easily translated into voltage, which is proportional to the T_e, increase ΔT_e, according to

$$\Delta V = I \left(\frac{dR}{dT} \right) \Delta T_e , \tag{7}$$

where I and R are the bridge current bias and resistance, respectively, and the derivative should be evaluated at $T_s = 94\,\mathrm{K}$, where $R(T)$ is linear. As we have illustrated in Fig. 1, the rise time of the transient in Fig. 9 corresponds to τ_{ct}, while the fall time is governed by τ_{e-ph}. In our simulations, we have taken the electron thermalization time into account by assuming that the electron system responds to the incident optical pulse with a broadened Gaussian shape. The input pulse width was adjusted until a least–squares fit to the rising edge of the transient was achieved, rendering $\tau_{et} = 0.56\,\mathrm{ps}$. Similarly, a least–squares fit was applied to the falling edge of the transient, resulting in $\tau_{e-ph} = 1.1\,\mathrm{ps}$.

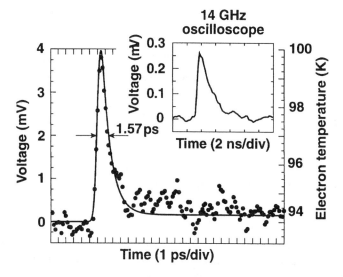

Fig. 9. Measured voltage transient (dots) and the fitted nonequilibrium electron temperature (solid line), when the bridge was biased in the resistive hot–spot state. The bias conditions are shown in Fig. 5. The inset shows the bolometric part of the photoresponse, registered with the help of a 14 GHz–bandwidth oscilloscope

The electron–heating signal was followed by a nanosecond–long bolometric component, which was directly observed on the fast oscilloscope, and shown in the inset in Fig. 9. The signal exhibited a few–ns–long fall time, consistent with the phonon escape time [24] and the $\sim 250\,\mu V$ amplitude predicted by the 2–T model. The direct comparison between the nonequilibrium (Fig. 9) and bolometric (inset in Fig. 9) compounds allowed us to determine that the bolometric signal corresponded to the 0.3 K increase of the phonon (lattice) temperature, and the ratio $C_{ph}/C_e = 38$ at T_s. The corresponding $\tau_{ph-e} = 42\,ps$.

We note that $\tau_{et} \approx 0.6\,ps$ is substantially longer than $\tau_{e-e} < 100\,fs$. As the electrons cool down, the scattering time increases, apparently, due to the increased e–ph contribution to thermalization increases. In our case, the energy of each incident 3 eV photon is sufficient for breaking an equivalent of at least 100 Cooper pairs, creating a very broadband distribution of excited electrons. Furthermore, T_e is a macroscopic quantity. The electron subsystem cannot be assigned a temperature until it has regained a Fermi–Dirac–like distribution, which, in turn, can be achieved after a very large number of scattering events with different initial electron energies [2]. One can also note that due to the very large value of C_{ph}/C_e (almost an order of magnitude larger than in conventional superconductors), we have $\tau_{ph-e} \gg \tau_{e-ph}$. Thus, secondary, thermal phonons essentially do not couple back to the electron subsystem. This latter observation is crucial in understanding the QP recombination process discussed in the next subsection.

5.2   Superconducting–State Photoresponse

Figures 10 and 11 present the photoresponse transients (dots) when the bridge was biased in the superconducting state at two different temperatures. The main, bipolar feature was characteristic for the nonequilibrium kinetic–inductive response (see also Sect. 2 and Fig. 8), with the positive part representing the process of breaking Cooper pairs and the negative part corresponding to the pair recombination. In Fig. 10, the solid lines represent the theoretical fit obtained finding f_{sc} from the 2–T model {(4)} and substituting it into (2) and (3), while in Fig. 11 the solid line is based on the R–T equations. We note that both models describe the main oscillatory feature of the transient quite well, but the fit to the negative part of the transient is much better in the case of the R–T model. At the same time, neither model can explain the post–pulse oscillations, most clearly visible in Fig. 11 (a). We also note that the main transient (without oscillations) lasts *only* about 2 ps with the thermalization part (positive signal) almost twice longer than $\tau_{et} = 0.56\,ps$ found in the measurements performed in the resistive state. The negative, recombination part is essentially as fast as the pair breaking; thus, there is clearly no phonon–trapping effect present in the photoresponse of YBCO. This observation is contrary to the standard response of a low-T_c thin film, but is in direct agreement with our earlier finding (see Sect. 5.1) that τ_{ph-e} in YBCO is very long and much longer than either τ_{e-ph} or τ_{ph-ph}, thus, phonons do not participate in the secondary Cooper–pair breaking and the experimental QP lifetime is the real QP recombination time τ_R. The above observation also agrees with the fact that in YBCO photoresponse time does not depend on the film thickness and optically thick films exhibit single–ps signals [31]. According to the 2–T model fit, $\tau_R = 1.8\,\mathrm{ps}$ and does not depend significantly on temperature. The least–squares fit to experimental transients in Fig. 11 (based on the R–T model), revealed that $\tau_B \approx 2\,\mathrm{ps}$, and $\mathcal{R} = (2.8 \pm 0.2) \times 10^{-18}\,\mathrm{ps}^{-1}\mathrm{cm}^3$ and again they do not depend strongly on temperature. Taking that $\tau_R = 1/(\mathcal{R}N_0)$ for weak perturbations, one gets extremely short, approximately $0.02\,\mathrm{ps}$ value, again much shorter than the phonon time τ_B. However, the QP–recombination process in YBCO with its lack of the phonon–trapping effect may not be properly described by the R–T model and the fit presented in Fig. 11 with the corresponding $\tau_R \ll 1\,\mathrm{ps}$ might be nonphysical.

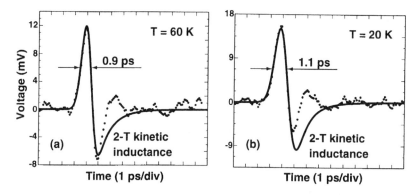

Fig. 10. Measured voltage transients (*dots*) and the fitted nonequilibrium kinetic–inductive voltages based on the 2–T model (*solid lines*), when the bridge was biased in the superconducting state; (**a**) $T = 60\,\mathrm{K}$ and (**b**) $T = 20\,\mathrm{K}$

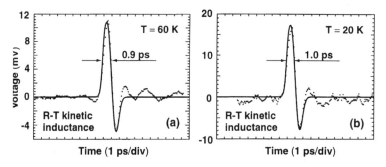

Fig. 11. Measured voltage transient (*dots*) and the fitted nonequilibrium kinetic–inductive voltages based on the R–T model (*solid lines*), when the bridge was biased in the superconducting state; (**a**) $T = 60\,\mathrm{K}$ and (**b**) $T = 20\,\mathrm{K}$

The nonexistence of the phonon–trapping effect can be attributed to the d–wave–pairing mechanism and the existence of allowed states at the gap nodes, where QPs can recombine. Sergeev and Reizer calculated in [46] the total inelastic scattering time in a pure d–wave superconductor and found only a small temperature–independent correction with respect to the scattering rate in the normal state. Consequently, their theory predicts τ_R of a few picoseconds near T_c, with the temperature dependence below T_c according to the power law $\tau_R \propto T^{-n}$, where n is an integer. On the other hand, a conventional s–wave superconductor is well known to exhibit an exponential increase of τ_R with the temperature decrease. Direct photoimpedance response and photoinduced mm–wave absorption experiments [50] were together implemented to measure the quasiparticle recombination time in NbN films. Both techniques jointly indicated that the magnitude and temperature dependencies of τ_R followed the behavior expected for s–wave superconductors. Our experimental evidence seems to support Sergeev and Reizer prediction. From the 2–T model, τ_R was estimated to be very close to τ_{e-ph} in the resistive state and no significant temperature dependence of τ_R was observed. We also measured a significant increase of τ_{et} below T_c, as it is predicted in [46]. It must be stressed, however, that our studies were restricted to the temperature range between $0.2T_c$ and $0.6T_c$, and the collected experimental data is far from being exhaustive.

Finally, the trailing oscillations are the new feature of the HTS photoresponse in the superconducting state and cannot be accounted for by the nonequilibrium kinetic inductance mechanism based solely on either 2–T or R–T models. One possible interpretation of this effect is to recognize that in our case, experimental τ_R is much shorter than τ_{e-ph} [46]. Thus, we may have concurrently existing in the bridge generation and recombination processes, leading to an oscillatory behavior superimposed on the simple exponential QP relaxation [34]. Clearly, more theoretical studies and further experiments are needed to fully resolve the origin of oscillations in the nonequilibrium quasiparticle relaxation in YBCO.

6   Conclusions

We have demonstrated that electrical transients of single–picosecond duration can be generated by a YBCO microbridge biased in either the resistive or the superconducting state. Nonequilibrium conditions are responsible for the observed photoresponse and can be explained with the help of the nonequilibrium electron excitation model. In the switched state, the microbridge is resistive and the resistance of the bridge increases due to the raised electron temperature, so a bias current produces a voltage spike. In the superconducting state, pair–breaking followed by fast QP recombination leads to a rapid change of the electron superfluid density, which in the presence of a bias current gives rise to an oscillatory transient due to nonequilibrium kinetic inductance. From our experiments we were able to extract the characteristic time constants $\tau_{et} = 0.56$ ps and $\tau_{e-ph} = 1.1$ ps in the switched state, and $\tau_{et} = 0.9 \pm 0.1$ ps in the superconducting state, which must be regarded as the intrinsic speed limitations for YBCO hot–electron photodetectors and mixers. We have also shown that while the response in the resistive state can be very well described by the 2–T model, the superconducting photoresponse in YBCO differs substantially from the behavior observed in ordinary, low-T_c superconductors. In the superconducting state, pair breaking is followed by very fast QP recombination and no phonon–trapping effect is observed. While the waveforms measured by us can be qualitatively reproduced by both the 2–T and R–T models, physical processes responsible for the lack of phonon trapping and the highly oscillatory, nonlinear transients are difficult to pinpoint at this time.

Our studies have also showed the high–speed capabilities of YBCO photodetectors, opening the door to many interesting applications in the area of superconducting optoelectronics. The measured time constants demonstrate that such photodetectors can operate in digital applications requiring data rates exceeding 300 Gbit/s, while mixers can reach an IF bandwidth greater than 100 GHz. The mechanism for the photoresponse is spectrally very broadband, and detection of all wavelengths from ultraviolet to 10 μm should be possible. HTS receivers for fiber optic communication with Gbit/s rates are an attractive possibility, especially if the operating wavelength is shifted toward 2 to 3 μm, where ultralow losses in fibers are expected. Ultrafast optical–to–electrical transducers for digital electronics using rapid single–flux–quantum (RSFQ) circuits and optical fibers for high–speed data transmission into the cryogenic environment are another area of high–speed applications of YBCO.

Acknowledgments

The author thanks A. Semenov from the Moscow State Pedagogical University, and all his friends and colleagues from the University of Rochester Superconducting Electronics and Ultrafast Optoelectronics Laboratories for their help and contribution. This research was supported by the U.S. Office of Naval Research grant N00014-98-1-0080. Additional support was provided by the Polish Government (KBN) grant 2.P03B.148.14.

References

1. Pals J.A., Weiss K., van Attekum P.M.T.M., Horstman R.E., Wolter J. (1982) Non–Equilibrium Superconductivity in Homogeneous Thin Films, Phys Repts 89:323 and references therein
2. Elesin V.F., Kopaev Yu.V. (1981) Superconductors with Excess Quasiparticles, Usp Fiz Nauk 133:259 [Sov Phys – Usp 24:116] and references therein

3. Testardi L.R. (1971) Destruction of Superconductivity by Laser Light, Phys Rev B 4:2189

4. Sobolewski R., Butler D.P., Hsiang T.Y., Stancampiano C.V., Mourou G.A. (1986) Dynamics of the Intermediate State in Nonequilibrium Superconductors, Phys Rev B 33:4604

5. Sergeev A.V., Semenov A.D., Kouminov P., Trifonov V., Goghidze I.G., Karasik B.S., Gol'tsman G.N., Gershenzon E.M. (1994) Transparency of a $YBa_2Cu_3O_7$–Film/Substrate Interface for Thermal Phonons Measured by Means of Voltage Response to Radiation, Phys Rev B 49:9091

6. Gershenzon E.M., Gershenzon M.E., Gol'tsman G.N., Semenov A.D., Sergeev A.V. (1984) Nonselective Effect of Electromagnetic Radiation on a Superconducting Film in the Resistive State, Sov Phys – JETP 59:442; Gousev Yu.P., Semenov A.D., Gol'tsman G.N., Sergeev A.V., Gershenzon E.M. (1994) Electron–Phonon Interaction in Disordered NbN Films, Physica B 194–196:1355

7. Chwalek J.M., Uher C., Whitaker J.F., Mourou G.A., Agostinelli J., Lelental M. (1990) Femtosecond Optical Absorption Studies of Nonequilibrium Electronic Processes in High T_c Superconductors, Appl Phys Lett 57:1696

8. Reitze D.H., Weiner A.M., Inam A., Etemad S. (1992) Fermi–Level Dependence of Femtosecond Response in Nonequilibrium High-T_c Superconductors, Phys Rev B 46:14 309

9. Gong T., Zheng L.X., Xiong W., Kula W., Kostoulas Y., Sobolewski R., Fauchet P.M. (1993) Femtosecond Optical Response of Y-Ba-Cu-O Thin Films: The Dependence on Optical Frequency, Excitation Intensity, and Electric Current, Phys Rev B 47:14 495

10. Shi L., Gong T., Xiong W., Weng X., Kostoulas Y., Sobolewski R., Fauchet P.M. (1994) Femtosecond Reflectivity of 60 K Y-Ba-Cu-O Thin Films, Appl Phys Lett 64:1150

11. Sobolewski R., Shi L., Gong T., Xiong W., Weng X., Kostoulas Y., Fauchet P.M. (1994) Femtosecond Optical Response of Y-Ba-Cu-O Films and Their Applications in Optoelectronics. In: Nahum M., Villegier J.- C. (Eds.) High–Temperature Superconducting Detectors: Bolometric and Nonbolometric, Proc SPIE 2159:110

12. Xiong W. (1995) Fabrication and Optoelectronic Properties of Y-Ba-Cu-O Thin Films with Different Oxygen Content, Ph D Thesis, University of Rochester

13. Stevens C.J., Smith D., Chen C., Ryan J.F., Podobnik B., Mihailovic D., Wagner G.A., Evetts J.E. (1997) Evidence for Two–Component High–Temperature Superconductivity in the Femtosecond Optical Response of $YBa_2Cu_3O_{7-x}$, Phys Rev Lett 78:2212

14. Eesley G.L. (1986) Generation of Nonequilibrium Electron and Lattice Temperatures in Copper by Picosecond Laser Pulses, Phys Rev B 33:2144

15. Elsayed-Ali H.E., Norris T.B., Pessot M.A., Mourou G.A. (1987) Time–Resolved Observation of Electron–Phonon Relaxation in Copper, Phys Rev Lett 58:1212

16. Schoenlein R.W., Lin W.Z., Fujimoto J.G., Easley G.L. (1987) Femtosecond Studies of Nonequilibrium Electronic Processes in Metals, Phys Rev Lett 58:1680

17. Kabanov V.V., Demsar J., Podobnik B., Mihailovic D. (1999) Quasiparticle Relaxation Dynamics in Superconductors with Different Gap Structures: Theory and Experiments on YBa$_2$Cu$_3$O$_{7-\delta}$, Phys Rev B 59:1497–1506

18. Semenov A.D., Nebosis R.S., Gousev Yu.P., Heusinger M.A., Renk K.F. (1995) Analysis of the Nonequilibrium Photoresponse of Superconducting Films to Pulsed Radiation by Use of a Two–Temperature Model, Phys Rev B 52:581

19. Heusinger M.A., Semenov A.D., Gousev Y.P., Kus O., Renk K.F. (1995) Temperature Dependence of the Photoresponse of Superconducting Kinetic Inductance Detectors. In: Dew-Hughes D. (Ed.) Applied Superconductivity, IOP Conf Ser No 148(v 2):749-752

20. Semenov A.D., Heusinger M.A., Hoffmann J.H., Renk K.F. (1996) Autocorrelation Study of Quasiparticle Relaxation in a YBa$_2$Cu$_3$O$_{7-x}$ Film, J Low Temp Phys 105:305

21. See, e.g.:
    Richards P.L. (1994) Bolometers for Infrared and Millimeter Waves, J Appl Phys 76:1

22. Frenkel A., Saifi M.A., Venkatesan T., Chinlon Lin, Wu X.D., Inam A. (1989) Observations of Fast Nonbolometric Optical Response of Nongranular High-T_c YBa$_2$Cu$_3$O$_{7-x}$ Superconducting Thin Films, Appl Phys Lett 54:1594

23. Kwok H.S., Zheng J.P., Ying Q.Y., Rao R. (1989) Nonthermal Optical Response of Y-Ba-Cu-O Thin Films, Appl Phys Lett 54:2473

24. Hegmann F.A., Preston J.S. (1993) Origin of the Fast Photoresponse of Epitaxial YBa$_2$Cu$_3$O$_{7-x}$ Thin Films, Phys Rev B 48:16 023

25. Carr G.L., Quijada M., Tanner D.B., Hirschmugl C.J., Williams G.P., Etemad S., Dutta B., De Rosa F., Inam A., Venkatesan T., Xi X. (1990) Fast Bolometric Response by High-T_c Detectors Measured with Subnanosecond Synchrotron Radiation, Appl Phys Lett 57:2725

26. Bluzer N. (1991) Temporal Relaxation of Nonequilibrium in Y-Ba-Cu-O Measured from Transient Photoimpedance Response, Phys Rev B 44:10 222

27. Ghis A., Villegier J.C., Pfister S., Nail M., Gibert P. (1993) Electrical Picosecond Measurements of the Photoresponse in YBa$_2$Cu$_3$O$_{7-x}$, Appl Phys Lett 63:551

28. Hegmann F.A., Hughes R.A., Preston J.S. (1994) Picosecond Photoresponse of Epitaxial YBa$_2$Cu$_3$O$_{7-x}$ Thin Films, Appl Phys Lett 64:3172

29. Donaldson W.R., Kadin A.M., Ballentine P.H., Sobolewski R. (1989) Interaction of Picosecond Optical Pulses with High T_c Superconducting Films, Appl Phys Lett 54:2470

30. Johnson M. (1991) Nonbolometric Photoresponse of YBa$_2$Cu$_3$O$_{7-x}$ Films, Appl Phys Lett 59:1371

31. Hegmann F.A., Jacobs–Perkins D., Wang C.–C., Moffat S.H., Hughes R.A., Preston J.S., Currie M., Fauchet P.M., Hsiang T.Y., Sobolewski R. (1995) Electro–Optic Sampling of 1.5 ps Photoresponse Signal from YBa$_2$Cu$_3$O$_{7-x}$ Thin Films, Appl Phys Lett 67:285

32. Lindgren M., Currie M., Williams C.A., Hsiang T.Y., Fauchet P.M., Sobolewski R., Moffat S.H., Hughes R.A., Preston J.S., Hegmann F.A. (1997) Ultrafast Photoresponse in Microbridges and Pulse Propagation in Transmission Lines Made from High-T_c Superconducting Y–Ba–Cu–O Thin Films, J Select Topics Quant Electron 2:668

33. Lindgren M., Currie M., Williams C., Hsiang T.Y., Fauchet P.M., Sobolewski R., Moffat S.H., Hughes R.A., Preston J.S., Hegmann F.A. (1999) Intrinsic Picosecond Response Times of Y-Ba-Cu-O Superconducting Photodetectors, Appl Phys Lett 74:853-855

34. Sobolewski R. (1999) Ultrafast Dynamics of Nonequilibrium Quasiparticles in High–Temperature Superconductors. In: Pavuna D., Bozovic I. (Eds.) Superconducting and Related Oxides, Physics and Nanoengineering III Proc SPIE 3481:480-491

35. Sobolewski R. (1992) Prospects for High-T_c Superconducting Optoelectronics. In: Superconductivity and Its Applications, AIP Conf Proc 251:659-670; — (1991) Application of High-T_c Superconductors in Optoelectronics. In: Singer S. (Ed.) Infrared and Optoelectronic Materials and Devices, Proc SPIE 1501:14-17

36. Il'in K.S., Lindgren M., Currie M., Sobolewski R., Cherednichenko S.I., Gol'tsman G.N., Gershenzon E.M. (1999) Picosecond Response and Gigahertz Bandwidth Characterization of Superconducting NbN Hot–Electron Photodetectors, Appl Phys Lett: to be published

37. Chi C.C., Loy M.M.T., Cronemeyer D.C. (1981) Transient Responses of Superconducting Lead Films Measured with Picosecond Laser Pulses, Phys Rev B 23:124

38. Il'in K.S., Milostnaya I.I., Verevkin A.A., Gol'tsman G.N., Gershenzon E.M., Sobolewski R. (1998) Ultimate Quantum Efficiency of a Superconducting Hot-Electron Photodetector, Appl Phys Lett 73:3938

39. Quinlan S.M., Scalapino D.J., Bulut N. (1994) Superconducting Quasiparticle Lifetimes Due to Spin–Fluctuation Scattering, Phys Rev B 49:1470

40. Ambrosch-Draxl C., Abt R. (1995) Anharmonic Properties of High-T_c Cuprates. In: Mihailovic D., Ruani G., Kaldis E., Müeller K.A. (Ed.) World Sci., Singapore,

41. Friedl B., Thomsen C., Cardona M. (1990) Determination of the Superconducting Gap in $RBa_2Cu_3O_{7-x}$, Phys Rev Lett 65:915

42. Semenov A.D. (1999) Physica C to be published

43. Anisimov S.I., Kapeliovich B.L., Perelman T.L. (1974) Emission of Electrons from the Surface of Metals Induced by Ultrashort Laser Pulses, Zh Éksp Teor Fiz 66:776 [— Sov Phys – JETP 39:375]

44. Gusev V.E., Wright O.B. (1998) Ultrafast Nonequilibrium Dynamics of Electrons in Metals, Phys Rev B 57:2878

45. Rothwarf A., Taylor B.N. (1967) Measurement of Recombination Lifetimes in Superconductors, Phys Rev Lett 19:27

46. Sergeev A.V., Reizer M.Yu. (1996) Photoresponse Mechanisms of Thin Superconducting Films and Superconducting Detectors, Int J Mod Phys 10:635 and references therein

47. Poulin G.D., Lachapelle J., Moffat S.H., Hegmann F.A., Preston J.S., (1995) Current–Voltage Characteristics of dc Voltage Biased High Temperature Superconducting Microbridges, Appl Phys Lett 66:2576

48. Phillips N.E., Fisher R.A., Gordon J.E. (1992) The Specific Heat of High-T_c Superconductors. In: Brewer D.F. (Ed.) Progress in Low Temperature Physics vol 13, Ch 5, Elsevier, Amsterdam, 267

49. Loram J.W., Mirza K.A., Cooper J.R., Liang W.Y. (1993) Electronic Specific Heat of $YBa_2Cu_3O_{6+x}$ from 1.8 to 300 K, Phys Rev Lett 71:1740

50. Semenov A.D., Gousev Y.P., Renk K.F., Voronov B.M., Gol'tsman G.N., Gershenzon E.M., Schwaab G.W., Feinaugle R. (1997) Noise Characteristics of a NbN Hot–Electron Mixer at 2.5 THz, IEEE Trans Appl Supercond 7:3083

A Semiphenomenological Approach
for Description of Quasiparticles
in High Temperature Superconductors

Z. Szotek[1], B. L. Gyorffy[2], W. M. Temmerman[1]
O. K. Andersen[3], and O. Jepsen[3]

[1] Daresbury Laboratory, Daresbury, Warrington, WA4 4AD, England, UK
[2] H H Wills Physics Laboratory, University of Bristol, Tyndall Avenue,
    Bristol, BS8 1TL, England, UK
[3] Max–Planck–Institut für Festkörperforschung, Postfach 800 665,
    D–70506 Stuttgart, Germany

Abstract. We present a semiphenomenological approach to calculating the quasiparticle spectra of High Temperature Superconductors (HTSC's). It is based on a particularly efficient parametrization of the effective electron–electron interaction afforded by the Density Functional Theory for superconductors and a Tight–Binding Linearized–Muffin–Tin–Orbital scheme for solving the corresponding Kohn–Sham–Bogoliubov–de Gennes equations. We illustrate the method by investigating a number of site and orbital specific, but otherwise phenomenological models of pairing in quantitative detail. We compare our results for the anisotropy of the gap function on the Fermi surface with those deduced from photoemission experiments on single crystals of $YBa_2Cu_3O_7$. We also compare our predictions for the low temperature dependence of the specific heat with measurements. We investigate the doping dependence of the superconducting transition temperature, T_c. We present new evidence that the Van Hove–like scenario is an essential feature of superconductivity in these materials. Since our description of pairing is semiphenomenological, we shed new light on the physical mechanism of pairing only indirectly and conclude, provisionally, that the dominant pairing interaction operates between electrons of opposite spin, on nearest neighbour Cu sites in $d_{x^2-y^2}$ orbitals.

1   Introduction

As is well known, the phenomenon of superconductivity arises when in a metal electrons pair up and occupy a single quantum state. Since electrons normally repel each other, one of the principal questions in the case of any superconductor is "why do such Cooper pairs form"? While for the conventional superconductors the answer is that the attraction is due to the electron–phonon coupling, in the case of the new, high temperature superconductors (HTSC) [1] the physical cause of the pairing remains a mystery [2]. In this paper, rather than speculating on the microscopic nature of the pairing, we develop a semiphenomenological strategy, whose aim is to determine which local orbitals the electrons occupy when they experience the attraction. This approach combines the first–principles local density approximation (LDA) electronic structure with a phenomenological representation of pairing potential. It has been formulated within the density functional theory (DFT) of superconductivity [3,4] which is briefly outlined in the next section of this paper. Of course, because our description of pairing is semiphenomenological, we shed new light on the physical mechanism of pairing only indirectly. As will be shown, this can be accomplished, in principle, by a particularly efficient representation of the electron–electron interaction afforded by the density functional theory of superconductivity and, in practice, by the development of powerful numerical methods, like linear muffin–tin orbitals (LMTO) method [5,6], for solving the corresponding Kohn–Sham–Bogoliubov–de Gennes (KS-BdG) equation [7,8].

The density functional theory for superconductors is not the usual way of doing superconductivity. In most cases one would rather apply the Eliashberg theory [9] or the conventional BCS microscopic theory [10]. Nevertheless, DFT, as originally derived by Hohenberg and Kohn [11] and Kohn and Sham [12], has proven very successful for studying systems with broken symmetry as for example in case of magnetism [13,14], and thus it may be expected that, eventually, it will be also useful in the context of the superconductors. In general, the density functional theory provides an exact mapping of a many–body electron problem occuring in solids onto a one–electron problem in an effective potential. DFT expresses this one–electron problem in terms of an electron charge density and an universal exchange–correlation functional of density, containing all information on many–body interactions in the system. A task of solving this many–body problem is thus reduced to finding sufficiently accurate expressions for the exchange–correlation functional and solving the relevant Schrödinger equation with an effective potential of which the exchange–correlation potential is a prominent part. This theory is in principle exact before any approximations for the exchange–correlation functional are made. LDA is one of such approximations, based on the local approximation for homogeneous but interacting electron gas, which has been widely and successfuly used for systems with moderate correlations.

The density functional theory for superconductors has been formulated in complete analogy to the density functional theory for spin polarised systems where the magnetisation density is the order parameter of the theory [13,14]. In the DFT for superconductors, it is the pairing amplitude, often called anomalous charge, $\chi(r, r')$, that is the order parameter. However, a central feature of the DFT for superconductors is an *electron–electron interaction kernel* $K(r_1, r_1'; r_2, r_2')$ which, when corresponding to an attractive interaction, leads to superconductivity. As shown in Sect. 3, this attractive interaction is parametrized by a set of interaction constants $K_{RL,R'L'}$, where R, R' and L, L' refer to the positions and orbital character, respectively, of the two electrons.

Given the success of the LMTO method in describing the electronic structures of the HTSC's in the normal state without adjustable parameters, and yet, in material specific quantitative detail [15,16], the above approach can be considered to be built on solid foundations. In particular, we expect that even though we make a local approximation and treat the expansion coefficients $K_{RL,R'L'}$ as the adjustable parameters of the theory, they remain physically meaningful. Namely, if a specific coefficient $K_{RL,R'L'}$ with all the others set equal to zero, can be adjusted to give a good quantitative account of the quasiparticle spectrum in the superconducting state of a particular superconductor, then we shall conclude, with appropriate caution, that the attraction operates between electrons in orbitals RL and $R'L'$.

In the application to $YBa_2Cu_3O_7$ (YBCO) we use the so–called eight–band orthonormal nearest neighbour tight–binding Hamiltonian [17], derived from the self–consistent, full LMTO band structure, to reflect the generic features of the crystal structure of the HTSC's. Details of this model and the quantitative insights to the band structure that it provides are discussed in Sect. 4, while in Sect. 5 different pairing scenarios are illustrated. Since the purpose of this work is to study the consequences of this approach in confronting experimental evidence, in Sect. 6 we present our results for the gap anisotropy, in comparison with the angular photoemission measurements, low temperature specific heat, and T_c versus doping. We summarize the paper in Sect. 7.

2   Density Functional Theory for Superconductors

In its simplest version, as derived by Oliveira, Gross and Kohn [3,4,18], the density functional theory for superconductors constitutes a relatively new formulation of the theory of superconductivity. It deals only with the case of the instantaneous electron–electron interaction and singlet pairing, but with this restriction, it is in principle exact and is fully equivalent to possible strong–coupling theories based on canonical perturbation theory [19]. However, the conventional, retarded electron–phonon mechanism of attraction can only be treated to the extent that it can be represented by a static, effective electron–electron interaction potential. On the other hand, the theory is fully applicable to pairing mechanisms due to electron–electron correlations such as, for example, spin fluctuations [20].

As already mentioned, the DFT for superconductors is formulated in a close analogy to the very successful spin–density–functional theory of magnetism [13]. The magnetisation density $m(r)$, the order parameter of the spin–density–functional theory, gets replaced by the superconducting order parameter, the *pairing amplitude*,

$$\chi(r, r') = \langle \psi_\uparrow(r), \psi_\downarrow(r') \rangle , \qquad (1)$$

with $\psi_\uparrow(r)$ and $\psi_\downarrow(r')$ being the electron annihilation operators for respectively a spin–up electron at r and a spin–down electron at r', and the theory of an equilibrium state with spontaneously broken gauge symmetry unfolds with minor modifications. Specifically, one defines the grand potential functional $\Omega[n, \chi]$ of the electron density $n(r)$ and pairing amplitude $\chi(r, r')$, which is minimized by

$$n(r) = 2 \sum_j [1 - f(E_j)]|v_j(r)|^2 + f(E_j)|u_j(r)|^2 \qquad (2)$$

and

$$\chi(r, r') = \sum_j [1 - f(E_j)]u_j(r)v_j^*(r') - f(E_j)u_j(r')v_j^*(r) , \qquad (3)$$

where $u_j(r)$ and $v_j(r)$ are respectively the electron– and hole–components of a Kohn–Sham like eigenvalue problem with the corresponding eigenvalues E_j. Here $f(E) \equiv \{1+\exp[E/k_BT]\}^{-1}$ is the Fermi function, and the normalization of the eigenfunctions is such that

$$\int |u_j(r)|^2\, d^3r + \int |v_j(r)|^2\, d^3r = 1 . \qquad (4)$$

The resulting Kohn–Sham like eigenvalue problem is of the Bogoliubov–de Gennes (BdG) form [3,4]

$$(-\tfrac{1}{2}\nabla^2 + V(r) - \mu)u_j(r) + \int \Delta(r, r')v_j(r')\, d^3r' = E_j u_j(r) ,$$

$$-\left(-\tfrac{1}{2}\nabla^2 + V(r) - \mu\right) v_j(r) + \int \Delta^*(r, r')u_j(r')\, d^3r' = E_j v_j(r) , \qquad (5)$$

and has to be solved self–consistently with respect to the chemical potential μ, the one–electron effective potential $V(r)$, and the pairing potential $\Delta(r, r')$. Like in other density functional theories, the solutions of this eigenvalue problem, i.e. the amplitudes $u_j(r)$ and $v_j(r)$, and the eigenvalues E_j, are strictly speaking only auxiliary quantities whose sole purpose is to provide a representation for $n(r)$ and $\chi(r, r')$, from which other physical observables, like the thermodynamic grand potential Ω_0, namely the minimum value of $\Omega[n, \chi]$, can be calculated. Nevertheless, experience with LDA calculations in the normal state suggests, that it may be useful to interpret provisionally the solutions of the BdG equation as descriptions of elementary excitations of the superconducting state and regard $u_j(r)$ as the amplitude for such an excitation being a quasiparticle and $v_j(r)$ as being that for a quasihole. The effective quasiparticle spectrum E_j of the BdG equation may be thought of as the normal–state single–electron spectrum ε_i, doubled up by folding around the chemical potential μ, and subsequently split by the pairing potential Δ. Therefore, the solutions come in pairs: a set of positive eigenvalues and a set of negative eigenvalues that are exactly symmetrical to each other about the chemical potential. Only the positive eigenvalues are physically meaningful but, for the sake of calculations, one can use the negative eigenvalues and the corresponding eigenvectors and interpret the results without loss of generality. Note that if the pairing potential were equal to zero, the BdG eigenvalue problem would get decoupled into two separate equations, one for particles and one for holes.

The *effective pairing potential* $\Delta(r, r')$ is defined as the functional derivative with respect to the pairing amplitude of the exchange–correlation contribution to the grand potential functional $\Omega_{xc}[n, \chi]$, i.e.

$$\Delta(r, r') = \frac{\delta \Omega_{xc}[n, \chi]}{\delta \chi(r, r')} \, , \tag{6}$$

and therefore it is electronic in nature. The *effective single–electron potential*

$$V(r) = V_{ext}(r) + \int \frac{n(r')}{|r - r'|} \, d^3 r' + \frac{\delta \Omega_{xc}[n, \chi]}{\delta n(r)} \tag{7}$$

is assumed independent of spin, and $V_{ext}(r)$ is the external potential, e.g. the Coulomb attraction from the protons.

Although exact, the above theory is useless until an explicit, approximate form for the functional $\Omega_{xc}[n, \chi]$ has been selected. To accomplish this we rewrite – without loss of generality [7] – (6) as

$$\Delta(r_1, r_1') = \iint K(r_1, r_1'; r_2, r_2') \chi(r_2, r_2') \, d^3 r_2 \, d^3 r_2' \, , \tag{8}$$

where

$$K(r_1, r_1'; r_2, r_2') \equiv \frac{\delta^2 \Omega_{xc}[n, \chi]}{\delta \chi(r_1, r_1') \delta \chi(r_2, r_2')} \, .$$

The pair–interaction kernel $K(r_1, r_1'; r_2, r_2')$, which in most general case is also a functional of $n(r)$ and $\chi(r, r')$, describes the scattering of an $(\uparrow, \downarrow)$–pair at (r_2, r_2') into an $(\uparrow, \downarrow)$–pair at (r_1, r_1'), and also the scattering of a $(\downarrow, \uparrow)$–pair at (r_2, r_2') into a $(\downarrow, \uparrow)$–pair at (r_1, r_1'). Since the microscopic pairing mechanism is not known, it is not possible to evaluate the pair–interaction kernel from first principles. However, the DFT for superconductors provides an efficient way of parameterizing it, by expanding in terms of local orbitals of the theory, and this is elaborated upon in the next section. Concerning the one electron effective potential, the usual local density approximation to the DFT [13] has been used for $\delta \Omega_{xc}[n, \chi]/(\delta n(r))$.

For the sake of comparison, in the spin density functional theory of collinear magnetism the analogous relation to (8) is:

$$V_{\uparrow}(r_1) - V_{\downarrow}(r_1) = \int I(r_1, r_2) m(r_2) \, d^3 r_2 \approx I(r_1) m(r_1) \, ,$$

which links the exchange potential to the magnetization *via* the exchange–interaction $I(r_1, r_2)$. The last approximation is the local spin–density (LSD) approximation, $I(r_1, r_2) \approx \delta(r_1 - r_2) I(r_1)$, which reduces the relation to the Stoner form.

3   Computational Details

To solve the BdG eigenvalue problem we need to employ a realistic band structure method that can be generalized to incorporate the pairing aspect of the theory. In the present study we use the tight–binding linear muffin tin orbitals method where the muffin tin orbitals provide a basis set $\{\varphi\}$ in terms of which we expand the single particle wave function

$$\begin{pmatrix} u_j(r) \\ v_j(r) \end{pmatrix} = \sum_{RL} \varphi_L(r - R) \begin{pmatrix} u_{RLj} \\ v_{RLj} \end{pmatrix} .$$

Here R labels the site ($\boldsymbol{R}$) and L the shape (e.g. atom– and angular–momentum type) of the orbital. Consequently, in the tight–binding representation the BdG eigenvalue problem acquires the following matrix form

$$\sum_{RL} \left[\begin{matrix} H_{R'L',RL} - (\mu + E_j)O_{R'L',RL} & \Delta_{R'L',RL} \\ \Delta^*_{RL,R'L'} & -H_{R'L',RL} + (\mu - E_j)O_{R'L',RL} \end{matrix} \right] .$$

$$\begin{bmatrix} u_{RL,j} \\ v_{RL,j} \end{bmatrix} = \begin{bmatrix} 0 \\ 0 \end{bmatrix} , \tag{9}$$

where

$$H_{R'L',RL} \equiv \int \varphi^*_{L'}(r - R') \left[-\frac{1}{2}\nabla^2 + V(r) \right] \varphi_L(r - R) \, \mathrm{d}^3 r$$

is an element of the (LMTO) Hamiltonian,

$$O_{R'L',RL} \equiv \int \varphi^*_{L'}(r - R')\varphi_L(r - R) \, \mathrm{d}^3 r$$

is an element of the overlap matrix, and

$$\Delta_{R'L',RL} \equiv \iint \varphi^*_{L'}(r' - R') \, \Delta(r',r) \, \varphi_L(r - R) \, \mathrm{d}^3 r' \, \mathrm{d}^3 r \tag{10}$$

is a matrix element of the pairing potential which couples to the relevant matrix elements of the pairing amplitude through the following matrix equation

$$\Delta_{R_1L_1,R'_1L'_1} = \sum_{R_2L_2} \sum_{R'_2L'_2} K_{R_1L_1,R'_1L'_1;\, R_2L_2,R'_2L'_2} \, \chi_{R_2L_2,R'_2L'_2} . \tag{11}$$

Until this stage, the expansion coefficients $K_{R_1L_1,\, R'_1L'_1;\, R_2L_2,\, R'_2L'_2}$ of the pair–interaction kernel have been unknown functionals of $n(r)$ and $\chi(r,r')$. At this point we make an assumption that these coefficients are relatively simple functions of the site and orbital indices and may therefore be approximated with relatively modest adverse consequences. As a result, first we assume that the pair interaction is *local* in the sense that it vanishes unless the spin–up electrons are in the *same* orbital, and similarly for the spin–down electrons, namely

$$K_{R_1 L_1, R_1' L_1'; R_2 L_2, R_2' L_2'} = \delta_{R_1 R_2} \, \delta_{L_1 L_2} \, \delta_{R_1' R_2'} \, \delta_{L_1' L_2'} \, K_{R_1 L_1, R_1' L_1'} \,. \tag{12}$$

Note that the quality of this local approximation and the values of the coefficients depend on the choice of local orbitals. Nevertheless, this approximation is expected to be relatively harmless for the HTSC's, on the account of the short coherence length which is of the order of the lattice parameter. By making use of (12), the relation (11) between the pairing potential and the pairing amplitude takes a simple form

$$\Delta_{RL,R'L'} = K_{RL,R'L'} \, \chi_{RL,R'L'} \,, \tag{13}$$

which states that, in the local–orbital representation, each matrix element of the pairing potential is proportional to the corresponding component of the pairing amplitude and is independent of the other components. The spin density functional analogue to (13) would be the 'Stoner relation': $V_\uparrow - V_\downarrow = Im$. Although the relation (13) has a very simple form, the expansion coefficients $K_{RL,R'L'}$ are still unknown. Therefore the next step is to abandon the hope of determining these coefficients as functionals of $n(r)$ and $\chi(r, r')$ from first principles (by specifying the mechanism of pairing), and use them instead as adjustable parameters. Clearly, this is what makes our theory phenomenological. However, since for the effective one–electron potential $V(r)$ we use the fully first–principles prescription of the LDA, we shall refer to our theory as semi–phenomenological.

There are two general reasons for regarding the above procedure promising. The first is that the effective electron–electron interaction kernel, $K(r_1, r_1'; r_2, r_2')$, like the direct correlation function in the theory of classical liquids [21], may very well turn out to be a relatively mundane function of its variables. This could mean that a small number of coefficients $K_{RL,R'L'}$ would suffice to fit a large amount of data. For instance, if we pick only one of these to be non–zero, and fix it by requiring that the transition temperature T_c agrees with experiments then, given the first–principles aspect of the theory, all other superconductive properties are predictions of the theory and can be calculated without further adjustable parameters. A number of such scenarios will be discussed in Sect. 5. The second reason for the expressed optimism is the physical meaning that can be attached to the coefficients $K_{RL,R'L'}$. Namely, if a particular one of these fits the experimental facts then we may conclude that, whatever its nature, the pairing mechanism operates between electrons occupying orbital L on site R for the one spin–direction, and orbital L' on site R' for the other spin–direction. As mentioned in the introduction, our entire strategy is based upon this possibility.

Since we are dealing with periodic solids, as a general consequence of Bloch theorem, the solutions of the BdG equation may be labeled by Bloch vector k and band index j. In this case, the BdG eigenvalue problem takes the form

$$\sum_L \begin{bmatrix} H_{L'L}^k - (\mu + E_j^k)O_{L'L}^k & \Delta_{L'L}^k \\ \Delta_{LL'}^{k*} & -H_{L'L}^k + (\mu - E_j^k)O_{L'L}^k \end{bmatrix} \begin{bmatrix} u_{L,j}^k \\ v_{L,j}^k \end{bmatrix} = \begin{bmatrix} 0 \\ 0 \end{bmatrix}, \quad (14)$$

which is very much the same as in the tight–binding representation, but the subscript RL has been replaced by L and all matrix elements have acquired a common superscript k. Explicitly, the matrix element of the pairing potential is:

$$\Delta_{L'L}^k \equiv \sum_T \Delta_{OL',TL} \, \exp[i\boldsymbol{k} \cdot \boldsymbol{T}] \qquad (15)$$

$$= \sum_T \exp[i\boldsymbol{k} \cdot \boldsymbol{T}] \int\int \varphi_{L'}^*(\boldsymbol{r}' - \boldsymbol{Q}_{L'}) \, \Delta(\boldsymbol{r}'\boldsymbol{r}) \, \varphi_L(\boldsymbol{r} - \boldsymbol{Q}_L - \boldsymbol{T}) \, d^3r' \, d^3r \,,$$

where we have used the fact that $\Delta_{T'L',TL} = \Delta_{L'L}(\boldsymbol{T} - \boldsymbol{T}')$ depends only on $\boldsymbol{T} - \boldsymbol{T}'$ because the pairing potential is invariant under lattice translations $\boldsymbol{T}$, $\Delta(\boldsymbol{r}'+\boldsymbol{T}, \boldsymbol{r}+\boldsymbol{T}) = \Delta(\boldsymbol{r}'\boldsymbol{r})$. Similarly for the Hamiltonian and overlap matrices we get: $H_{L'L}^k \equiv \sum_T H_{OL',TL} \exp[i\boldsymbol{k} \cdot \boldsymbol{T}]$ and $O_{L'L}^k \equiv \sum_T O_{OL',TL} \exp[i\boldsymbol{k} \cdot \boldsymbol{T}]$. The solutions of the above BdG eigenvalue problem, namely

$$\begin{bmatrix} u_{L,j}^k \\ v_{L,j}^k \end{bmatrix}$$

are the expansion coefficients of the wave function in a basis set of the Bloch–symmetrized orbitals

$$\varphi_L^k(\boldsymbol{r}) \equiv \sum_T \varphi_L(\boldsymbol{r} - \boldsymbol{Q}_L - \boldsymbol{T}) \, \exp[i\boldsymbol{k} \cdot \boldsymbol{T}] \,,$$

with k being the Bloch–vector, and $\boldsymbol{Q}_L$ the site of orbital φ_L in the cell at the origin. For neatness of notation we have written E_j^k for E_{kj}.

Finally, a few details are due on the computational aspects of the theory. In principle, for a chosen effective pair–interaction described by a set of $K_{RL,R'L'}$'s we should solve the BdG equation (14) self–consistently by two interlocking iterative procedures. Firstly, for a fixed number of electrons per cell, n, and fixed $H_{L'L}^k$ and $O_{L'L}^k$ we should iterate the chemical potential and the matrix elements of the pairing potential and the pairing amplitude to self–consistency. Then we should evaluate the charge density and recalculate $H_{L'L}^k$ and $O_{L'L}^k$ using the usual LMTO formulæ for the next cycle of iterations. This second loop corresponds to strong coupling effects and may very well be important in many applications. However, in the present work we simplified the calculations by using $H_{L'L}^k$ and $O_{L'L}^k$ determined by the self–consistent calculation for the *normal* state [17], which makes our semi–phenomenological approach *a weak coupling theory*. Of course, the whole self–consistency procedure has to be completed at each temperature, T, separately. Starting from

$T = 0$, the lowest temperature for which the pairing amplitude converges to zero is the transition temperature T_c.

Note, that in what follows, the local pair interaction parameters may be referred to in a few equivalent ways, namely

$$K_{RL,R'L'} \equiv K_{TL,T'L'} \equiv K_{LL'}(T' - T), \qquad (16)$$

however, the latter will usually be the most preferred option.

4   Eight–Band Model

Although the BdG equation, with the parameterized kernel K, can be solved self–consistently for Nb, using the $k-$ and band–dependent formulation of the formalism, and the full LMTO Hamiltonian [7] for evaluating $H^k_{L'L}$ and $O^k_{L'L}$, the interesting cases of the HTSC's still represent a challenge with currently available computers. Thus, to explore the potential of the formalism,

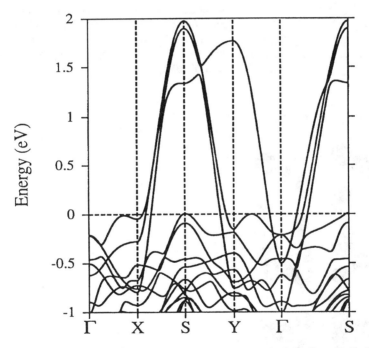

Fig. 1. The electronic structure of $YBa_2Cu_3O_7$ between the $k_\parallel$ points $(0,0)$, $(\pi/a,0)$, $(\pi/a,\pi/b)$, $(0,\pi/b)$, $(0,0)$, $(\pi/a,\pi/b)$ in the $k_z = 0$ plane of the orthorhombic Brillouin zone as calculated by the full–potential LMTO method and using the local density functional theory

for $YBa_2Cu_3O_7$ we adopted a computationally more tractable, albeit less accurate approach; we used the eight–band model [17] to represent the LDA energy bands ε_{ki} and wave functions $\psi_{ki}(r)$. In order to fully appreciate the eight–band model Hamiltonian of $YBa_2Cu_3O_7$, we start by illustrating the main features of the first–principles LDA band structure and the corresponding Fermi surface of this system. Since these features are fairly independent of the band structure method used in the calculations, here we limit ourselves to the results of the LMTO method [22]. The full LDA band structure of YBCO, along $\mathbf{\Gamma XSY\Gamma S}$ directions in the basal plane ($k_z = 0$), is presented in Fig. 1. The corresponding Fermi surface cross–sections for the basal plane are shown in Fig. 2. Since it is of importance to understand the relationship between

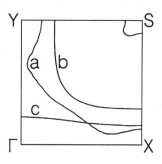

Fig. 2. Intersection of the Fermi surface of $YBa_2Cu_3O_7$ with the $k_z = 0$ plane of the orthorhombic Brillouin zone as calculated by the full–potential LMTO method using the local density functional theory. The coordinates are for $\mathbf{\Gamma}$: $(0,0)$, $\mathbf{X}$: $(\pi/a,0)$, $\mathbf{S}$: $(\pi/a,\pi/b)$, and $\mathbf{Y}$: $(0,\pi/b)$

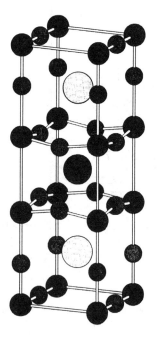

Fig. 3. The crystal structure of $YBa_2Cu_3O_7$. Y atom (*big black*) in the middle of the orthorhombic unit cell separates the two copper–oxygen planes (copper is *black* and oxygen is *dark gray*). The copper–oxygen chains run along the edges of the unit cell. Ba (*light gray*) lie between the planes and the chains

the specific features of the band structure and crystal structure of YBCO, the latter is given in Fig. 3. As can be seen, at the optimal doping, YBCO crystallizes in the orthorhombic structure whose most prominent feature is a set of CuO_2–planes which are dimpled, meaning that the oxygens occupy slightly out of plane positions. The CuO_2–planes are arranged in pairs (bilayers), with the second CuO_2–plane being the mirror image with respect to the yttrium [001]–plane of the first CuO_2–plane. The CuO_2–planes are a generic feature of high T_c materials and, as such, are of particular interest for the present study. Another important feature of YBCO is a CuO_3–chain which, however, is not a generic feature of HTSC's. All these structural features are reflected in the band structure (Fig. 1) and topology of the Fermi surface (Fig. 2). The band structure features an anti–bonding (of CuO_2–planes) $pd\sigma$–band(s), with a total width of about 10 eV, and a chain–related band, crossing the Fermi level, marked by energy zero. The strongly hybridized anti–bonding $pd\sigma$–band has maximum at the **S**–point and minimum at the Γ–point. It exhibits a set of bifurcated saddle points near **X** and **Y**. These bifurcated saddle points are the most surprising LDA prediction for HTSC's with dimpled CuO_2–planes. They give rise to a bifurcated Van Hove singularity (VHS) in the density of states, which is logarithmic. It is not clear how the bifurcated saddle points close to the Fermi level are related to the extended saddle points observed by the angle–resolved photoemission spectroscopy (ARPES) [23]. ARPES provides by far the most direct experimental data on band dispersion and Fermi surface shape in cuprates. These studies, like all LDA calculations, have found that Fermi level in these systems is very close to VHS, at the composition of optimal doping. However, LDA calculations have been unable to produce any structure in the cuprates which leads to extended saddle points. Saddle points are points at which Fermi surface first touches the Brillouin zone boundary. For an ordinary saddle point the curvature of the energy bands has the opposite sign in two orthogonal directions. The magnitude of these curvatures can take any two independent values. In an extended saddle point one of the curvatures goes as k^4, and thus in one direction the energy band is extremely flat, while in the other it disperses as k^2. This leads to a divergence in the density of states which is square–root rather than logarithmic. However, it remains a challenge for the theory to understand where the extended VHS comes from.

The Fermi surface of YBCO consists of four sheets, associated with carriers on several different layers, and there have been reports [24] that all four pieces of the Fermi surface are close to the calculated shapes and sizes (Fig. 2). The main cross–sections of the Fermi surface, marked by letters a and b in Fig. 2 originate from the CuO_2–planes, while the sheet marked by c is chain–related. The smallest element of the Fermi surface, situated around the **S**–point, the so–called stick, is due to BaO–layers [15].

Recently, Shen and Dessau [25] reviewed photoemission studies on cuprates. The experimentally derived dispersion of the CuO_2 anti–bonding band(s) for a series of cuprates [25] show a remarkable similarity to one another and good agreement with LDA calculations at optimal doping. Therefore, the motivation for deriving the eight–band model was to represent faithfully the generic features of the electronic structure of cuprates, i.e. those originating from CuO_2–planes, with a minimal set of orbitals.

As has been shown in Ref. [16], the LDA bands originating from the generic structural element of the HTSCs, the CuO_2–plane, may be accurately described within ± 1 eV around the Fermi level by the *eight–band, orthonormal, nearest neighbor, tight–binding Hamiltonian*, $\mathcal{H}^k_{L'L} = \delta_{L'L}\varepsilon_L + \sum_T t_{OL',TL} \exp[i\mathbf{k} \cdot \mathbf{T}]$ with $O^k_{L'L} = \delta_{L'L}$. This Hamiltonian has been first derived for the stoichiometrically doped, bilayered, 92 K superconductor $YBa_2Cu_3O_7$, by downfolding the LDA energy bands using the LMTO method [17]. In the eight–band model, bands originating from non–generic structural elements separating the CuO_2 layers, such as the chain and BaO–layers, have been removed. As a consequence, the eight–band model is two–dimensional. It features eight orthonormal local orbitals φ_L ($L = 1$ to 8) per CuO_2–plane (layer). The first four of these are the σ–orbitals: $O2\,p_x$, $Cu\,d_{x^2-y^2}$, $O3\,p_y$, and $Cu\,s$. The others are the π–orbitals: $O2\,p_z$, $Cu\,d_{xz}$, $O3\,p_z$, and $Cu\,d_{yz}$. The centers of the Cu orbitals form a slightly orthorhombic planar lattice, and the oxygen atoms occupy slightly out of the plane positions. Since YBCO has two CuO_2–layers (bilayer) the full unit cell contains 2×8 orbitals. With respect to the symmetry of the wave functions for a bilayer we search for even and odd solutions corresponding to even and odd bands, with respect to the plane where yttrium atoms would be situated in a real crystal structure.

The three first–mentioned σ–orbitals, $O2\,p_x$, $Cu\,d_{x^2-y^2}$, and $O3\,p_y$, are nearly degenerate ($\varepsilon_p - \varepsilon_d \sim 1$ eV; $\varepsilon_p \equiv \varepsilon_x \approx \varepsilon_y$ $\varepsilon_d \equiv \varepsilon_{x^2-y^2}$). They form the standard set which gives rise to a bonding (between copper and oxygen), a non–bonding (purely oxygen–like), and an anti–bonding $pd\sigma$–band with a total width of about 10 eV ($t_{pd} \approx 1.6$ eV). The anti–bonding band, somewhat hybridized with the remaining five orbitals, is the conduction band of the hole–doped HTSCs and, on electron scale, it is somewhat less than half full. As may be seen from the uppermost dashed– or dotted–bands in Fig. 4a, the conduction band has its minimum at $\mathbf{\Gamma}$, saddle–points near $\mathbf{X}$ and $\mathbf{Y}$, and maximum at $\mathbf{S}$, as in the case of the full LDA band structure (see Fig. 1). With merely these three orbitals, and limiting their interaction to nearest neighbors, as well as neglecting the slight orthorhombicity of the bands, the saddle–points would be at $\mathbf{X}$ and $\mathbf{Y}$, and the constant–energy contour passing through these points would be the two sets of $\mathbf{XY}$–lines crossing at right angles (chequer–board). This contour, which is perfectly nested with nesting–vector $\mathbf{q} = \mathbf{S}$, would be the Fermi surface for half filling (no hole doping); the nearest neighbor three–orbital model, in fact, has perfect electron–hole symmetry with $\mathbf{q} = \mathbf{S}$ at half filling. This type of saddle–point with $\partial^2\varepsilon/\partial k_x^2 = -\partial^2\varepsilon/\partial k_y^2$ has been called ordinary or isotropic [26].

The Cu s orbital, whose energy ε_s is about $5\,\text{eV}$ above the Fermi level, repels the conduction band ($t_{sp} \approx 2.3\,\text{eV}$ towards lower energy near $\mathbf{X}$ and $\mathbf{Y}$, but not along $\mathbf{\Gamma S}$ where there can be no hybridization between Cu s and Cu $d_{x^2-y^2}$–like symmetries [16]. The saddle–point energies are thus lowered and the constant–energy contour passing through them corresponds to finite hole–doping. The repulsion causes the angle between the two sheets of the constant–energy contour which passes through the saddle–points to be acute towards $\mathbf{\Gamma}$ and obtuse towards $\mathbf{S}$ whereby the nesting with $\mathbf{q}$ in the [11]–direction deteriorates (anisotropic saddle point). This effect is usually included in the three–band model by allowing for O–O hopping and it corresponds to folding the Cu s character into the tails of the neighboring O2 p_x and O3 p_y orbitals. Keeping the Cu s orbital explicitly in the basis set (and limiting the hopping to nearest neighbors), however, leads to further insights [16]. Since the Cu s orbital is the most diffuse, it provides most of the hopping perpendicular to the layer and, hence, is mostly responsible for the splitting of the two conduction bands in a bilayered material into even and odd bands (a bilayer has 16 orbitals). This is the reason why the splitting of the bilayer bands is strongest near $\mathbf{X}$ and $\mathbf{Y}$, and almost vanishes along $\mathbf{\Gamma S}$, as may be seen in Fig. 4. Secondly, ε_s is the band parameter which exhibits the strongest variation between HTSC materials; it is for instance high in $\text{La}_{2-x}\text{Sr}_x\text{CuO}_4$ and falls through the sequence from YBCO–, bismuth–, thallium–, Hg–, to infinite–layer compounds. This trend can easily be understood when noting that the Cu s orbital in fact includes some Cu d_{3z^2-1} and apical oxygen p_z–character. Therefore, the weaker the interaction between apical–oxygen p_z and plane–copper, the lower the energy ε_s of the effective Cu s orbital. Such

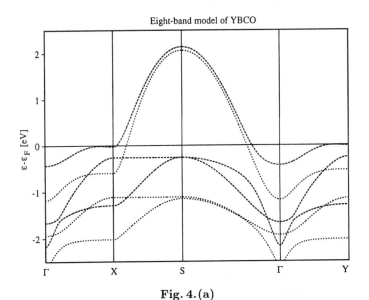

Fig. 4. (a)

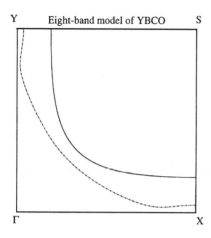

Y   Eight-band model of YBCO   S

Γ   X

Fig. 4. (a) *(previous page)* – The energy bands of the eight–band model of a CuO$_2$ bilayer of YBa$_2$Cu$_3$O$_7$ along the symmetry lines in the two–dimensional Brillouin zone. The dashed curves refer to the eight odd (anti–bonding between the layers) bands and the dotted curves to the eight even (bonding) bands. **(b)** *(left)* – The odd (chain curve) and even (full curve) sheets of the Fermi surface of a CuO$_2$ bilayer of YBa$_2$Cu$_3$O$_7$ in the irreducible part of the two–dimensional Brillouin zone. It is this part of the electronic structure which the eight–band model is designed to reproduce accurately

weak interaction can be caused by absence of apical oxygen, long distance from apical oxygen to the plane, or strong interaction of apical–oxygen p_z with another orbital, such as chain copper $d_{z^2-y^2}$;

With ε_s low, the four π–orbitals may come into play, because a low ε_s depresses the saddle–points of the conduction band to near the top of the anti–bonding O2 $p_z -$ Cu d_{xz} band at **X** and the O3 $p_z -$ Cu d_{yz} band at **Y**. If the CuO$_2$–layer is flat, the σ and π bands cannot hybridize, but if it is dimpled like in YBa$_2$Cu$_3$O$_7$ or buckled, as found theoretically for the infinite–layer compound [27], these bands repel. Since the relevant matrix element vanishes right at **X** and **Y**, this repulsion, if it exceeds a certain critical value [17], will make the saddle–point at **X** *bifurcate* away from **X** towards Γ, so that **X** is now a minimum. Similarly for the saddle–point originally at **Y**. As found from accurate LDA calculations [15] (Fig. 1 and 2), and as may be seen in Fig. 4, this indeed is the case for the odd bands in YBa$_2$Cu$_3$O$_7$. Close to a bifurcated saddle–point a fraction of the Cu $d_{x^2-y^2}$ character is transferred to oxygen p_z–character. If the $\sigma - \pi$ interaction is exactly at its critical value, e.g. if the dimpling angle is exactly at its critical value, the saddle–point will be *extended*, that is, it will have k^4–dispersion towards Γ and k^2–dispersion towards **S**. The two sheets of constant–energy contour passing through an extended saddle–point will *touch* with the common tangent passing through Γ. The often–noted fact that the (LDA) Fermi surface is [11]–oriented in La$_{2-x}$Sr$_x$CuO$_4$ but [10]–oriented in YBa$_2$Cu$_3$O$_7$ is the combined result of the ε_s–lowering and the dimpling–induced $\sigma - \pi$ coupling. In YBa$_2$Cu$_3$O$_7$ the Fermi level is only a few meV above the bifurcated saddle–point of the odd conduction band.

The *Van Hove singularity* in the density of states for a two–dimensional *isotropic or anisotropic* saddle–point is logarithmic, $N(\varepsilon) \propto -\ln|\varepsilon - \varepsilon_{\mathbf{X}}|$, and is little influenced by the degree of anisotropy because the $\mathbf{k}$–space angle where the band is flat (towards Γ) is smaller than the angle where the band is steep (towards $\mathbf{S}$). For an *extended* two–dimensional saddle–point, $N(\varepsilon) \propto |\varepsilon - \varepsilon_{\mathbf{X}}|^{-1/4}$, which is a stronger singularity. Finally, for a *bifurcated* saddle–point, the singularity is, as already mentioned, logarithmic, but the prefactor is larger than if there were no $\sigma - \pi$ coupling, that is, if the layer were flat. Moreover, since the $\mathbf{X}$–point is now a (shallow) band–minimum, $N(\varepsilon)$ jumps at $\varepsilon_{\mathbf{X}}$.

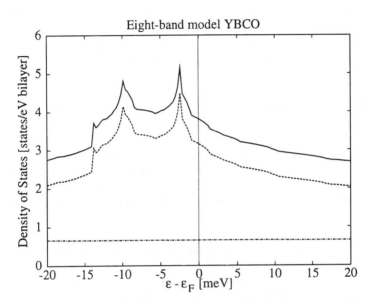

Fig. 5. The normal–state density of states in the neighborhood of the Fermi level for the eight odd (dashed curve) and eight even (chain curve) plane–bands of YBa$_2$Cu$_3$O$_7$, as well as their sum (full curve). The two logarithmic van Hove singularities are due to the saddle–points of the odd plane–band near respectively $\mathbf{X}$ and $\mathbf{Y}$

In Figure 5 the dashed curve results from a numerical calculation of the normal–state density of states, $N(\varepsilon)$, for the odd band in YBa$_2$Cu$_3$O$_7$. Merely 2 meV below the Fermi level we see the logarithmic singularity caused by the saddle–point bifurcated away from $\mathbf{X}$; the jump due to the band–minimum at $\mathbf{X}$ is seen 14 meV below. The slight orthorhombicity of the bands (for a discussion see [16]), splits the degeneracy of the saddle–points bifurcated away from respectively $\mathbf{X}$ and $\mathbf{Y}$ by 8 meV. On the fine energy scale of the figure, the density of states for the even band is constant (chain curve) because its saddle–points are more than 0.5 eV below the Fermi level. It is obvious that the value of $N(\varepsilon_{\mathrm{F}})$ depends critically on the position of ε_{F} for such a two–dimensional model. It is, however, reassuring that our value, ~ 2 states per spin, eV, and bilayer, for the total density of states (full curve) compares reasonably well with the value 2.5 states per spin, eV, and bilayer, obtained in accurate LDA calculations [15] for three–dimensional $YBa_2Cu_3O_7$.

In order to calculate quasiparticle spectra, in the superconducting state, we have generalized the eight–band model by incorporating a phenomenological electron–electron attraction, using the formal structure of density functional theory for superconductors [8,28]. So, we have replaced the full LMTO Hamiltonian in (14) by the eight–band model Hamiltonian featuring eight orbitals per layer of a bilayer material. The advantage of using the eight–band model over the one– or three–band models lies in that we are dealing with physical orbitals to which we can assign electrons. This way we can study a variety of pairing scenarios that can hopefully provide us with further insights to a possible origin of the pairing mechanism.

5   Pairing Scenarios

In this section we concentrate on a number of pairing scenarios [28] leading to various symmetries of the calculated superconducting gap. In Figure 6 we summarize the most interesting cases out of about 20 that we have studied. The convention of these schematic diagrams is that they explicitly indicate the orbitals occupied by members of a singlet electron pair when the attractive interaction, whose strength is measured by $K_{LL'}(T)$, is operative. The numerical values of $K_{LL'}(T)$ for each scenario are those which yield $T_c \cong 92 \, K$ when deployed as the free parameter in our self–consistent solution of the BdG equation for the 2×8–band model. Note that these interaction parameters are not dimensionless and are surprisingly large, $\sim 0.7 - 14 \, eV$. To make contact with the dimensionless coupling constant λ, one should rather use the Fermi surface average of $K_{LL'}(T)$ coupling coefficient and relate the average to the band width which in the present case is about $10 \, eV$. This leads to λ's which are, as they should be, less than or of the order of one for all the scenarios studied here.

In the case of conventional superconductors the effective electron–electron interaction induced by exchange of phonons is short ranged. Therefore, because the phonon Green's–function falls off rapidly in real space the dominant contribution comes from the physical situation when the two electrons are on the same atomic site, albeit at different times. In our methodology this would be described by $K_{LL'}(0)$ where L and L' refer to orbitals on the same site.

The three scenarios on the left of Fig. 6 describe such cases to which we shall refer to as 'on–site'. The scenarios on the right–hand side of the figure are the 'off–site' scenarios, or as we shall refer to them, the intralayer nearest neighbor scenarios. For the latter both orbitals belong to different but nearest neighbor Cu atoms.

That even the 'on–site' pairing can lead to a very anisotropic gap is shown in Fig. 7, where we plot the gap $\Delta_k(T = 0)$ as a function of Fermi

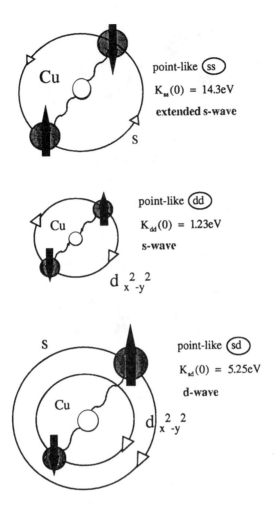

Fig. 6. The pairing interaction: $T_c = 92\,\mathrm{K}$

Fermi surface length for all scenarios given in Fig. 6. The reason for this is that the k–dependence of the pairing potential can substantially differ from the k–dependence of its expectation value in a band state. In case of the Cu on–site s – Cu s scenario the corresponding gap is highly anisotropic (Fig. 7a). It vanishes where the Fermi surface crosses the ΓS line and it rises to $2\Delta^k(0) \sim 40$ meV near the **X** and **Y** points. Since it approaches ΓS with zero slope, the gap can be interpreted to have the full point group symmetry (C_{4v}) with respect to rotations about the z–axis perpendicular to the bilayer. Such symmetry is colloquially referred to as s–type. But to differentiate it from the conventional symmetric gap states, as is customary, we shall refer to a gap displaying the above features as having extended s–wave symmetry. In

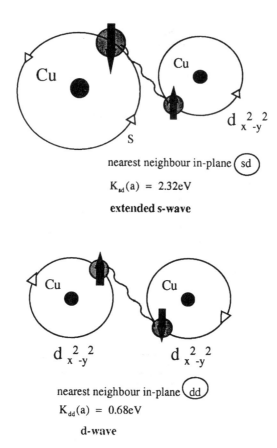

$d_{x^2-y^2}$

S

nearest neighbour in-plane (sd)

$K_{sd}(a) = 2.32eV$

extended s-wave

$d_{x^2-y^2}$ $d_{x^2-y^2}$

nearest neighbour in-plane (dd)

$K_{dd}(a) = 0.68eV$

d-wave

Fig. 6. ...continued. Schematic diagrams depicting studied scenarios of various pairing interactions, operating between $Cu\,s$ and $Cu\,d_{x^2-y^2}$ orbitals in the CuO_2 layers. Details are given in the text

the present case the interesting gap–anisotropy arises solely from the peculiar k–dependence of the $Cu\,s$ hybridization across the Fermi surface. For the BdG solution we found that the proper order parameter $\chi(r,r')$, or actually, the s–component of its lattice Fourier transform, has the same 'extended–s–wave' symmetry as the gap and, hence, the present scenario is an example of a very anisotropic s–wave pairing due to on–site interaction. However, due to a rather large value of $K_{ss}(0) = 14.3\,eV$, this scenario will be highly unlike to materialize in reality.

The Cu on–site $d_{x^2-y^2} - d_{x^2-y^2}$ scenario, with $K_{dd}(0) = 1.23\,eV$, leads to a fairly constant gap over the Fermi surface as is also observed to be the case in conventional superconductors. It is illustrated in Fig. 7b. The LDA

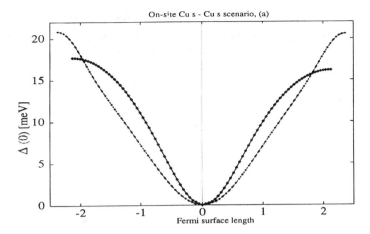

Fig. 7.a

The calculated anisotropy of (half) the gap for the on–site $Cu\,s$–$Cu\,s$ (**a**), the on–site $Cu\,d_{x^2-y^2}$–$Cu\,d_{x^2-y^2}$ (**b**), and the on–site $Cu\,s$–$Cu\,d_{x^2-y^2}$ (**c**), the intralayer nearest neighbor $Cu\,s$–$Cu\,d_{x^2-y^2}$ (**d**), and the intralayer nearest neighbor $Cu\,d_{x^2-y^2}$–$Cu\,d_{x^2-y^2}$ (**e**) scenarios, as a function of the Fermi–surface length measured from the crossing with the $\mathbf{\Gamma S}$–line and in units of the inverse lattice constant, for the odd (crosses) and even (diamonds) sheets of the Fermi surface

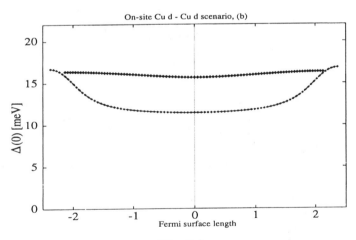

Fig. 7. b

calculations indicate that the Cu $d_{x^2-y^2}$ character is nearly constant along the Fermi surface, and since in this case the Fourier transform of the pairing potential does not contribute to the k–dependence of the gap, the resulting gap is fairly isotropic.

As in the case of the Cu on–site $s-s$ scenario the on–site Cu s–Cu $d_{x^2-y^2}$ scenario, governed by $K_{sd}(0) = 5.25$ eV, also gives a very anisotropic gap. However, as shown in Fig. 7c, it is of 'd–wave' symmetry, indicated by the characteristic cusp at the origin. The k–dependence in this case is due to the

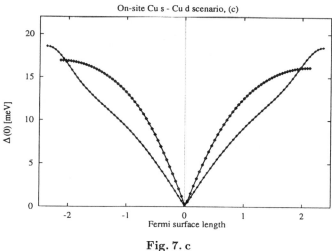

Fig. 7. c

fact that the Cu s character changes substantially across the Fermi surface, while the Cu $d_{x^2-y^2}$ character remains fairly constant. However, since $K_{sd}(0)$ is rather large, this scenario, like the Cu on–site $s-s$ scenario would probably be hard to realize in reality.

Concerning the nearest neighbor interactions, if a pair of electrons has to be on nearest neighbor sites in order to attract each other, their lowest energy state is likely to have a large amplitude for them to be so situated. Then, if the order parameter for all the nearest neighbors to a site has the same sign, using the somewhat confusing terminology derived from symmetry under continuous rotations, it is said to have extended 's–wave' symmetry. If it oscillates in sign as one follows around the nearest neighbors surrounding a site, the symmetry is referred to as 'p','d','f','g', etc... Thus, as opposed to the 'on–site' scenarios, the nearest neighbor scenarios are expected to give rise to extended 's' or exotic, non–s–wave, pairing. As indicated in Figs. 7d and e, this indeed turns out to be the case.

First we discuss the gap along the Fermi surface for the nearest neighbor intralayer Cu s–Cu $d_{x^2-y^2}$ coupling characterized by $K_{sd}(a) = 2.32$ eV (Fig. 7d). Here a denotes the vector distances to the *four nearest* copper

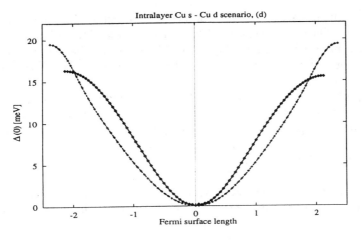

Fig. 7. d

neighbors, $(\pm a, 0)$ and $(0, \pm b)$, in the same layer. The rather low value of the coupling constant makes this an attractive candidate but the symmetry is extended $s-$ and not d-wave. That is to say, the order parameter does go to zero on the (π, π) and $(\pi, -\pi)$ lines in the Brillouin zone but it rises through positive values on either sides of these lines. In the k–dependence of the gap this manifests itself in the fact that it approaches the (π, π) line with zero slope from both sides. If experimental evidence for 'extended– s–wave' pairing emerges this would be a useful model to investigate further [29]. In this case the k–dependence of the gap is due to both the lattice Fourier transform of the pairing potential and the k–dependence of the Cu s.

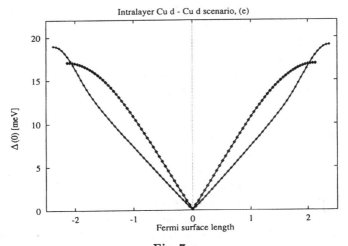

Fig. 7. e

The final scenario we want to discuss is the intralayer $Cu\,d_{x^2-y^2}-Cu\,d_{x^2-y^2}$ scenario. If for a fixed T_c $(= 92\,K)$, a low–$K_{LL'}(T)$ is regarded more likely than a high–$K_{LL'}(T)$ scenario, then clearly the most interesting case is the $Cu\,d_{x^2-y^2}-Cu\,d_{x^2-y^2}$ interaction for which we found $K_{dd}(a) = 0.68\,eV$. This is the smallest pair–interaction parameter $K_{LL'}(T)$ among all cases we have studied. The corresponding gap on the odd and even sheets of the Fermi surface shown in Fig. 7e features the cusp at the origin which implies that the gap has pure $d_{x^2-y^2}$–wave symmetry. As it turns out, so does the order parameter. Surprisingly, the variation of the gap along the Fermi surface is well described by the simple, '$d_{x^2-y^2}$–wave', function $(\cos ak_x - \cos bk_y)/2$. With a $d-d$ off–site interaction kernel, and because the $Cu\,d_{x^2-y^2}$ character is fairly constant across the Fermi surface, it is now the Fourier sum which provides the $\boldsymbol{k}$–dependence of the gap.

6   Results

6.1   Gap Anisotropy and Quasiparticle Density of States

Having identified the nearest neighbor intralayer $Cu\,d_{x^2-y^2}-Cu\,d_{x^2-y^2}$ scenario as the most probable one, on the account that it leads to the gap of d–wave symmetry and is associated with the smallest coupling coefficient $K_{LL'}(T)$, here we compare its gap anisotropy to the measurements of Schabel et al. [30], who studied four different samples of YBCO, all corresponding to different T_c's. Although the agreement of our calculations was reasonable for all samples [28], in Fig. 8 we only present detailed comparison with the measurements for sample XVII, whose T_c is equal to the value for which the calculations were performed. Considering the fact that we did not fit our one free parameter, $K_{dd}(a)$, to any feature of the gap measured by Schabel et al. [30], but only to obtain T_c of 92 K, the nearly quantitative agreement between theory and experiment for $\Delta^{\boldsymbol{k}}(0)$ can be taken as evidence that the relation between the attractive force, represented by $K_{dd}(a)$, and the gap in the quasiparticle spectrum, is correctly described by our BdG equation with the eight–band model Hamiltonian. Note that the value of $\sim 21\,meV$ for the half of the measured maximum gap compares favorably with the calculated value of $\sim 19\,meV$. Also, the calculated BCS–like ratio $2\Delta(0)k_BT_c = 4.8$ is in good agreement with the experimental value of 5.4 obtained for the sample XVII. That the $Cu\,d_{x^2-y^2}-Cu\,d_{x^2-y^2}$ intralayer nearest neighbor interaction is the preferred 'd–wave' scenario is illustrated in Fig. 9, where we compare the respective gaps of this scenario, along both even and odd sheets of the Fermi surface, with the corresponding gaps of another 'd–wave' scenario, specifically, where the interaction operates between $Cu\,s$ and $Cu\,d_{x^2-y^2}$ orbitals on the same site. As can be seen, the latter scenario leads to substantially different dependence of the gap as a function of $|\cos(ak_x) - \cos(bk_y)|/2$.

After exploring the anisotropy of the gap in the quasiparticle spectrum, it is also of interest to look at the elementary excitations in full energy range

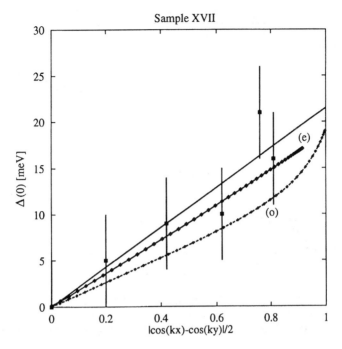

Fig. 8. A comparison of the calculated gap for the intralayer nearest neighbor $\mathrm{Cu}\,d_{x^2-y^2}-\mathrm{Cu}\,d_{x^2-y^2}$, scenario with $K_{dd}(a) = 0.68\,\mathrm{eV}$ for the even (e) and odd (o) sheets of the Fermi surface, with the experimental data deduced by Schabel *et al.* [30], from photoemission measurements on their sample XVII

for both d–wave scenarios studied in this paper. To make this possible we have calculated the quasiparticle density of states

$$N_S(E) = \sum_{kj} \delta(E_j^k - E) \,, \tag{17}$$

where, in this case, j runs over both positive– and negative–energy solutions. As will be seen in the next section, this is a useful quantity for calculations of the thermodynamic properties. To evaluate the sum over the Brillouin zone in (17) we used the same very accurate tetrahedron method as in the calculation of the normal density of states. Nevertheless, considerable extra effort was necessary to establish the linear dependence of $N_S(E)$ at the low energy end of the spectrum, a result which is a general consequence of the d–wave pairing. The need for that can be understood while inspecting $N_S(E)$ for the intralayer nearest neighbor $\mathrm{Cu}\,d_{x^2-y^2}-\mathrm{Cu}\,d_{x^2-y^2}$ scenario, as shown in Fig. 10. The resulting quasiparticle density of states was evaluated at each energy E using 256×256 points in the full two dimensional Brillouin zone, of size $\frac{2\pi}{a} \times \frac{2\pi}{b}$. Although there is a well–defined pseudo gap, $2\Delta_{ps}(0) \sim 30\,\mathrm{meV}$, at the low–energy end of the spectrum, there are states within this gap down

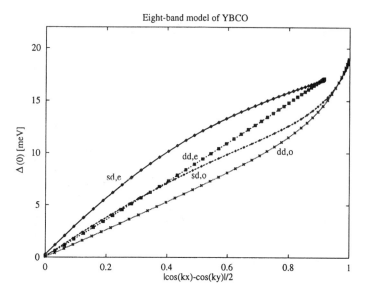

Fig. 9. A comparison of the calculated gap anisotropies for two different 'd–wave' pairing scenarios: The intralayer nearest neighbor $Cu\,d_{x^2-y^2}-Cu\,d_{x^2-y^2}$ scenario with $K_{dd}(a) = 0.68\,eV$ and the on–site $Cu\,s-Cu\,d_{x^2-y^2}$ scenario, with $K_{sd}(0) = 5.25\,eV$

to 1 meV. Evidently, the states below $2\Delta_{ps}(0)$ arise from the points in the two–dimensional Brillouin zone where the gap vanishes. Once this is realized, it is clear that the lower limit to such states below 1 meV is a numerical artefact caused by the finite number of k–points in our integration scheme. Namely, there is always an E_j^k which falls closest to zero and below which there are no states. Focusing only on the regions of the Brillouin zone where the gap goes to zero and increasing the density of k–points, while at the same time reducing the area of integration, we can show that $N_S(E)$ goes linearly to $E = 0$, as should be the case for d–wave pairing. This is illustrated in Fig. 11, where we compare the slopes of the linear variation of $N_S(E)$ for two different d–wave scenarios. Since the slopes depend on the specific k–dependence of the gap function near its nodes, one can see that although both scenarios correspond to the same T_c, their respective quasiparticle spectra are very different. This can be verified experimentally, because the slope determines the low temperature specific heat as a function of temperature. In the next subsection we shall make such contact with available experiments.

6.2   Specific Heat

Given that the BdG equation provides a fairly complete numerical description of the quasiparticle spectra for each scenario, one of the specific heat $C_v(T)$

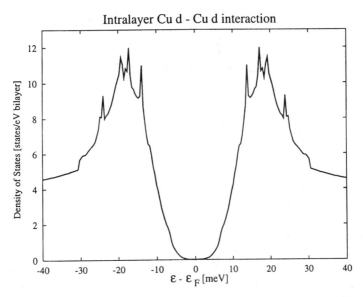

Fig. 10. The total quasi–particle density of states of the eight–band model of a CuO$_2$ bilayer of YBa$_2$Cu$_3$O$_7$, in superconducting state, for the intralayer nearest neighbor Cu$d_{x^2-y^2}$–Cu$d_{x^2-y^2}$ scenario, with $K_{dd}(a) = 0.68$ eV

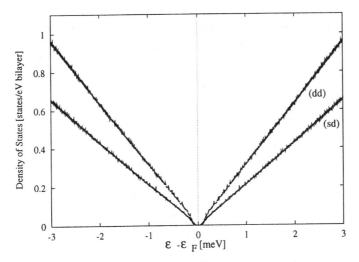

Fig. 11. A comparison of the linear behavior of the total quasi–particle density of states in the vicinity of the chemical potential for two different 'd–wave' pairing scenarios: The intralayer nearest neighbor Cu$d_{x^2-y^2}$–Cu$d_{x^2-y^2}$ scenario with $K_{dd}(a) = 0.68$ eV and the on–site Cus–Cu$d_{x^2-y^2}$ scenario with $K_{sd}(0) = 5.25$ eV

at constant volume, which in case of independent fermions is

$$C_v^s(T) - \sum_{kj} \frac{\beta}{2T} \left[E_j^k + \beta \frac{\partial E_j^k}{\partial \beta} \right] \frac{E_j^k}{\cosh^2 \left(\beta E_j^k / 2 \right)} , \qquad (18)$$

with $\beta = 1/k_B T$.

As shown in Fig. 12, there are two features of $C_v^S(T)$ in the superconducting state that are of general interest: one is the size and nature of the jump, $[C_v^s(T_c) - C_v(T_c)]/C_v(T_c)$, and the other is the temperature dependence as T goes to zero. Here C_v^s and C_v stand for specific heats of superconducting and normal states, respectively. The results presented in Fig. 12 reflect the specific heat calculated for the one–band model with two different paring scenarios [28]. The full line corresponds to 's–wave' pairing, with the usual constant gap of BCS type [19], and the dotted line was calculated using a single band 'd–wave' pairing model. Concerning the temperature dependence of the gap, it was taken from Mühlschlegel [31]. Note that in these calculations the jump for the 'd–wave' case is bigger than the universal 's–wave' value. Since, however, it is not universal, in principle, it could be used to differentiate between different semi–phenomenological scenarios. Unfortunately, our eight–band model calculations are very difficult near T_c where the gap is very small. Also, fluctuation effects, which are large for high T_c superconductors, on account of the short coherence length, $\xi_0 \sim a$, make the interpretation of the experiments in terms of a jump problematic [32,33]. Therefore, we did not pursue our computations of $C_v^S(T)$ near T_c.

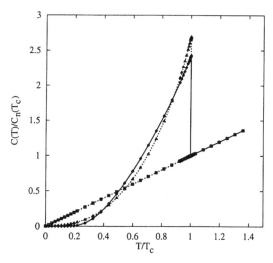

Fig. 12. A comparison of the normal state electronic specific heat (squares) with the electronic specific heat for 's–wave' (diamonds) and 'd–wave' (triangles) pairing interactions, calculated for a one–band model with $T_c = 92\,\mathrm{K}$

The second noteworthy feature of $C_v^S(T)$ is its approach to zero at low temperatures. As illustrated in Fig. 12, in the usual 's–wave' case this is an exponential decay while for 'd–wave' superconductors it is a power law. In our two–dimensional calculations the latter is T^2. Since the *coefficient* of the T^2–term depends on the quantitative details of the electronic structure, it can serve as a suitable testing ground for our computations. Indeed, for the investigated 'd–wave' scenarios, the linear energy dependence of $N_S(E)$ translates directly to a T^2–contribution to $C_v^S(T)$ and a quantitative prediction for the coefficients of T^2.

In Figure 13 we display our results calculated from the quasiparticle spectra corresponding to two different d–wave scenarios [28]. In the case of $\mathrm{Cu}\,d_{x^2-y^2}-\mathrm{Cu}\,d_{x^2-y^2}$ intralayer nearest neighbor interaction we have evaluated the full Brillouin zone integral in (17). Reassuringly, $C_v^S(T)$ vs. T^2 curves are convincing straight lines as befits a 'd–wave' pairing scenario. For the $\mathrm{Cu}\,s-\mathrm{Cu}\,d_{x^2-y^2}$ on–site interaction we used only the linear part of $N_S(E)$ depicted in Fig. 11 and, hence, the T^2 behavior was a forgone conclusion. As expected, the slopes are significantly different indicating that while both scenarios predict the same $T_c = 92\,\mathrm{K}$ their respective quasi–particle spectra are different in detail. In particular, as our discussion in the previous section indicates, this difference is due to the different rate at which $\Delta^k(T)$ rises from its node in k–space.

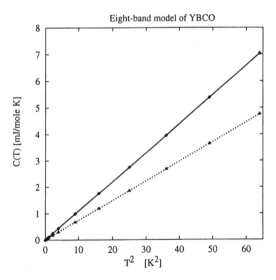

Fig. 13. A comparison of the calculated low temperature electronic specific heat for two different 'd–wave' scenarios featuring the intralayer nearest neighbor $\mathrm{Cu}\,d_{x^2-y^2}-\mathrm{Cu}\,d_{x^2-y^2}$ coupling with $K_{dd}(a) = 0.68\,\mathrm{eV}$ (*full line*) and the on–site $\mathrm{Cu}\,s-\mathrm{Cu}\,d_{x^2-y^2}$ coupling with $K_{sd}(0) = 5.25\,\mathrm{eV}$ (*dotted line*)

Unfortunately, also the interpretation of the low temperature specific heat measurements is fraught with difficulties [34,25]. Nevertheless, Moler *et al.* [35] were able to extract a T^2 contribution from their measurements on very high quality YBCO crystals and found $C_v^S(T)^{\mathrm{exp}} = 0.95(T/T_c)^2$ (J/mole K). This is remarkably close to our result, $0.93\,(T/T_c)^2$ (J/mole K) for the Cu $d_{x^2-y^2}$–Cu $d_{x^2-y^2}$ intralayer nearest neighbor scenario, and thus favors this scenario to its rival, involving 's' and 'd' orbitals on the same Cu site.

6.3 T_c versus Doping and Van Hove Scenario

In this section we want to elucidate, in quantitative detail, the relationship between the Van Hove singularities and such superconducting properties as the transition temperature T_c. In particular, we reproduce the rise and fall of T_c with doping in quantitative agreement with experiments [36].

In general, there are two principal facts responsible for the persistence of the belief that a Van Hove scenario, implying that many of the special properties of the cuprate superconductors are due to the presence of saddle points in the band structure close to the Fermi level, is relevant to the high T_c problem. One is that it provides a natural explanation of the variation of T_c with the electron per atom ratio (e/a), namely doping. Specifically, one readily finds that when ε_F is at the Van Hove singularity, T_c is maximum and it falls as e/a moves away from optimal doping as observed in experiments [24,37,38]. The second is that most first principles calculations of the electronic structure of high T_c cuprates do find Van Hove–like features near ε_F [17,39,40].

Of course, there are also many objections to the above suggestions [24,41]. In general, these are based on the expectation that a Van Hove singularity, being a delicate feature of the electronic structure, will be rendered ineffective as soon as realistic band structure, strong coupling, or disorder is introduced into the model. Here we want to demonstrate that the Van Hove–like singularities predicted by first principles local density approximation (LDA) calculations in the electronic structure of high T_c materials can support the above phenomenon of T_c enhancement.

Assuming that the Hamiltonian matrix $\mathcal{H}_{L'L}^{k}$ and the interaction $K_{LL'}^{q}$ do not depend on the band filling (doping) n, we have solved the BdG equations for two different scenarios: the on–site Cu $d_{x^2-y^2}$–Cu $d_{x^2-y^2}$ scenario (s–wave symmetry) and the intralayer nearest neighbor Cu $d_{x^2-y^2}$–Cu $d_{x^2-y^2}$ scenario (d–wave symmetry), at several temperatures for each value of the deviation $\delta n = n - n_c$, where n_c is the band filling characteristic of YBa$_2$Cu$_3$O$_7$. We have determined the transition temperature T_c as the lowest temperature for which the order parameter $\chi_{LL'}^{k}$ goes to zero. In Fig. 14 we show the Fermi surfaces for selected values of δn. The Fermi energies, and therefore the corresponding deviations δn from optimal doping are indicated in Fig. 15, where we plot the density of states for the normal state. Clearly, in the optimally doped material, ε_F is slightly above the orthorhombicity–split Van Hove singularities, caused by the bifurcated saddle–points, and moves through them as it is lowered to increase the number of holes. The above calculations were repeated for many other band fillings and both scenarios mentioned earlier. The corresponding transition temperatures T_c are plotted in Fig. 16 against the deviation of the number of holes from that at optimal doping, that is to say $-\delta n$.

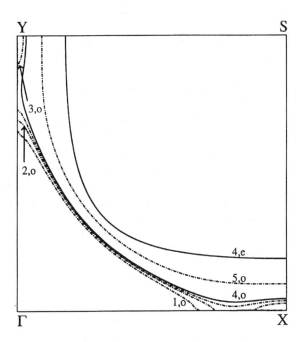

Fig. 14. The odd (o) and even (e) sheets of the Fermi surface of a CuO_2 bilayer of $YBa_2Cu_3O_7$, in the irreducible part of the two–dimensional Brillouin zone. Here $\Gamma = (0,0)$, $\mathbf{X} = (\pi/a,0)$, $\mathbf{Y} = (0,\pi/b)$, and $\mathbf{S} = (\pi/a,\pi/b)$. Moreover, 'odd' and 'even' refer to the symmetry of the wave functions with respect to the mirror plane between the two CuO_2 layers of the bilayer. The two *solid curves* labeled '4,o' and '4,e' are the odd and even Fermi surface sheets, respectively, of the optimally doped compound with T_c of 92 K. The other odd sheets, labeled by '5,o', '3,o', '2,o', and '1,o' correspond to [other] under– and over–doped compounds with the following T_c's and approximate hole dopings beyond optimal doping: (43 K, -0.11), (86 K, 0.01), (73 K, 0.03), and (28 K, 0.05). The saddle–points are bifurcated from $\mathbf{X}$ and $\mathbf{Y}$ towards Γ due to dimpling of the plane and they are split in energy due to chain–induced orthorhombicity. The relevant even sheets have not been presented here, because they are very much the same as the one for the optimally doped compound labeled by '4,e'. In Figure 15, where the density of states for the normal–state bilayer of $YBa_2Cu_3O_7$ is shown, the corresponding Fermi energies are marked by arrows labeled from '5' to '1'

The first thing to note about these curves is that T_c rises and falls with increasing number of holes as in a conventional d–wave Van Hove scenario [38]. Thus we have demonstrated that the bifurcated saddle points of the eight–band model can readily give rise to an enhancement of T_c by a factor of three or so. In fact the asymmetric shape of the curve is very similar to that due to an extended saddle–point [43], and this finding in itself is of considerable general interest. The second point of interest is the width of the T_c versus $-\delta n$ curve. As indicated in Fig. 16, our calculations for the intralayer nearest neighbor $Cu\, d_{x^2-y^2}-Cu\, d_{x^2-y^2}$ scenario imply a width at half maximum, $\delta n = 0.14$. This is remarkably close to the universal T_c versus $-\delta n$ curve deduced from experiments by Markiewicz [24]. In particular, he finds $\delta n = 0.15$. In this connection it is important to stress that for simple 2D models featuring logarithmic Van Hove singularities, the width is a function of the average gap and pair interaction constant K. For instance, the width is influenced by the anisotropy of the gap: for d–wave pairing with the lobes along the Cu–O bond, the relevant density of states is $N_d(\varepsilon) \equiv \sum_{\boldsymbol{k}} \frac{1}{2}[\cos(ak_x) - \cos(bk_y)]^2 \delta[\varepsilon(\boldsymbol{k}) - \varepsilon]$, and this sharpens up the peak because the $\mathbf{X}$– and $\mathbf{Y}$–points have maximal– and the $\mathbf{\Gamma S}$–line no weight. As a result, $N_d(\varepsilon) \lesssim N(\varepsilon)$ near the Van Hove singularities, as long as the bifurcation is not too large, and it vanishes quadratically at the band edges [26]. Hence, the fact that our calculation, in which the only adjustable parameter $K_{dd}(a)$ was chosen to fit $k_B T_c$, yields a width in agreement with experiment can be taken as a strong indication that the superconductivity, if nothing else, of $YBa_2Cu_3O_7$ can be described by a fairly simple BCS–like model, combined with a realistic band structure which supports Van Hove–like singularities.

It is important that the experimentally observed rapid fall of T_c with overdoping is well reproduced by the theory. In our calculations, the reason for the fall–off is that, with increasing overdoping, the Fermi surface recedes from those parts of $\boldsymbol{k}$–space which have a strong $[\cos(ak_x) - \cos(bk_y)]^2$ weight (see Fig. 14). Bifurcation of the saddle–points further enhances this effect, because the $\mathbf{X}$– and $\mathbf{Y}$–centered electron pockets, created by the bifurcation, are lost at lower energies and, as seen in Fig. 15, this causes discontinuous drops in $N(\varepsilon)$ and, with full weight, in $N_d(\varepsilon)$. Moreover, we observe that when ε_F is some $100\,\text{meV}$ above the Van Hove singularities T_c is still a substantial $50\,\text{K}$. This surprising insensitivity of T_c to the distance of ε_F above the Van Hove singularities may be the explanation for the ubiquitous occurrence of high T_c's in the cuprates. Conveniently, it makes the somewhat mysterious pinning of ε_F to such specific features of the density of states unnecessary. Another important point to make here is that the orthorhombic $10\,\text{meV}$ splitting (see Ref. [17]) of the Van Hove singularity does not produce a two–peaked structure in Fig. 16. Clearly, this is due to the fact that the splitting is smaller than the average gap.

Leaving this section, we recall that the strategy behind investigating various interaction scenarios specified by the one, non–zero, interaction constant

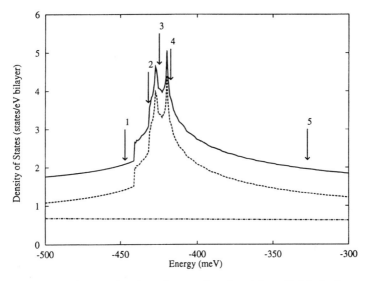

Fig. 15. The normal–state density of states for the eight odd (*dashed curve*) and eight even (*chain curve*) plane–bands of $YBa_2Cu_3O_7$, as well as their sum (*full curve*). The arrows labeled from '1' to '5' mark the corresponding Fermi levels of several over– and under–doped compounds as described in Figs. 14 and 16. The logarithmic Van Hove singularities between markers 4 and 3 and between 3 and 2 are due to the saddle–points of the odd plane–band near **Y** and **X**, respectively. The two discontinuous drops of the density of states near marker 2 and between markers 2 and 1 are caused by the disappearances of the electron pockets at respectively **Y** and **X**. Such a logarithmic singularity followed by a discontinuous drop is the characteristic of a bifurcated saddle point in two–dimensions

$K_{RL,R'L'}$, which is fitted to give the observed T_c, is that by calculating different superconducting properties and comparing the results with experiments, we can eliminate some of them in favor of the others. From Fig. 16 it is clear that the width corresponding to the on–site $Cu\,d_{x^2-y^2}-Cu\,d_{x^2-y^2}$ scenario is too large, and the case of the intralayer, nearest neighbor, $Cu\,d_{x^2-y^2}-Cu\,d_{x^2-y^2}$ scenario is to be preferred.

7   Conclusions

We have presented a semi–phenomenological approach for calculating the quasiparticle spectra of high T_c materials in the superconducting state. We have illustrated it on the example of $YBa_2Cu_3O_7$ by calculating some of its superconducting properties. The principle virtue of this approach is that it is able to make contact with the experimental data in quantitative detail while avoiding the proliferation of adjustable parameters. The approach is based on an effective parameterization of the electron–electron interaction

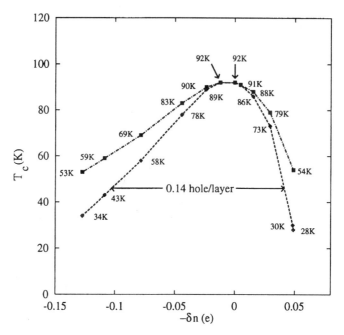

Fig. 16. T_c versus deviation of the number of holes from that at optimal doping for the CuO_2 bilayer of $YBa_2Cu_3O_7$. The dashed curve corresponds to the intralayer nearest neighbor $Cu\,d_{x^2-y^2}-Cu\,d_{x^2-y^2}$ scenario, while the dash–dotted curve represents the on–site $Cu\,d_{x^2-y^2}-Cu\,d_{x^2-y^2}$ scenario. On both curves T_c's corresponding to different hole concentrations are marked

with the density functional description of the superconducting state and the TB-LMTO method for solving the Kohn–Sham–Bogoliubov–de Gennes equations. As indicated by comparison with experimental data for various quantities studied here, the favorable pairing scenario is the one where the attractive force operates between electrons of opposite spins sitting in $d_{x^2-y^2}$ orbitals on nearest neighbor Cu–sites of the CuO_2 layers. It should be stressed that d–wave pairing has not been assumed here, but is the direct consequence of this particular interaction and the electronic structure described by the eight–band model. Moreover, the fact that T_c, the quasiparticle spectra, and the low temperature specific heat are all described quantitatively using only one adjustable parameter $K_{dd}(a)$, suggest that irrespective of the nature of the pairing interaction, these properties are related to each other as dictated by the structure of the simplest BdG equations, that is to say, a simple BCS–like theory.

References

1. Bednorz J.G., Müller K.A. (1986) Z Phys B 64:189
2. Annett J.F., Goldenfeld N., Leggett A.J. (1996) In: Ginsberg D.M. (Ed.) Physical Properties of High Temperature Superconductors vol 5, World Sci., Singapore
3. Oliveira L.N., Gross E.K.U., Kohn W. (1988) Phys Rev Lett 60:2430
4. Kohn W., Gross E.K.U., Oliveira L.N. (1989) J Phys (Paris) 50:2601
5. Andersen O.K. (1975) Phys Rev B 12:3060
6. Skriver H.L. (1984) The LMTO Method, Springer, Berlin
7. Suvasini M.B., Temmerman W.M., Gyorffy B.L. (1993) Phys Rev B 48:1202
8. Temmerman W.M., Szotek Z., Gyorffy B.L., Andersen O.K., Jepsen O. (1996) Phys Rev Lett 76:307
9. Eliashberg G.M. (1960) Sov Phys – JETP 11:696
10. Bardeen J., Cooper L.N., Schrieffer J.R. (1957) Phys Rev 108:1175
11. Hohenberg P., Kohn W. (1964) Phys Rev 136:864
12. Kohn W., Sham L.J. (1965) Phys Rev 140:A1133
13. Dreizler R.M., Gross E.K.U. (1990) Density Functional Theory, Springer, Berlin New York
14. Jones R.O., Gunnarsson O. (1989) Rev Mod Phys 61:689
15. Andersen O.K., Liechtenstein A.I., Rodriguez C.O., Mazin I.I., Jepsen O., Antropov V.P., Gunnarsson O., Gopalan S. (1991) Physica C 185–189:147
16. Andersen O.K., Liechtenstein A.I., Jepsen O., Paulsen F. (1995) J Phys Chem Solids 56:1573
17. Andersen O.K., Jepsen O., Liechtenstein A.I., Mazin I.I. (1994) Phys Rev B 49:4145
18. Gross E.K.U., Kurth S., Capelle Klaus, Lüders M. (1995) In: Gross E.K.U., Dreitler R.M. (Eds.) Density Functional Theory, vol 337 of NATO ASI Series B, Plenum Press, New York
19. Fetter A.L., Walecka J.D. (1971) Quantum Theory of Many–Particle Systems, McGraw Hill, New York;
    Abrikosov A.A., Gorkov L.P., Dzaloshinskii I.E. (1965) Quantum Field Theoretical Methods in Statistical Physics, Pergamon Press, Oxford
20. Monthoux P., Balatsky A., Pines D. (1989) Phys Rev Lett 62:961;
    Pines D., Monthoux P.J. (1995) J Phys Chem Solids 56:1651
21. Chaikin P.M., Lubensky T.C. (1995) Principles of Condensed Matter Physics, Cambridge at the University Press
22. Andersen O.K., Liechtenstein A.I., Rodriguez C.O., Mazin I.I., Jepsen O., Antropov V.P., Gunnarsson O., Gopalan S. (1991) Physica C 185–189:147–155
23. Tobin J.G., Olson C.G., Gu C., Liu J.Z., Solal F.R., Fluss M.J., Howell R.H., O'Brien J.C., Radousky H.B., Sterne P.A. (1992) Phys Rev B 45:5563;
    Gofron K., Campuzano J.C., Ding H., Gu C., Liu R., Dabrowski B., Veal B.W., Cramer W., Jennings G. (1993) J Phys Chem Solids 54:1193;
    — (1994) Phys Rev Lett 73:3302;
    Dessau D.S., Shen Z.-X., King D.M., Marshall D., Lombardo L.W., Dickenson P., Di Carlo J., Park C.H., Loeser A.G., Kapitulnik A., Spicer W.E. (1993) Phys Rev Lett 71:2781;
    Abrikosov A.A., Campuzano J.C., Gofron K. (1993) Physica C 214:73
24. Markiewicz R.S. (1997) J Phys Chem Solids 58:1179–1310

25. Shen Z.-X., Dessau D.S. (1995) Phys Rept 253:1
26. Andersen O.K., Savrasov S.Y., Jepsen O., Liechtenstein A.I. (1996)
    J Low Temp Phys 105:285
27. Savrasov S.Y., Andersen O.K. (1996) Phys Rev Lett 21:4430
28. Gyorffy B.L., Szotek Z., Temmerman W.M., Andersen O.K., Jepsen O. (1998)
    Phys Rev B 58:1025
29. Sun A.G., Gajewski D.A., Maple M.B., Dynes R.C. (1994)
    Phys Rev Lett 72:2267
30. Schabel M.C., Park C.H., Matsuura A., Shen Z.-X., Bonn D.A.,
    Liang Ruixing, Hardy W.N. (1997) Phys Rev B 55:2796
31. Mühlschlegel B. (1959) Z Phys 155:313
32. Junod A. (1990) In: Ginsburg D.M. (Ed.) Physical Properties of High
    Temperature Superconductors II, World Sci., Singapore, 13
33. Loram J.W., Mirza K.A., Cooper J.R. (1998) IRC (Cambridge), Research
    Review
34. Redcliffe J.W., Loram J.W., Wade J.M., Witschek G., Talon J.L. (1996)
    J Low Temp Phys 105:903;
    Junod A., (1996) In: Narlikar A.V., (Ed.) Studies in High Temperature
    Superconductors, Chapt 15, Nova Sci., Commack, New York
35. Moler K.A., Baar D.J., Urbach J.S., Liang Ruixing, Hardy W.N.,
    Kapitulnik A. (1994) Phys Rev Lett 73:2744
36. Szotek Z., Gyorffy B.L., Temmerman W.M., Andersen O.K. (1998)
    Phys Rev B 58:522
37. Friedel J. (1989) J Phys Cond Matt 1:7757
38. Newns D.M., Tsuei C.C., Pattnaik P.C. (1995) Phys Rev B 52:13 611
39. Novikov D.L., Freeman A.J. (1996) In: Klamut J., Veal B.W.,
    Dabrowski B.M., Klamut P.W., Kazimierski M. (Eds.) Recent Developments
    in High Temperature Superconductivity, Springer, Berlin Heidelberg, 17–35
40. Pickett W.E. (1997) Physica C 289:51
41. Radtke R.J., Levin K., Schuttler H–B., Norman M.R. (1993)
    Phys Rev B 48:15 957;
    Levin K., (1992) In: Ashkenazi J., Vezzoli G. (Eds.) Electronic Structure and
    Mechanisms for High Temperature Superconductivity, Plenum, New York, 481
42. Chandrasekhar B.S., Einzel D. (1993) Ann Phys 2:535–546
43. Dasgupta I., Andersen O.K., Jepsen O. (1998) private communication
44. Szotek Z., Gyorffy B.L., Temmerman W.M. (1999) to be published
45. Bonn D.A., Kamal S., Zhang Kuan, Liang Ruixing, Baar D.J., Klein E.,
    Hardy W.N. (1994) Phys Rev B 50:4051

Superconductivity and Magnetic Properties of Spin–Ladder Compounds

H. Szymczak[1], R. Szymczak[1], M. Baran[1], L. Leonyuk[2], G.- J. Babonas[3], and V. Maltsev[2]

[1] Institute of Physics, Polish Academy of Sciences, Al. Lotników 32/46, PL-02-668 Warsaw, Poland
[2] Moscow State University, Moscow, Russia
[3] Semiconductor Physics Institute, 2600 Vilnius, Lithuania

Abstract. A review is given of the structural, magnetic and transport studies on two–leg spin–ladder $(M_2Cu_2O_3)_m(CuO_2)_n$ systems (where M are divalent or/and trivalent cations). The crystals belonging to these systems consist of two interpenetrated subsystems. The first subsystem $[M_2Cu_2O_3]$ is composed of (Cu_2O_3) two–leg ladder planes and M ions coordinated to them. The second subsystem consists of CuO_2 1D–chains. In these materials, the superconductivity was discovered for $m/n = 1/1$, $5/7$, $7/10$. The intrinsic and extrinsic superconducting properties of spin–ladder systems are presented and discussed in detail. The important role played by the hole transfer from CuO_2 planes to the spin–ladder planes is stressed. It is shown that the superconductivity in this new family of high–temperature superconductors should be described as an extreme type II limit.

1 Introduction

In the past decade, the study of cuprate compounds, especially high T_c superconductors, has been one of the most attractive subjects in the field of condensed matter physics. In addition to high-T_c superconductors, the variety of low–dimensional cuprates became the subject of extensive investigations and discussions. The low dimensional cuprates exhibit unusual magnetic phenomena such as the spin–Peierls transition in $CuGeO_3$ [1] and the Haldane gap in Y_2BaNiO_5 [2]. The ladder systems formed by n–spin chains coupled side by side are of particular interest since for these systems a possibility to observe the superconductivity has been suggested [3]. The quantum effects characteristic of low dimensional systems lead to a dramatic dependence of the properties of ladder systems on the number of legs (n) [4]. Ladders with an odd number of legs are characterized by a gapless spin excitation spectrum and a power–law falloff of the spin–spin correlations. Ladders with an even number of chains have singlet ground states, separated from the lowest triplet states by a finite gap. The spin–spin correlations in this case fall as an exponential function of the distance. These theoretical predictions have been verified experimentally in materials having planes with weakly coupled arrays of ladders (see [4] and references therein).

Among various spin–ladder compounds the $(M_2Cu_2O_3)_m(CuO_2)_n$ cuprates (where M are divalent and trivalent cations) have attracted the largest attention because they comprise both simple chains and two–leg ladders of copper ions. This system consists of the Cu_2O_3 spin–ladder planes as well as CuO_2 chains.

The orthorhombic structure of $(M_2Cu_2O_3)_m(CuO_2)_n$ compounds is very often of incommensurate–type due to the presence of two sublattices $(M_2Cu_2O_3)$ and (CuO_2) (see Fig. 1). The sublattices possess very close lattice parameters along the a– and b–axes. The lattice parameters along the c–axis in two sublattices are incommensurate because the corresponding structural units are composed of the CuO–squares sharing corners or edges. The commensurate version of the $(M_2Cu_2O_3)_m(CuO_2)_n$–type structure means that (in addition to the parameters c_1 and c_2 corresponding to two sublattices) the reflexes ascribed to the common lattice (corresponding to $mc_1 = nc_2 = c$) are observed in the XRD pattern. At present, several groups of $(M_2Cu_2O_3)_m(CuO_2)_n$ compounds have been successfully synthesized with the m/n–values equal to 1/1, 5/7, 7/10 and 9/13.

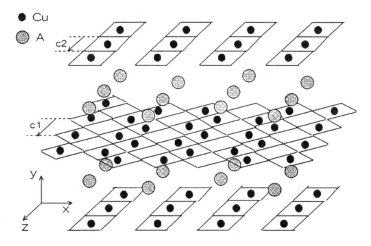

Fig. 1. Model of the crystallographic structure of the $(M_2Cu_2O_3)_m(CuO_2)_n$ compounds

The ladder–type compounds contain the Cu_2O_3 plane which has been predicted [5] to manifest the high-T_c superconductivity. The superconductivity was indicated experimentally [6,7] in several $(M_2Cu_2O_3)_m(CuO_2)_n$–type compounds. However, a surprisingly large difference in the T_c–values of various $(M_2Cu_2O_3)_m(CuO_2)_n$ samples is noticeable. The transition to a superconducting (SC) state was indicated at ambient pressure at $80 - 85\,K$ in the single crystals of the $(M_2Cu_2O_3)_m(CuO_2)_n$ (where $M =$ Ca, Sr, Bi, Y) with $m/n = 1/1$ [8] and 5/7 [9] while the samples $Sr_{0.4}Ca_{13.6}Cu_{24}O_{41.84}$ corresponding to $m/n = 7/10$ were superconducting at 3-4.5 GPa with T_c of about 10 K [7].

In this paper, we review the recent results on several groups of $(M_2Cu_2O_3)_m(CuO_2)_n$ compounds with various m/n parameters (1/1, 5/7, 7/10). The experimental results will be analyzed in order to reveal some correlations between the crystal structure and the manifestation of super-conductivity in these materials. The main attention in this review is devoted to the system with $m/n = 5/7$, in which the T_c as high as about 80 K was obtained, and for which detailed investigations have been done both for non-superconducting (NSC) as well as for SC samples.

2   Spin–Ladder System $(M_2Cu_2O_3)(CuO_2)$

The structure of $(M_2Cu_2O_3)_m(CuO_2)_n$ compounds can be efficiently ana-lyzed [10] in the fragment approximation. From this point of view, the $(M_2Cu_2O_3)_m(CuO_2)_n$–type compound is considered as a polysome mAnB in the polysomatic series with end members MCu_2O_3 (A) and $M_{1-x}CuO_2$ (B). The structure of polysome is represented by the combination of mA and nB slabs characteristic of two end members A and B.

The two–legs ladder–type structure MCu_2O_3 is most characteristic for the case $M = Sr$, in which Sr is coordinated by 8 oxygen atoms. The $M_{1-x}CuO_2$ for $M = Sr$ is realized at $1 - x = 0.73$ [11] with the same Sr–coordination polyhedron. In the case of $M = Ca$, the $M_{1-x}CuO_2$–type structure is stable at $1 - x = 0.83$. However, for $CaCu_2O_3$ the ladder–plane is conjugated and the Ca–polyhedron is octahedron. So, the combination of the slabs A and B for $M = Ca$ into the structural type $(M_2Cu_2O_3)_m(CuO_2)_n$ is not stable and the corresponding compound can be obtained only at high pressures. Thus, the $(M_2Cu_2O_3)_m(CuO_2)_n = (MCu_2O_3)_m(M_{m/n}CuO_2)_n$–type structure is most stable for $M = Sr$ at $m/n = 7/10$. The substitution of cations with various values of ionic charge and radius into M–sites allows one to realize the stable compounds with different m/n–values.

When one slab is highly deficient, the corresponding reflexes in the struc-tural determination are relatively weak. As a result, in the commensur-ate description of the $(M_2Cu_2O_3)_m(CuO_2)_n$–type structure only the para-meter (c_1 or c_2) of a more complete slab is recognized as a common para-meter c for the whole structure. Usually, the reflexes due to the parameter $c_1 \sim 3.9$ Å of the ladder–plane are better developed [8,12–14]. However, we have also grown the single crystals in which the common parameter of the $(M_2Cu_2O_3)_m(CuO_2)_n$–type structure corresponds to the Cu–O chains with characteristic value $c_2 \sim 2.75$ Å. It should be emphasized that parameters $a \sim 11$ Å and $b \sim 13$ Å of the $(M_2Cu_2O_3)_m(CuO_2)_n$–type structure were clearly indicated in all the cases mentioned above.

These considerations show that the $(M_2Cu_2O_3)_m(CuO_2)_n$–type com-pounds with $m/n = 1/1$, in which one fragment is more developed than another, are not stable. Small deviations of the melt composition during the process of a crystal growth can lead to relatively large variations in the chemical composition and the distribution of cations over the M–sites along the sample. However, at the same time the detailed investigation of the structural features along such non–homogeneous samples gives one a possibility to develop a model of the formation of the $(M_2Cu_2O_3)_m(CuO_2)_n$–type crystals and to reveal the correlation between the structure and superconductivity in incommensurate phases. Such a situation was found in a relatively long needle–like sample of 8 mm in length which was cut into three pieces. For all three samples, detailed structural analysis was performed. The results of studies performed on this unique set of samples predict the following scheme for the formation of $(M_2Cu_2O_3)_m(CuO_2)_n$ $(m/n = 1/1)$ samples. First it should be emphasized that the a– and b–parameters typical of the $(M_2Cu_2O_3)_m(CuO_2)_n$ crystals were fixed, indicating that both fragments were presented in the series of the samples under consideration. At early stage of crystal growth, sample 1 was formed with well–developed planes of Cu–O chains ($a = 11.294$ Å, $b = 12.720$ Å, $c = 2.769$ Å). The samples formed at the next stage represent the case of the most unstable structures with a possible domination of the phase with c–parameter close to 3.9 Å (sample 2). At later stages, the $1/1$–type structure can be transformed to the structure of incommensurate type with various m/n–values which are caused by the characteristics of M–cations. The instability of structure can be illustrated by various sets of the lattice parameters determined from different numbers of reflexes in the XRD as in the sample 3. The sets $a = 11.348$ Å, $6b = 12.915$ Å, $c = 74.25$ Å; $a = 11.399(3)$ Å, $b = 12.886$ Å, $c = 19.499(4)$ Å; and $a = 22.811$ Å, $b = 25.735(5)$ Å, $c = 39.02(1)$ Å were obtained for 25, 12 786 and 103 892 reflexes, respectively. Other samples from the same batch were smaller, and as usual with reduction of dimensions, they were characterized by a better shape and smooth shiny metallic natural faces. Such samples were characterized by a high perfection as the structural refinement by single crystal XRD was performed with a low R–factor value [8]. These samples (one of them is referred to hereafter as sample 4) have the lattice constants $a = 11.361(4)$ Å, $b = 12.906(7)$ Å, $c = 3.9067(8)$ Å (space group $Fmmm$). It should be noted that sample 4 could be also described in the commensurate approximation with large m– and n–values ($m/n = 31/44$).

Figure 2 presents the zero field cooled (ZFC) magnetic susceptibility data for the samples described above. It is worth noting that T_c–values are similar and close to 82 K while the shielding effect varies significantly. It should be mentioned that the values of shielding effect were determined taking into account demagnetizing factors of the samples. The observed differences in the shielding effect are significantly larger than those caused by any possible errors resulting from improper determination of demagnetizing factors. One can see from Fig. 2 that the shielding for sample 1 is the lowest and close to that of sample 4. At the same time this last sample was shown [8] to possess highly perfect structure. The largest shielding was observed for sample 2, the most unstable from structural point of view.

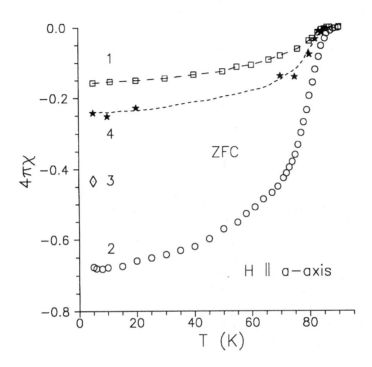

Fig. 2. Temperature dependences of ZFC susceptibility for several crystals of $(M_2Cu_2O_3)_m(CuO_2)_n$ $(m/n = 1/1)$ system

It is worth mentioning that the manifestation of superconductivity in the $(M_2Cu_2O_3)_m(CuO_2)_n$, $m/n = 1/1$–type phases depends on the type of M–cations [8].

3 Spin–Ladder System $(M_2Cu_2O_3)_5(CuO_2)_7$

In contrast to the non–deficient phases from the group $m/n = 7/10$ [12,15,16], the incommensurate $m/n = 5/7$ phases are characterized by a small defi-ciency of Cu–sites distributed approximately equally in both the ladder–type Cu_2O_3 plane and the CuO_2–plane of the chains. Superconductivity was found in the crystals of following compositions and unit cell parameters:
$(Sr_{3.5}Y_{0.1}Ca_{5.9}Al_{0.1}Bi_{0.3}Pb_{0.1})Cu_{15.1}O_{29}$
$(a = 11.319(2)$ Å, $b = 12.763(2)$ Å, $c = 19.49(1)$ Å, $T_c = 80$ K);
$(Sr_{3.1}Ca_{5.9}Bi_{0.4}Y_{0.3}Pb_{0.3})Cu_{16.7}O_{29}$
$(a = 11.32$ Å, $b = 12.75$ Å, $c = 19.49$ Å, $T_c = 80$ K);

$(Sr_{4.02}Ca_{5.84}Bi_{0.14})Cu_{15.84}O_{29}$
$(a = 11.346(1)$ Å, $b = 12.809(3)$ Å, $c = 19.52(1)$ Å, $T_c = 82$ K);
$(Sr_{4.42}Ca_{4.86}Bi_{0.05})Cu_{17}O_{29}$ (SC1)
$(a = 11.349(7)$ Å, $b = 12.896(5)$ Å, $c = 19.49(3)$ Å, $T_c = 84$ K).
As an example of NSC composition, the following single crystal can be presented: $(Sr_{6.1}Ca_{3.3}Bi_{0.1}Y_{0.3})Cu_{16.4}O_{29}$ $(a = 11.346(3)$ Å, $b = 12.996(3)$ Å, $c = 19.586(9)$ Å).

On the basis of the XRD single crystal structural determination we have revealed the correlation between the structural features and the occurrence of superconductivity. The structure of SC and NSC samples differs by the coordination polyhedra for Cu in the ladder–planes. In NSC samples the 5–apexes polyhedron was fixed with the Cu–O distances $2.58 - 2.79$ Å and $1.95 - 2.17$ Å for out–of–plane and in–plane oxygen atoms, respectively. In SC samples the distance from Cu to out–of–plane oxygen $(2.85 - 3.14$ Å) is significantly larger than that $(1.88 - 2.00$ Å) for in–plane oxygen. The situation is similar to that for SC and NSC Y-123–type compounds for the distances of Cu, in the CuO_2–plane, to in–plane oxygen atoms and to the apical oxygen. The difference in the Cu-O distances in the plane of CuO_2 chains of SC and NSC $(M_2Cu_2O_3)_m(CuO_2)_n$–type phases with $m/n = 5/7$ was also indicated. The CuO–squares are smaller in the SC samples. The difference in the optical data of the SC and NSC compounds can be interpreted on the basis on the structural changes mentioned above [17].

Two features of the manifestation of superconductivity can be distinguished in the $(M_2Cu_2O_3)_m(CuO_2)_n$–type phases with $m/n = 1/1$ and $m/n = 5/7$. The M–cation radius varies in the range $1.26 - 1.28$ Å and the formal Cu–valence is higher than 2.30. The first feature is related to the value of the b–parameter, which is distinctly smaller than ~ 13.0 Å in the SC samples. A smaller b–value favors the charge transfer from the plane of the CuO_2 chains to the ladder plane. Along with the first feature, the second one characterizes the cation composition of the superconducting sample and is directly related to the concentration of free carriers. The concentration of free carriers is higher in superconducting samples and the difference is clearly seen in the optical spectra [17].

A very effective method to study pure and doped spin–ladder compounds is based on the measurements of magnetic properties of the system. Using this method we have been successful in determining the carrier distribution between spin–chains and spin–ladders. The analysis and discussion of properties of the $(M_2Cu_2O_3)_5(CuO_2)_7$ system presented in this part of the paper is based mainly on the results of magnetic measurements which have been presented in [18] and [19].

Generally, one should expect that the critical temperature T_c is dependent not only upon the concentration of the carriers, but also on the charge carrier density in the ladder plane which is responsible for superconductivity in this group of superconducting compounds. The transfer of carriers from chains to ladders has been determined for NSC and SC samples analyzing the temperature dependences of magnetic susceptibility $\chi(T)$. Fig. 3a presents the reciprocal magnetic susceptibility versus temperature of a typical NSC sample measured in magnetic field applied along three main crystallographic axes. One can see that the magnetic susceptibility χ is highly anisotropic: $\chi_b > \chi_c \approx \chi_a$.

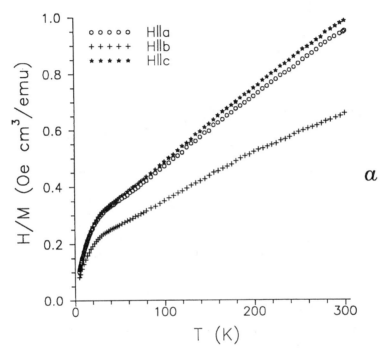

Fig. 3. a Temperature dependences of the inverse volume susceptibility for the nonsuperconducting sample, in an external magnetic field $H = 5\,\mathrm{kOe}$ applied along the main crystallographic axes a, b and c

The susceptibility $\chi(T)$ shows a strong upturn in the low temperature region (see Fig. 3b for NSC sample with composition $Sr_{6.1}Ca_{3.3}Bi_{0.1}Y_{0.3}Cu_{16.4}O_{29}$) due to a Curie–Weiss contribution. This contribution may arise from finite length chains with odd number of spins [20], isolated spins, and/or other paramagnetic impurities. The detailed analysis of the temperature dependence of magnetic susceptibility $\chi(T)$ [21] has shown that χ should be described as a sum of a temperature–independent term χ_0, a Curie–Weiss term χ_{cw}, and a dimer term related to the CuO_2 chains, χ_{ch}:

$$\chi(T) = \chi_0 + \chi_{cw}(T) + \chi_{ch}(T) , \qquad (1)$$

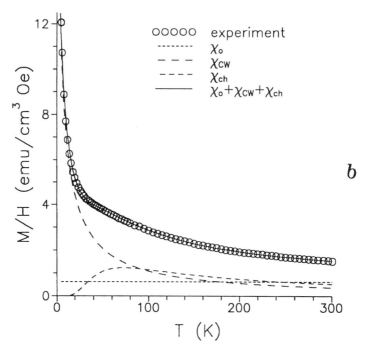

Fig. 3. b Volume susceptibility of NSC sample vs temperature for $H \parallel b$–axis. *Solid line* – fitting of (2) to experimental points (*open squares*), *dashed lines* – components χ_0, χ_{cw} and χ_{ch} of susceptibility

where $\chi_{cw} = C_g/(T - \Theta)$ with $C_g = n_f g^2 \beta^2 \mathcal{N}_A/(4kM_{mol})$,
$\chi_{ch} = D/(T(3 + \exp[\Delta/kT]))$ with $D = 2n_d g^2 \beta^2 \mathcal{N}_A/(kM_{mol})$,
n_f is number of free Cu^{2+} ions ($S = \frac{1}{2}$) per f.u., β – the Bohr magneton, $\mathcal{N}_A$ – the Avogadro number, k – the Boltzmann constant, M_{mol} – the molar mass, n_d – the number of dimers per f.u., and Δ is the energy gap for dimers. The g–factors used in these expressions are usually determined from ESR experiments. The ESR experiments [22] performed on NSC and SC samples revealed strong coupling of spins to carriers. This dynamic coupling leads to the shift and averaging of the Cu^{2+} spectra. To avoid this effect (the magnitude of which is difficult to determine), in the further consideration the g–factors were taken the same as determined for Cu^{2+} in chains in the nonmetallic $Sr_{14}Cu_{24}O_{41}$ single crystal [23]:

$$g_a = 2.05, \quad g_b = 2.26, \quad g_c = 2.04.$$

Since free Cu^{2+} ions are attributed to unpaired ions located in chains [23] the above values of g–factors were also used to calculate n_f parameters. The susceptibility given by (1) was fitted to experimental data. The results of fitting for three cases of magnetic field applied along a, b and c axes are given in the Table 1. For $H \parallel b$ they are shown in Fig. 3b as dotted lines for three components, χ_0, $\chi_{cw}(T)$, $\chi_{ch}(T)$, the total susceptibility $\chi(T)$ being shown as solid line. Standard errors of n_f and n_d values are estimated to be in the limit of 1.6% and 3.2% of these values, respectively. The obtained differences in values of both n_f and n_d for three crystallographic directions (which are less than 10%), seem to be mainly due to unknown true values of g–factors for Cu ions in chains and ladders.

Table 1. The set of fitting parameters describing the paramagnetic susceptibility of NSC sample according to (2) (C_g and D in emuK/gOe; Θ and Δ/k in K)

Axis	$C_g \cdot 10^4$	Θ	n_f	$D \cdot 10^4$	Δ/k	n_d
a	1.44 ± 0.02	-3.5 ± 0.1	0.834	9.4 ± 0.2	92 ± 2	0.680
b	1.88 ± 0.03	-3.5 ± 0.1	0.896	1.24 ± 0.04	97 ± 2	0.738
c	1.47 ± 0.02	-3.6 ± 0.1	0.860	9.7 ± 0.3	96.6 ± 1.5	0.710

In the above analysis, the χ_{ch} was taken in the approach of noninteracting dimers [24]. This approach, applied to our case, gives a rather good fit as shown in Fig. 3b. This means that the interdimer interactions in this case are rather small and may be neglected. Spontaneously dimerized Cu^{2+} ions in antiferromagnetic chains resemble a spin–Peierls instability. The dimerization reported here seems to be of different origin from that of standard spin–Peierls transitions found in $CuGeO_3$ [25] where the dimerization occurs as a result of strong spin–lattice coupling leading to a spontaneous lattice distortion and a structural phase transition. In our case, the dimerized state is rather a purely quantum effect originating from the competition between nearest–neighbor and next–nearest–neighbor exchange interactions and does not involve any structural phase transitions. The dimerization may also be connected with charge ordering expected to exist in CuO_2 chains. Since the formal valency of copper ions in $(M_2Cu_2O_3)_5(CuO_2)_7$ is about +2.24 this means the existence of 4 holes/f.u. In NSC samples, these holes are expected to be localized in the chains and are coupled with the copper spins to form the Zhang–Rice singlet [26]. Therefore the dimerized state of the chain may be expressed as $\langle +0-\rangle$ (+ and − represent the copper spins, 0 represents the Zhang–Rice singlet) charge ordering state. It cannot be excluded that this charge ordering effect is responsible for a lattice dimerization. From magnetic susceptibility measurements, it is apparent that the total spin density in spin chains ($n_f + 2n_d$) is about 2.3, which is smaller than expected for the ideal $M_{10}Cu_{17}O_{29}$ stoichiometric compound. But this difference is small (0.7) and may be due to non–stoichiometry.

Calculating the nominal number of holes/f.u. expected for a NSC sample of composition $(Sr_{6.1}Ca_{3.3}Bi_{0.1}Y_{0.3})Cu_{16.4}O_{29}$ it is easy to show that the formal number of holes is 5.2/f.u. Since ($n_f + 2n_d$) = 2.3 one should expect

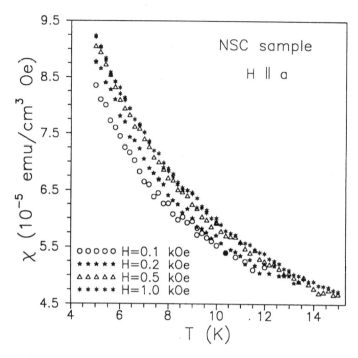

Fig. 4. ZFC volume susceptibility vs temperature for the nonsuperconducting sample, for various intensities of magnetic field parallel to the a–axis

(assuming ideal oxygen stoichiometry) that 0.5 holes/f.u. is transferred to the ladder plane where holes are expected to be delocalized. This transfer is too small to make the sample superconducting. But unexpectedly, the existence of sufficient mobile carriers concentration (at least in some small regions) has been confirmed studying the dependence of the magnetic susceptibility $\chi(T)$ as a function of magnetic field H. It was found that the magnetic susceptibility $\chi(T)$ depends on the magnetic field for $H < 1\,$kOe. This is shown in Fig. 4, where $\chi(T)$ measured in several magnetic fields is presented. In the higher field region (1–10 kOe), the $\chi(T)$ is magnetic field independent. Then in fields from 10 kOe to 50 kOe, it slightly decreases with increasing field. In Fig. 5, the magnetization data $M(H)$ at $T = 5\,$K is shown for $H < 800\,$Oe. One can see in the low field region a deviation of M to values lower than those of the linear relation $M(H)$ extrapolated from the field range of $1 - 10\,$kOe. The net deviation is also shown in the figure and in the enlarged scale in the inset. The results, shown in Figs. 4 and 5, together with a small but distinct difference between ZFC and FC susceptibilities observed in $H = 200\,$Oe (these results are not presented here) could indicate an appearance of the traces of superconductivity. The only way to explain these results is to assume the concept developed by Nagaev (see review paper

[27] for references) and Hiznyakov and Sigmund [28]. Recently, we have used this concept to describe the magnetic properties of $YBa_2Cu_3O_x$ [29] and La_2CuO_{4+x} [30,31]. According to this model, the holes introduced into CuO_2 planes create small spin polarized clusters (magnetic polarons) which, with an increasing hole concentration, start to overlap and build up a percolation network. From this point of view, the percolation network can exist also in the NSC sample, but in this case the percolation threshold is not reached. The external magnetic field (of magnitude higher than first penetration field) suppresses traces of superconductivity existing in the hole–rich region making magnetization of the sample field–dependent. A magnetic field of about 1 kOe is high enough to suppress the superconductivity in superconducting regions (superconducting drops) inside the sample. For higher magnetic fields, the magnetic susceptibility becomes field independent.

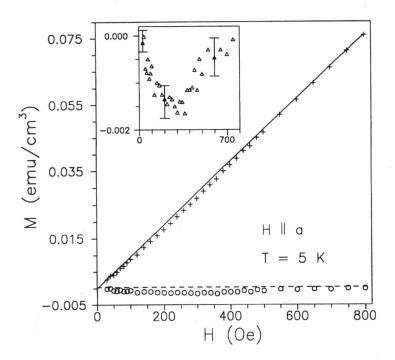

Fig. 5. Magnetization curve $M(H)$ at $T = 5\,K$ presented as crosses and the solid line extrapolated from a linear part of $M(H)$ $(1 - 10\,kOe)$. *Open circles* at the bottom of the figure present the difference between experimental and extrapolated data. The same difference is shown in more convenient scale in the *inset* (M and H are expressed in the inset in the same units as in the main figure)

In the above analysis of the properties of NSC samples, the important point was a possibility to determine n_d and n_f values from the temperature dependence of magnetic susceptibility $\chi(T)$. Such procedures cannot be directly applied to SC samples because of diamagnetic contribution below T_c. In order to suppress this contribution, the magnetization $M(T)$ for SC samples has been measured in a relatively high field of about 50 kOe. Results for the field cooled regime are visually quite similar to those presented in Figs. 3a and 3b. In the case of SC samples, similarly as for NSC samples, the relation $\chi_b > \chi_a \approx \chi_c$ is also fulfilled. For SC samples, the paramagnetism also contributes a significant amount to the susceptibility. So neglecting the contribution of superconductivity and applying the same fitting procedure as for NSC samples, we obtained the parameters presented in Table 2. Standard errors of n_f and n_d are in this case lower than those obtained for the NSC sample (being in the limit of 0.7% and 1.4%, respectively). However, since the applied experimental procedure could not suppress the superconductivity entirely, we should assume that these parameters, as compared with those for NSC samples, are determined with the same or even larger errors. In the case of SC samples, the values of g–factor components were taken the same for free Cu^{2+} ions and for Cu^{2+} ions in chains and equal to those taken for NSC samples.

Table 2. The set of fitting parameters describing the paramagnetic susceptibility of SC sample according to (2) (C_g and D in emuK/gOe; Θ and Δ/k in K)

Axis	$C_g \cdot 10^4$	Θ	n_f	$D \cdot 10^4$	Δ/k	n_d
a	1.942 ± 0.014	-12.3 ± 0.1	1.053	1.178 ± 0.016	138 ± 0.85	0.798
b	2.22 ± 0.01	-9.8 ± 0.06	0.990	1.402 ± 0.014	128.5 ± 0.6	0.782
c	1.677 ± 0.007	-8.89 ± 0.06	0.918	1.047 ± 0.01	124 ± 0.6	0.716

For the above studied SC sample, the sum $(n_f + 2n_d)$ is higher than that obtained for the NSC sample and equal to about 2.6. Formally calculated number of holes/f.u. for the SC sample is considerably higher (about 7/f.u.) than in the case of the NSC sample. This fact, together with relatively high number $(n_f + 2n_d)$, unambiguously indicates a large number of holes transferred from chains to ladders. This transfer is responsible for superconductivity in spin–ladder compounds. The large number of nominally determined holes may indicate that the measured sample is in an overdoped regime. In order to check this possibility, the relation between critical temperature T_c and nominal concentration of holes has been determined [32] taking samples with different T_c and different compositions. It is seen from Fig. 6 that the curve describing the dependence of T_c on the calculated number of holes/f.u. (n_h) is "bell shaped," as is the case for many other superconducting cuprate systems.

It is worth noting that the temperature dependence of electrical resistivity $\rho(T)$ for highly overdoped sample (Fig.7) has semiconductor–like character

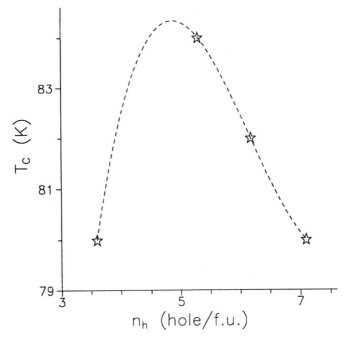

Fig. 6. Dependence of T_c on a calculated number of holes per f.u. The fitted line is only a guide to the eye

above T_c. A similar semiconductor–like temperature dependence of $\rho(T)$ was observed in Y-123, $(La,Sr)_2CuO_4$ and Bi-2212 superconductors, with various carrier concentration for out–of–plane resistivity [33–35].

The results of dc susceptibility measurements performed for the zero field cooled (ZFC) sample in the magnetic field $H = 10$ Oe applied along three crystallographic axes are presented in Fig. 8. Two features are well demonstrated in the Figure:

- the ZFC susceptibility (corrected for demagnetizing factor) is less than that of an ideal diamagnet;
- the susceptibility is highly anisotropic (it also concerns the FC susceptibility) but in contrast to the paramagnetic region, in the diamagnetic region the following dependence is observed: $|\chi_c| < |\chi_b| < |\chi_a|$.

The observed deviation of the magnetic susceptibility from the ideal $-1/4\pi$ value in ZFC regime is due to the very low value of the lower critical field H_{c1} which seems to be characteristic of this new family of high-T_c compounds. The anisotropy of high and low field magnetic susceptibility $\chi(T)$ is due to the anisotropy of H_{c1} and consequently due to the anisotropy of the carrier effective mass.

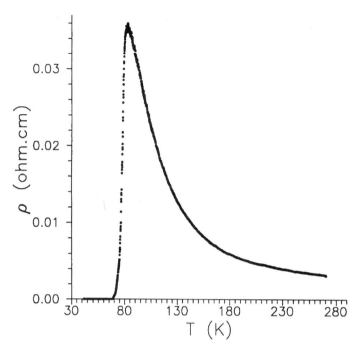

Fig. 7. Dependence of electrical resistivity ρ on temperature

In measurements of hysteresis loops, a considerable anisotropy has also been observed in the entire irreversible region of magnetization ($|M_c| <$ $|M_b| \leq |M_a|$). The width of the hysteresis loop, ΔM, defined as a difference of magnetization for increasing and decreasing magnetic field, for all investigated samples (in entire magnetic field range) was distinctly lower for $\boldsymbol{H} \parallel \boldsymbol{c}$ than for $\boldsymbol{H} \parallel \boldsymbol{b}$ and $\boldsymbol{H} \parallel \boldsymbol{a}$. The ratio $\Delta M_a/\Delta M_b$ depending on the sample (and on a field value) was between 1 and 2. The hysteresis loops for one of the samples, measured at 5 K for magnetic field applied along three crystallographic axes, are given in Fig. 9. The experimental points shown in Fig. 9 are corrected for the paramagnetic contribution taking the paramagnetic susceptibility described earlier in this section.

Taking the values of ΔM it was possible to evaluate the critical current density. The estimation was made using whole dimensions of crystals (neglecting any possible substructure as possible blocks or mesoscopic structure) and was treated as a lower limit of critical current density at 5 K and in $H = 0$. At first, we considered the case of an isotropic critical current using the Bean formula $J_c = 20\Delta M/a(1 - a/3b)$ ($a \leq b$) modified for a rectangular cross–section. The critical current density in $(\boldsymbol{a}, \boldsymbol{b})$ plane, corresponding to the smallest hysteresis loop, is in the limits $(3 - 5) \times 10^3$ A/cm^2, as estimated in such a way for several crystals. For $(\boldsymbol{a}, \boldsymbol{c})$ and $(\boldsymbol{b}, \boldsymbol{c})$ planes the

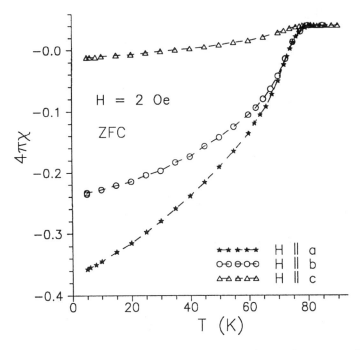

Fig. 8. ZFC susceptibility vs temperature for one of SC samples measured in $H = 2$ Oe applied along the a, b and c axes

estimated J_c values are several times higher. These results are indicative of the anisotropy of critical current densities and suggest that the critical currents probably fulfill a relation $J_{cc} > J_{cb} \geq J_{ca}$, where J_{ca}, J_{cb} and J_{cc} are critical currents along a, b and c axes, respectively. Taking into account the above we tried to apply the extended Bean model for a superconductor with anisotropic critical currents [36]. The results of such analysis of experimental data confirmed the relation between critical current densities along three crystallographic axes given above. For the crystal with data shown in Fig. 9, we estimated critical currents densities (expressed in 10^3 A/cm^2) as follows: $J_{ca} = 2$, $J_{cb} = 4$ and $J_{cc} = 15$. However, the anisotropy of J_c can be underestimated because of inconvenient shape of crystals and of weak ΔM dependence on the anisotropy [36].

Relatively low values of estimated critical currents could be characteristic of this new family of superconductors, but this fact can be also caused by some kind of granularity of the samples. The roughly estimated transport critical current along the c–axis being at least one order of magnitude smaller than magnetic critical current suggests the latter possibility. However, our X-ray measurements do not indicate on the existence of any texture or other large scale imperfectness of investigated single crystals, so one should assume

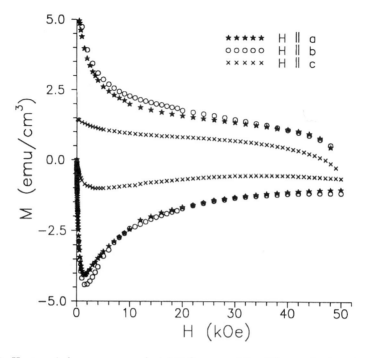

Fig. 9. Hysteresis loops measured at 5 K for one of the SC samples in a magnetic field applied along the main crystallographic axes. The experimental points shown in the figure are corrected for the paramagnetic contribution

the presence in the samples' specific granularity in the sense defined by Wohl-leben *et al.* [37]. According to [37], the crystal can be intersected by surfaces of molecular thickness and some of them are related to various structural defects. The defect surfaces can subdivide the crystal into superconducting blocks. The crucial point of the model [37] is the assumption that the size of superconducting blocks is of the order of the penetration depth. In frames of this "granular" or "mesoscopic defect" model it is very easy to explain low values of transport J_c (because of weak links between grains), underestima-tion of magnetic J_c, as well as the incomplete shielding. The existence of the weak links and/or Josephson junctions is to some extend confirmed by our ESR measurements. Taking into account the fact that grains or blocks have the same crystallographic orientations, and assuming that their dimensions are yet sufficient for existence of regular fluxon network and that supercon-ducting fraction in our crystals is close to 100%, we have made an attempt to estimate critical state parameters in frames of the anisotropic Ginzburg–Landau model.

According to the standard Ginzburg–Landau theory, the reversible magnetization depends on the magnetic field [38] as follows:

$$M(H,T) = \frac{\Phi_0}{32\pi^2\lambda^2} \ln\left[\frac{\eta H_{c2}}{eH}\right] , \qquad (2)$$

where λ is the penetration depth, Φ_0 ($= h/2e$) is the flux quantum and η is a constant of the order of unity. Equation (2) is commonly used to determine the value of the penetration depth λ and the upper critical field H_{c2}.

Anisotropic superconductors, such as high-T_c superconductors, are not well described by the isotropic Ginzburg–Landau theory. The simplest way to modify the theory for layered superconductors is to incorporate anisotropy *via* an effective mass tensor. Such a procedure was developed by Kogan and collaborators [38–41]. It is possible to extend the Kogan results to treat orthorhombic superconductors:

$$M(\boldsymbol{H} \parallel \boldsymbol{x}_i) = \frac{\Phi_0\sqrt{m_i}}{32\pi^2\lambda^2} \ln\left[\frac{\eta H_{c2}(\boldsymbol{H} \parallel \boldsymbol{x}_i)}{eH}\right] ,$$

where $\boldsymbol{x}_i$ axes correspond to principal crystallographic directions, $\boldsymbol{a}$, $\boldsymbol{b}$ and $\boldsymbol{c}$ for i equal 1, 2 and 3, respectively, m_i is a dimensionless normalized effective mass ($m_i = M_i/(M_1 M_2 M_3)^{1/3}$), M_i is the effective mass along the direction $\boldsymbol{x}_i$. The penetration depth λ_i associated with the component of the screening current flowing along $\boldsymbol{x}_i$ is $\lambda_i = \lambda(m_i)^{1/2}$. Since $m_1 m_2 m_3 = 1$

$$\lambda = \sqrt[3]{\lambda_1\lambda_2\lambda_3} .$$

It can be shown [42] that

$$H_{c2}(\boldsymbol{H} \parallel \boldsymbol{x}_1) = \frac{\Phi_0}{2\pi\xi_j\xi_k} , \qquad i \neq j \neq k ,$$

where $\xi_i = \xi(m_i)^{1/2}$ is the coherence length along $\boldsymbol{x}_i$ direction and $\xi = (\xi_1\xi_2\xi_3)^{1/3}$.

The values of λ and $H_{c2}(\boldsymbol{H} \parallel \boldsymbol{x}_i)$ were derived from measurements of reversible magnetization $M(H,T)$, for magnetic field along the principal crystallographic directions, in the temperature region $30 - 60$ K. This temperature range was chosen taking into account that from one side we would like to have a sufficiently large field range where $H > H_{irr}$, and from another side we would like to limit the influence of thermal fluctuations on $M(H)$ dependence. Both paramagnetic and superconducting contributions to the reversible magnetization, $M_{rev}(H)$, can be expressed in the form:

$$M_{rev}(H) = c_1 + c_2 H + c_3 \ln H , \qquad (3)$$

where c_1, c_2 and c_3 are constants for a fixed temperature, chemical composition, and orientation of the sample with respect to the applied magnetic field. The second term (c_2H) describes the paramagnetic contribution to the reversible magnetization while the others describe the behavior of a superconductor in the broad field domain $H_{c1} \ll H \ll H_{c2}$ [38]. By fitting (3) to experimental data it was possible to separate both components of magnetization. It is worth to mention that a difference between the paramagnetic susceptibility determined in this way and that determined from $M(T)$ measurements in 50 kOe does not exceed 5%. After subtraction of paramagnetic contribution from $M_{rev}(H)$, the remaining net superconducting component of magnetization can be presented as a linear relation of M versus $\ln H$. As an example, in Fig. 10 such dependences at various temperatures are presented for H along the a axis.

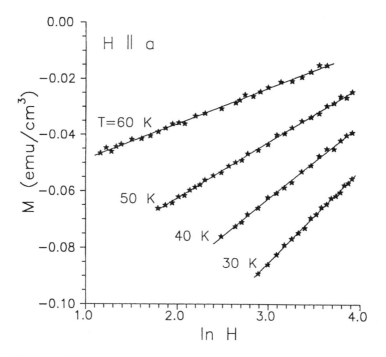

Fig. 10. Reversible magnetization vs $\ln H$ at several temperatures for H applied along a–axis

One should realize that such a large contribution from the paramagnetic component to the magnetization (as in the case of our samples) significantly limits the precision of λ and H_{c2} determination. We evaluate that the λ values are estimated with errors of 20% to 30%. The upper critical field was determined with considerably higher error than the penetration depth because of two following reasons: (i) the uncertainty in the value of η, which depends on the geometry of the vortex lattice and which can be determined only together with H_{c2} (we assume that $\eta = 1.4$); (ii) much stronger influence of errors in determination of parameters c_1, c_2, c_3 from (3) on H_{c2} than on λ.

The penetration depths $\lambda_a(T)$, $\lambda_b(T)$ and $\lambda_c(T)$, as derived from the experimental data, are plotted in Fig. 11 as $\lambda^{-1/2}$ *versus* $(1 - (T/T_c)^2)$. One can see that the function $\lambda_i(T) = \lambda_i(0)((1 - (T/T_c)^2)^{-1/2}$ satisfactorily describes the experimental data. From the fitting of above formula, we have obtained the following $\lambda_i(0)$ values: $\lambda_a \approx \lambda_b \approx 2.5\,\mu m$ and $\lambda_c \approx 0.6\,\mu m$. The T^2 dependence of λ_i^{-2} has been observed previously in $YBa_2Cu_3O_{6+x}$ fine particles [43] and in cobalt and zinc doped $YBa_2Cu_3O_{6+x}$ [44] but its physical meaning is yet unclear. It has been suggested that this dependence arises due to the strong temperature dependence of the mean free path of carriers leading to the additional temperature dependent factor contributing to $\lambda(T)$. The T^2 term may also arise as a consequence of unconventional pairing (see [45] for details and references). Such a mechanism seems to be very attractive, but to confirm its existence further experiments are needed.

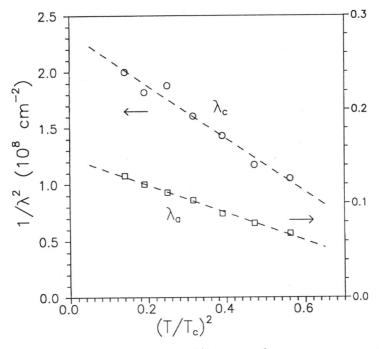

Fig. 11. Temperature dependences of $\lambda^{-1/2}$ vs $(T/T_c)^2$ for a and c axes. Values of $\lambda^{-1/2}$ for the b–axis are not presented (as being close to those for the a–axis) to make the figure more legible

Taking H_{c2} values estimated for series of temperatures with H along a, b and c axes, we estimated the values of $\xi_i(T)$. Assuming that the temperature dependence of the coherence length could also be described by the function describing $\lambda(T)$ (in order to make the Ginzburg–Landau parameter κ temperature independent) it was possible to roughly estimate $\xi_i(0)$ values, obtaining 20 Å, 13 Å and 45 Å for $i = a, b, c$, respectively. The fact that the ξ value for c axis is distinctly larger than those for a and b axes confirms the expectations that the observed superconductivity has quasi–one dimensional character and should be related to the ladders or chains.

Taking the values of $\lambda_i(0)$ and $\xi_i(0)$, and realizing that these last values are only roughly estimated, we have derived from them κ_i and $H_{c2}(T = 0, i)$ values. The obtained results are as follows: $\kappa_i = 600, 400, 1600$ and $H_{c2}(0, i) = 560\,\mathrm{kOe}, 370\,\mathrm{kOe}, 1300\,\mathrm{kOe}$, for $i = a, b, c$, respectively. The high values of κ indicate that the SC samples can be described as an extreme type II limit.

4   Spin–Ladder System $(M_2Cu_2O_3)_7(CuO_2)_{10}$

The spin–ladder compounds $(M_2Cu_2O_3)_7(CuO_2)_{10} = M_{14}Cu_{24}O_{41}$ (where $M = $ La, Sr, Ca) were initially reported by McCarron et al. [15] and Siegrist et al. [12]. These materials were intensively studied using various experimental methods. The existence of the spin gaps in both ladder and chain sites was confirmed on the base of magnetic measurements [23,46], inelastic neutron scattering [47–49], NMR [50–53] and ESR [23].

The $Sr_{14}Cu_{24}O_{41}$ is the parent compound for this family of spin–ladders. The average valency of Cu ions in the $Sr_{14}Cu_{24}O_{41}$ is +2.25, which means that holes are inherently doped. It was shown that the holes only enter the CuO_2 chains [24,46] and that they form there nonmagnetic CuO_2 units breaking up the chains. The inelastic neutron scattering experiments have shown that the residual Cu ions in the chains form the dimers, and that the dimers are formed between Cu ions which are separated by 2 and 4 times the distance between the nearest–neighbor copper ions [47]. Recently performed inelastic neutron scattering experiments on $Sr_{14}Cu_{24}O_{41}$ single crystals [54] have confirmed a model which has an interdimer distance of four Cu–Cu distances. The NMR experiments have confirmed the neutron studies showing Cu^{+2} ions in dimerized state and Cu^{+3}–like sites [55–57].

The substitution of Ca in the Sr place increases the conductivity of $Sr_{14-x}Ca_xCu_{24}O_{41}$ [24,46] although the average Cu valence does not change due to the isovalence of the Ca substitution. This is because the Ca doping leads to an increase in the chemical pressure inducing efficient transfer of holes from the chains to the ladders [46,56,57]. The holes in ladders become mobile and could take a part in formation of conditions necessary for an appearance of superconductivity [7]. The charge redistribution upon Ca doping has been confirmed by optical studies [58]. It follows from these measurements that the holes obtain itinerary only when transferred onto the ladders and that the electronic properties of the $Sr_{14-x}Ca_xCu_{24}O_{41}$ system are determined by the holes doped into ladders, while a substantial number of holes localized in the chains are dimerized. It has also been established that upon Ca doping, the spin gap for the chain is nearly independent of the Ca concentration, but the spin gap for the ladder decreases with doping [51].

Recently, superconductivity with T_c = 12 K has been found in $Sr_{0.4}Ca_{13.6}Cu_{24}O_{41.84}$ under pressure of 3 GPa [7]. The application of the hydrostatic pressure to this system has an effect similar to Ca substitution, increasing the hole transfer from the chains to the ladders [59]. The resistivity in this system shows remarkable anisotropy confirming the one–dimensional charge transport in the ladders. The electrical resistivity measurements performed on a single crystal of $Sr_{2.5}Ca_{11.5}Cu_{24}O_{41}$ under high hydrostatic pressure have shown a dimensional crossover in the charge dynamics from one to two suggesting that the superconductivity in the ladder compounds might be a phenomenon in a 2D anisotropic electronic system [60].

The Hall effect measurements performed on crystals in which superconductivity (under pressure) has been observed indicate that the carrier density increases with increasing pressure and that localization of hole carriers at low temperature is substantially suppressed by applying pressure [61]. Another important result has been obtained quite recently [53]. The first measurements of ^{17}O NMR in undoped and hole doped spin–ladders clearly demonstrate the evolution of spin excitations as a function of both temperature and hole doping. It has been shown, in particular, the existence of the crossover from the spin gap to the paramagnetic regime with increasing of temperature. The analysis of the electrical and magnetic anomalies in the spin–ladder cuprate $Sr_{14-x}Ca_xCu_{24}O_{41}$ leads to the conclusion that the holes in the ladder are paired and localized at low temperatures where almost all copper ions in the ladder complete to form spin–singlet pairs [62].

5   Conclusions

We have reviewed the structural, magnetic and transport properties of the new family of high–temperature superconductors. It has been shown that the superconductivity in this system is determined not only by the hole concentration but also by the effective hole transfer from the CuO_2 chains to the two–leg ladders. The effectiveness of this transfer depends strongly on the size of M ions. On the basis of magnetic measurements performed on the 5/7 system it has been shown that the spin–ladder superconductors should be described as an extreme type II limit.

Acknowledgements

The work was partly supported by Polish State Committee for Scientific Research (KBN) under grant No 2 P03B 102 14.

References

1. Hase M., Terasaka I., Uchinokura K. (1993) Phys Rev Lett 70:3651
2. Darriet J., Regnault L.P. (1993) Solid State Commun 86:409
3. Dagotto E., Riera J., Scalapino D. (1992) Phys Rev B 45:5744
4. Dagotto E., Rice T.M. (1996) Science 271:618
5. Rice T.M., Gopalan S., Sigrist M. (1993) Europhys Lett 23:445–449
6. Leonyuk L., Babonas G.–J., Vasil'ev A., Szymczak R., Reza A., Maltsev V., Ponomarenko L. (1996) Czech J Phys 46(S3):1457–1458
7. Uehara M., Nagata T., Akimitsu J., Takahashi H., Mori N., Kinoshita K. (1996) J Phys Soc Jpn 65:2764–2767
8. Leonyuk L., Rybakov V., Sokolova E., Maltsev V., Shvanskaya L., Babonas G.–J., Szymczak R., Szymczak H., Baran M. (1998) Z Krist 213:406–410
9. Leonyuk L., Babonas G.- J., Rybakov V., Sokolova E., Szymczak R., Maltsev V., Shvanskaya L. (1998) J Phys Chem Solids 59:1591
10. Leonyuk L., Maltsev V., Babonas G.J., Reza A., Szymczak R. (1998) Int J Mod Phys B 12:3110–3112
11. Karpinski J., Schwer H., Meijer G.I., Conder K., Kopnin E.M., Rossel C. (1997) Physica C 274:99–106
12. Siegrist T., Schneemeyer L.F., Sunshine S.A., Waszczak J.V., Roth R.S. (1988) Mat Res Bull 23:1429–1438
13. Slobodin B.V., Fotiev A.A., Kosminin A.S., Shtern G.E., Garkushin N.K., Balashov V.L., Trunin A.S. (1990) Sverkhprovodimost' 3:523–526 (in Russian)
14. Vallino M., Mazza D., Abbattista F., Brisi C., Lucco-Borlera M. (1989) Mater Chem Phys 22:523–529
15. McCarron E.M., Subramian M.A., Calabrese J.C., Harlow R.L. (1988) Mater Res Bull 23:1355
16. Arakcheeva A.V., Shlyapkina L.V., Leonyuk L.I., Lubman G.U. (1996) Kristallografiya 41:43 (in Russian)
17. Babonas G.–J., Leonyuk L., Galickas A., Reza A., Shvanskaya L., Dapkus L., Maltsev V. (1997) Supercond Sci Technol 10:496–501
18. Szymczak R., Szymczak H., Baran M., Mosiniewicz-Szablewska E., Leonyuk L., Babonas G.–J., Maltsev V., Shvanskaya L. (1999) Physica C 311:187–196
19. Szymczak H., Szymczak R., Baran M., Leonyuk L., Babonas G.- J., Maltsev V. (1999) In: Majchrowski A., Zieliński J. (Eds.) Single Crystal Growth, Characterization, and Applications, Proc SPIE 3724:22–32
20. Bonner J.C., Fisher M.E. (1964) Phys Rev 135:A640–A658
21. Szymczak R., Szymczak H., Leonyuk L., Babonas G.- J. (1998) J Magn Magn Mater 177–181:711
22. Mosiniewicz-Szablewska E., Szymczak H., Baran M., Szymczak R., Świątek K. (1999) Phys Rev B to be published
23. Matsuda M., Katsumata K. (1996) Phys Rev B 53:12 201–12 205
24. Carter S.A., Batlogg B., Cava R.J., Krajewski J.J., Peck W.F., Rice T.M. (1996) Phys Rev Lett 77:1378
25. Hase M., Terasaki I., Uchinokura K. (1993) Phys Rev Lett 70:3651–3654
26. Zhang F.C., Rice T.M. (1988) Phys Rev B 37:3759–3761
27. Nagaev E.L. (1995) Usp Fiz Nauk 165:529–554 (in Russian)

28. Hiznyakov V., Sigmund E. (1988) Physica C 156:655–666
29. Gnatchenko S.L., Ratner A.M., Baran M., Szymczak R., Szymczak H. (1997) Phys Rev B 55:3876–3885
30. Szymczak H., Szymczak R. (1996) Phase Trans B 57:59–68
31. Baran M., Szymczak H., Szymczak R. (1995) Europhys Lett 32:79–84
32. Baran M., Szymczak R., Szymczak H., Leonyuk L., Babonas G.-J. (1999) In: CIMTEC'98, 14-19 June 1998, Florence, Italy, in press
33. Ito T., Nakamura Y., Takagi H., Uchida S. (1991) Physica C 185–189:1267–1268
34. Heine G., Lang W., Wang X.L., Wang X.Z. (1997) Physica C 282–287:1561–1562
35. Yan Y.F., Matl P., Harris J.M., Ong N.P. (1995) Phys Rev B 52:R751–R754
36. Gyorgy E.M., van Dover R.B., Jackson K.A., Schneemeyer L.F., Waszczak J.V. (1989) Appl Phys Lett 55:283
37. Wohlleben D., Michels G., Ruppel S. (1991) Physica C 174:242
38. Kogan V.G., Fang M.M., Mitra S. (1988) Phys Rev B 38:11 958
39. Kogan V.G. (1981) Phys Rev B 24:1572
40. Campbell L.J., Doria M.M., Kogan V.G. (1988) Phys Rev B 38:2439
41. Kogan V.G. (1988) Phys Rev B 38:7049
42. Clem J.R. (1994) In: Bontemps N., Bruynseraede Y., Deutcher G., Kapitulnik A., (Eds.) The Vortex State, Kluwer, Dordrecht, 25
43. Fleisher V.G., Laiho R., Lähderanta E., Stepanov Y.P., Traito K.B. (1996) Physica C 264:295
44. Porch A., Cooper J.R., Zheng D.N., Waldram J.R., Campbell A.M., Freeman P.A. (1993) Physica C 214:350
45. Blazey K.W. (1990) In: Bednorz J.G., Müller K.A. (Eds.) Earlier and Recent Aspects of Superconductivity, Springer, Berlin Heidelberg, 262
46. Kato M., Shiota K., Koike Y. (1996) Physica C 258:284
47. Matsuda M., Katsumata K., Eisaki H., Motoyama N., Uchida S., Shapiro S.M., Shirane G. (1996) Phys Rev B 54:12 199
48. Eccleston R.S., Azuma M., Takano M. (1996) Phys Rev B 53:R14 721
49. Matsuda M. (1998) Physica B 241–243:758
50. Tsuji S., Kumagai K., Kato M., Koike Y. (1996) J Phys Soc Jpn 65:3474
51. Kumagai K., Tsuji S., Kato M., Koike Y. (1997) Phys Rev Lett 78:1992
52. Kitaoka Y., Magishi K., Matsumoto S., Ishida K., Ohsugi S., Asayama K., Uehara M., Nagata T., Akimitsu J. (1998) J Magn Magn Mater 177–181:487
53. Imai T., Thurber K.R., Shen K.M., Hunt A.W., Chou F.C. (1998) Phys Rev Lett 81:220
54. Eccleston R.S., Uehara M., Akimitsu J., Eisaki H., Motoyama M., Uchida S. (1998) Phys Rev Lett 81:1702
55. Takigawa M., Motoyama N., Eisaki H., Uchida S. (1998) Phys Rev B 57:1124
56. Magishi K., Matsumoto S., Kitaoka Y., Ishida K., Asayama K., Uechara M., Nagata T., Akimitsu J. (1998) Phys Rev B 57:11 533
57. Carretta P., Ghina P., Lascialfari A. (1998) Phys Rev B 57:11 545
58. Osafune T., Motoyama N., Eisaki H., Uchida S. (1997) Phys Rev Lett 78:1980
59. Motoyama N., Osafune T., Kakeshita T., Eisaki H., Uchida S. (1997) Phys Rev B 55:R3386
60. Nagata T., Uehara M., Goto J., Akimitsu J., Motoyama N., Eisaki H., Uchida S., Takahashi H., Nakanishi T., Mori N. (1998) Phys Rev Lett 81:1090

61. Nakanishi T., Mori N., Murayama C., Takahashi H., Nagata T., Uehara M., Akimitsu J., Kinoshita K., Motoyama N., Eisaki H., Uchida S. (1998) J Phys Soc Jpn 67:2408
62. Adachi T., Shiota K., Kato M., Noji T., Koike Y. (1998) Solid State Commun 105:639

Vortex Pinning and Critical Current Enhancements in High-T_c Superconductors with Fission–Generated Random Columnar Defects

J. R. Thompson[1,2], J. G. Ossandón[3], L. Krusin-Elbaum[4], K. J. Song[2], D. K. Christen[1], J. Z. Wu[5], and J. L. Ullmann[6]

[1] Oak Ridge National Laboratory, Oak Ridge, TN 37831–6061, USA
[2] Dept. of Physics, University of Tennessee, Knoxville, TN 37996–1200, USA
[3] Facultad de Ingeniería, Universidad de Talca, Curicó, Chile
[4] IBM Watson Research Center, Yorktown, NY 10598, USA
[5] Department of Physics, University of Kansas, Lawrence, KN 66045, USA
[6] Los Alamos National Laboratory, Los Alamos, NM 87545, USA

Abstract. Randomly oriented columnar defects can efficiently increase and stabilize the current density in many high-T_c superconductors. Irradiation with 0.8 GeV protons induces a prompt fission of heavy constituent nuclei, such as Hg, Tl, Pb, and Bi, in the cuprate compounds; then recoiling fragments generate randomly oriented columnar tracks. This markedly enhances the persistent current density J, elevates the irreversibility line to higher fields and temperatures, and reduces the temporal rate of current decay. The associated crystalline disorder depresses T_c by $\sim 0.1 - 1$ K per 1016 proton/cm^2. At optimal proton fluency, J is enhanced by one or more orders of magnitude (compared with unirradiated virgin materials) and the logarithmic decay rate $\mathrm{d}\ln(J)/\mathrm{d}\ln(t)$ is diminished. However, we show that while these vortex pins greatly reduce thermally activated current decay in the highly anisotropic material Bi-2212, significant temperature–independent current decay remains, due to quantum tunneling of vortices. An analysis of the thermally-induced current decay in a "Maley" framework provides the effective pinning energy $U(J,T)$ of irradiated materials, for comparison with the virgin superconductors. The influence of quantum decay on this analysis is shown.

1 Introduction

For many of the large scale applications envisioned for high-T_c superconductors (HTS), the materials must conduct high density currents with minimal dissipation, often in the presence of a large magnetic field. This combination of current and magnetic field generally produces a Lorentz force that tends to drive magnetic flux through the superconductor. However, movement of flux lines (vortices) through the effectively viscous medium dissipates energy – thus it is imperative that the magnetic flux be immobilized as well as possible. More specifically, the motion of vortices comes from a Lorentz–like driving force (per unit volume) of the form $\boldsymbol{f} = \boldsymbol{J} \times \boldsymbol{B}/c$, where $\boldsymbol{B} = \boldsymbol{H} + 4\pi\boldsymbol{M}$ is the magnetic induction, $\boldsymbol{J}$ the macroscopic current density, and c the speed of light. The displacement velocity of the vortices $\boldsymbol{v}$ is usually perpendicular to B, so that there is inside the sample an electric field $E = B \times v/c$, which is parallel to J. Consequently, a positive power dissipation $P = J \cdot E$ occurs unless v (and hence E) vanishes. Therefore, for a loss–free supercurrent to exist, the vortices must be anchored or "pinned" in the crystal. This anchoring is best achieved by creating defects of optimal morphology in the material [1], with *columnar defects* providing nearly optimal pinning of vortices [2]. In the first studies, the defects were formed by irradiation with heavy ions. The ions transfer considerable energy to the electronic system and thereby amorphize the material along the path of the ion. Typically the tracks are long, $\sim 10\,\mu m$ or more, have a diameter of 5–10 nm, and are nominally straight and parallel. Unfortunately, heavy ions generally have a limited range in matter of a few tens of μm's, so forming columnar defects through protective cladding or thick conductors is difficult.

An alternative method [3] for forming columnar tracks uses 0.8 GeV protons, which are very penetrating. These ions have a range of ~ 0.6 *meters* in a high-T_c material. In this process, an incident proton is absorbed by a heavy nucleus such as Bi, Hg, Tl, Pb,..., which are natural constituents of most (but not all) families of the highest-T_c superconductors. In these high-Z nuclei, the cross–section [4] $\sigma_f \sim 100$ mb for prompt fission is moderately large; σ_f increases approximately exponentially with the quantity Z^2/A, where Z is the atomic number and A is the atomic mass. Thus none of the components of YBaCuO (or the rare–earth–based counterparts) are directly suitable for this process. When absorbed by a heavy nucleus, the energetic proton first causes spallation of several nucleons, then the highly excited nucleus fissions into two fragments with approximately half of the mass and charge of the parent nucleus and energy of ~ 100 MeV. The recoiling fragment generates a columnar track deep within bulk material. As the fission process is nominally random in direction, the tracks are randomly distributed in angle, i.e. they are strongly splayed. Previously we have shown transmission electron micrographs of the resulting defect morphology [5]. Here we survey results from applying this methodology in several high-T_c families, including Bi-2212/Ag tapes [6,7] and other HTS materials [8], including a series of Hg–cuprates [9].

2   Experimental Aspects

Materials investigated include c–axis–oriented Bi-2212/Ag tapes [10], bulk polycrystalline [11] $(TlPb)(SrBa)_2Ca_2Cu_3O_x$; and epitaxial [12], high-J_c Hg-1223 thin films of $HgBa_2Ca_2Cu_3O_x$ (0.4 μm thick). With a thin film, an "amplifier foil" of heavy metal (Pb or Au) was sometimes placed in front of it to increase the areal density of columnar defects. For the Hg-1223 films, a 40 μm foil of Pb was used; fragments from fission of Au or Pb escape from the last 2–3 μm of foil and create additional splayed defects in the film. This produces a greater density of defects in the film for a given proton fluency. Characterization methods include magnetic studies using a SQUID–based magnetometer (Quantum Design MPMS-7) and some dc transport measurements. Precharacterized samples were irradiated in air at the WNR facility at LANSCE (Los Alamos Neutron Science Center) using 0.8 GeV protons. The ion fluency, with a typical beam density of $< 1\,\mu A/cm^2$, was obtained both from the integrated beam current and activation of Al dosimetry foils. The volume density of fission events is $\Phi_p\,\sigma_f\,(N/V)$, where Φ_p is the proton fluency; σ_f is the fission cross–section; N/V is the number density of fissionable nuclei. For comparison with the vortex density, it is convenient to reexpress this as an areal density of columnar defects, in units of flux density, B_Φ. Thus we estimate

$$B_\Phi \approx \Phi_p\,\sigma_f\,(N/V)\,\ell\,\varphi_0 , \qquad (1)$$

where φ_0 is the flux quantum and ℓ is either the track length in bulk material ($\sim 7\,\mu m$), the escape depth from an amplifier foil ($\sim 3\,\mu m$), or the thickness of a thin film without amplifier foil.

3   Influence on T_c

The columnar defects are amorphized regions in the material and this depresses somewhat the overall order parameter of the superconductor. Thus the transition temperature T_c is progressively reduced, as shown in Fig. 1 for polycrystalline Hg-1223 materials and textured Bi-2212/Ag tapes. In each case, the T_c decreases approximately linearly with proton fluency Φ_p, but at markedly different rates. Both features can be understood qualitatively, if one assumes that the decrease ΔT_c is proportional to the number density of fission events that damage the superconductor. Then the rate $-\Delta T_c/\Delta\Phi_p$ should be higher for Bi-2212 than for Hg-1223, as the density (N/V) is larger with two (rather than one) fissionable nucleus per unit cell. Furthermore, the cross–section σ_f for Bi is significantly larger than for Hg. Offsetting these differences is the fact that the creation rate for columnar defects increases with the same factors. To zero-th order, then, the depression in T_c is proportional to the matching field B_Φ (assuming that other factors such as the recoil energy and ℓ do not change significantly).

4   Enhancements in Current Density J

To illustrate the benefits of the randomly oriented columnar defects, Fig. 2 shows the transport current density J_c measured in a thin film of textured Tl-1223 superconductor that was deposited on polycrystalline yttria–stabilized zirconia. Results are shown as a function of magnetic field $H \parallel c$–axis for the same film, prior to irradiation (virgin) and after irradiation with 2×10^{16} protons/cm^2. This produced a density of defects $B_\Phi \approx 0.1\,T$; no amplifier foil was used. The critical current density was defined with the standard criterion of $1\,\mu V/cm$. As is evident, J_c increased markedly, with

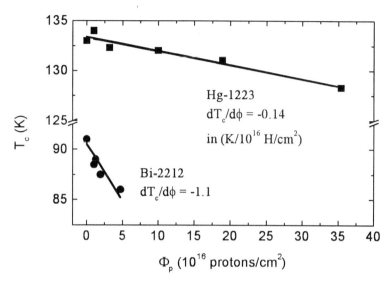

Fig. 1. The transition temperature T_c for Bi-2212 and Hg-1223 superconductors, irradiated with 0.8 GeV protons

much of the increase occurring at fields exceeding the nominal defect density B_Φ.

Let us next consider a bulk superconductor, in the form of polycrystalline $(TlPb)(SrBa)_2Ca_2Cu_3O_x$. This material was irradiated with 5.5×10^{16} p/cm², forming columnar defects with a matching field $B_\Phi \approx 0.7$ T. This fluency depressed the T_c by 1.7 K from the value of 117 K for the unirradiated (virgin) material. The defect density lies somewhat below, but near the optimal value for enhancing J_c, as discussed below. Figure 3 compares the magnetization $M(H)$ of the virgin and irradiated materials at $T = 100$ K. Irradiation–induced defects have clearly increased the magnetic hysteresis ΔM; as represented by the vertical arrow in the figure, this is the difference in M between the lower (increasing H) and upper (decreasing H) branches of the hysteresis curve. In the Bean critical state model [13], ΔM is directly proportional to the circulating persistent current density $J = 15 \, \Delta M/r$, where r is the radius of a cylindrical sample. With this polycrystalline material, r is mean grain radius of ~ 4 μm. (For other configurations of materials, we set $r = a/2 \sim 0.15$ cm for the roughly square thin films with edge length a; for the Bi-2212/Ag tapes, $r =$ platelet size ~ 100 μm). Thus the increased hysteresis ΔM is direct evidence for a substantially increased intra–granular current density J. In the present case, the critical state model gives that at $H = 1$ T, the irradiated material has $J = 1 \times 10^4$ A/cm², while it is immeasurably small (< 500 A/cm²) in the virgin superconductor. This also means that the irreversibility line (IL) $H_{irr}(T)$ is pushed upward by the random columnar defects. The result for this material is noteworthy, as the IL

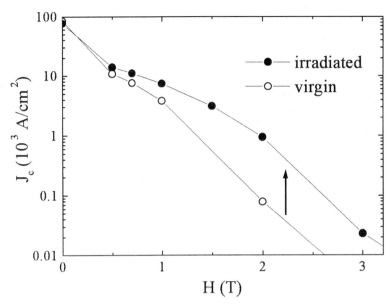

Fig. 2. Transport critical current density J_c versus magnetic field H, applied perpendicular to the plane of a textured film of Tl-1223 deposited on a polycrystalline YSZ substrate. The film was irradiated, with no "amplifier foil", with $\Phi_p = 2 \times 10^{16}\,\mathrm{H/cm^2}$, giving $B \approx 0.1\,\mathrm{T}$

in $(\mathrm{TlPb})(\mathrm{SrBa})_2\mathrm{Ca}_2\mathrm{Cu}_3\mathrm{O}_x$ is already higher than those obtained for either the isomorphic parent compound Tl-1223 or Tl(BaSr)-1223 [11]. Increases of the IL expand the useful $H - T$ region for potential applications.

Doping a material with columnar defects destroys the superconductivity locally, leading to vortex pinning. Qualitatively, a greater density of pins might be expected to support an even higher J. However, more defects also depress the overall order parameter and T_c, as noted above. Also, once there are enough defects to pin all the vortices ($B_\varphi > B$), the addition of more columnar tracks may even promote depinning, due to the proximity of nearby potential wells. Thus the optimization of fluency and pin density can depend on the range of magnetic field and temperature under consideration. To illustrate some of these features, Fig. 4 shows the dependence of J on proton fluency (i.e. defect density) for the series of $(\mathrm{TlPb})(\mathrm{SrBa})_2\mathrm{Ca}_2\mathrm{Cu}_3\mathrm{O}_x$ materials at temperatures of $70 - 100\,\mathrm{K}$, for $H = 1\,\mathrm{T}$. The addition of defects causes a rapid initial increase in J, followed by a broad maximum. At still higher fluencies, the persistent current density is expected to decrease for the reasons cited.

The next family to consider is the group of Hg–cuprate superconductors. For a different morphology, let us consider an epitaxial film of Hg-1223. The

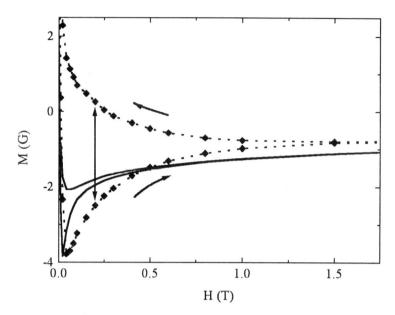

Fig. 3. Magnetization M versus field H for polycrystalline (TlPb)(SrBa)$_2$Ca$_2$Cu$_3$O$_x$ superconductor at T = 100 K, before (*solid line*) and after (*discrete symbols*) irradiation with 5.5×10^{16}/cm^2 protons of 0.8 GeV energy. The increased width of the hysteresis loop ΔM (represented by the vertical double arrow) shows that the persistent current density increased significantly in the material

film of 0.4 μm thickness was irradiated through a 40 μm Pb foil with a fluency of 1.7×10^{16} p/cm^2, giving $B_\varphi \approx 0.4$ T. Figure 5 shows the persistent current density J versus H at temperatures of 40 and 60 K, for a virgin film (open symbols; T_c = 126.5 K) and the same film irradiated (filled symbols; T_c = 118 K). In the virgin film, J falls off quickly and quasi–exponentially with field, due to very fast thermally activated flux creep. After irradiation, J is markedly increased (except at low fields) and exhibits increases by factors of 10–100 or even more near the virgin IL. The addition of defects reduced both the flux creep and the field dependence of J. In other studies of bulk Hg–based cuprate superconductors that were irradiated with 0.8 GeV protons, transmission electron microscopy revealed the presence of randomly oriented columnar defects with the expected number density. Corresponding studies with other defect morphologies, collision cascades from fast neutron irradiation [14] or point–like defects produced by MeV proton irradiations [15], show considerably smaller enhancements of the magnetic hysteresis ΔM and J. These comparisons give experimental evidence that these extended defects – randomly distributed fission–generated tracks – provide qualitatively better

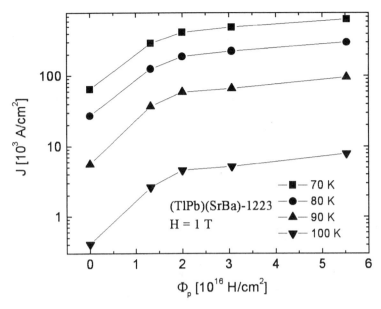

Fig. 4. The dependence of J on fluency of 0.8 GeV protons, for $(TlPb)(SrBa)_2Ca_2Cu_3O_x$ at 80 K in an applied field $H = 1$ T. At the highest fluency (where the matching field $B_\varphi \approx 0.7$ T), the current density is near its maximum

vortex pinning, particularly at elevated temperatures and in large magnetic fields.

The Hg–cuprate family of superconductors is scientifically interesting [16], in that the structure is tetragonal and the first three members are synthesized relatively easily, all with high T_c's. The superconductive anisotropy γ increases significantly as the number of adjacent CuO layers increases. The effect of anisotropy is to reduce the in–plane component of magnetic field H that is applied at some arbitrary angle, while leaving the normal component unaffected [17]; the rescaled field is shifted toward the c–axis. Qualitatively, the angularly dispersed columnar defects are similarly refocused, creating a narrowed cone of angles, and the extent of rescaling increases as γ increases. Thus a comparative study [18] of Hg-1201, Hg-1212, and Hg-1223 polycrystalline materials showed that high anisotropy effectively recreated an approximately parallel array of columnar defects in Hg-1223. Evidence for this rescaling came from studies of the vortex dynamics, which revealed "variable range hopping" of vortices, similar to the VRH dynamics observed with parallel columns in $YBa_2Cu_3O_{7-\delta}$ crystals. In both materials, there is an easy movement of flux lines that causes a rapid, thermally activated decay of J, until some other mechanism stabilizes the system of vortices.

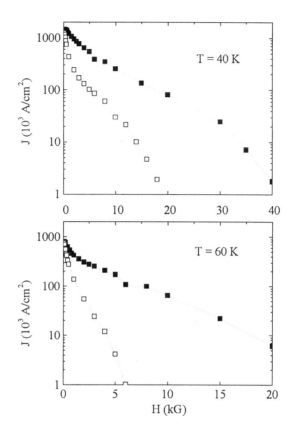

Fig. 5. The persistent current density J versus magnetic field H for a Hg-1223 film (a) at 40 K and (b) at 60 K. The film was irradiated with 1.7×10^{16} p/cm^2 through an amplifier foil of 40 μm Pb

5   Thermal and Quantum Tunneling of Vortices in Bi-2212

To illustrate more quantitatively the enhancement of vortex pinning, we next discuss Bi-2212 synthesized with its c–axis perpendicular to a silver substrate. The tapes with 3.4 μm of superconductor were irradiated, with no amplifier foil, to a maximum fluency of 5×10^{16} protons/cm^2, giving $B_\varphi \approx 1.4$ T. In addition to conventional hysteresis loops $M(H)$, we measured the decay of M and J with time t at fixed fields. An example is shown in Fig. 6, which traces the decay of current during 75 minutes of continual measurements. Qualitatively, the length of the trace at each temperature represents the fractional decay of J. This is quantified by determining the normalized decay rate $S = -\,\mathrm{d}\ln(J)/\,\mathrm{d}\ln(t)$. Results for S are shown in the inset of Fig. 6 for this sample and, for comparison, the corresponding measurements for an unirradiated portion of the tape. It is clear that the addition of columnar defects greatly reduced the thermally activated depinning of vortices. Interestingly, there remains a significant rate of nominally T–independent decay [19] in the region below $\sim 10\,\text{K}$. This arises from quantum tunneling of vortices between pinning sites [20]; experimentally, the phenomenon is more pronounced in lower fields and with higher defect densities where $B \ll B_\varphi$; in the case shown here, we have $B \cong B_\varphi$. Note, too, that extrapolating the roughly linear $S(T)$ for the virgin sample to $T = 0$ gives a quantum tunneling rate S_Q that is similar in magnitude to that observed with columnar defects.

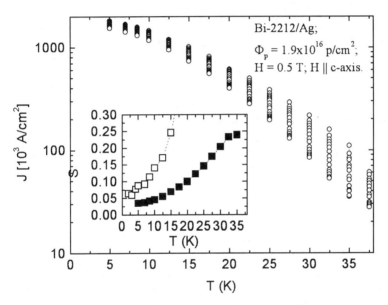

Fig. 6. Measurements of current J during a time period of 75 minutes at temperatures T. The vertical arrays of points trace the decay of J with time; the magnetic field $H = 0.5\,\text{T}$ was applied perpendicular to the plane of the textured Bi-2212/Ag tape. Inset: the corresponding normalized decay rate S for this sample (*closed symbols*) and for an unirradiated portion of the same tape (*open symbols*)

The time dependent data contain further information. Indeed, these "flux creep" experiments are analogous to measurements of the intragrain current density J versus electric field E, but with electric fields that are much smaller than typical in transport studies. A useful method for obtaining the effective vortex pinning energy $U(J,T)$ in the presence of thermally activated flux creep was devised by Maley *et al.* [21]. The procedure is based on the master rate equation [22]

$$\mathrm{d}J/\,\mathrm{d}t = -\frac{J_\mathrm{c}}{\tau} \exp\left[-\frac{U(J,T)}{T}\right] . \tag{2}$$

Here τ is the elementary attempt time for vortex hopping and $U(J,T)$ (expressed in units of Kelvin) corresponds to the instantaneous current J measured at time t. Inverting (2) provides U from the experimental data, by adjusting the remaining unknown $\ln(J_\mathrm{c}/\tau)$ so as to form a continuous curve. This procedure is well established for piecing together the thermal decay at different T's, but there is no provision for non–thermal decay processes. Applying this construction to the data in the range $T = 10 - 30\,\mathrm{K}$ gives the results shown in Fig. 7. In this range, the fundamental superconductive parameters vary little and the factor $\ln(J_\mathrm{c}/\tau)$ can be taken as constant. The 3D figure shows the explicit dependence on T as well as J. The usual presentation of U only as a function of J is shown in the projection of the data onto the $U - J$ plane. As the Maley analysis includes solely *thermally activated creep*, we exclude from consideration the low temperature data with pronounced quantum creep, when fixing the factor $\ln(J_\mathrm{c}/\tau)$. Quantum tunneling affects the Maley analysis and leads to the discontinuities in $U(J)$ observed at low temperatures and high currents; in effect, the J is smaller than it should be, based only on thermally activated decay. Quantum creep substantially reduces the observed current density at low temperatures, compared with the values that would be attained in its absence.

6   Conclusions

We have demonstrated that deeply penetrating 0.8 GeV protons can create columnar defects *via* a fission process, and have shown that they enhance the properties of Tl–, Hg– and Bi–based cuprate superconductors. A Pb amplifier foil increases the defect density in thin films. Finally, an analysis of the decay rate of the current density quantifies the increase of pinning energy upon irradiation of oriented Bi-2212 tapes and shows that quantum creep effects are significant and persist to surprisingly high temperatures.

Acknowledgments

We wish to acknowledge valuable discussions with M.P. Maley and H.R. Kerchner and the generous provision of materials by M. Paranthaman. Collaborative research between the University of Tennessee and IBM was supported in part by NSF grant DMR 95-10731. Work at the University of Talca was sponsored by the Chilean FONDECYT grant 1960316. The Weapons Neutron Research facility, LANSCE, Los Alamos National Laboratory is supported by USDOE contract W–7405–ENG–36. Research at the Oak Ridge National Laboratory is supported by the Division of Materials Sciences, U.S.

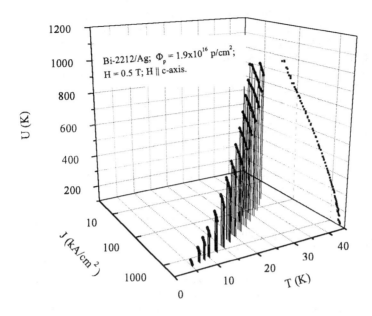

Fig. 7. The effective vortex pinning energy $U(J,T)$ for Bi-2212/Ag (same irradiated material as in Fig. 6), deduced using the construction of Maley

Department of Energy under contract number DE–AC05–96OR22464 with Lockheed Martin Energy Research Corp.

References

1. For a review, see:
   Blatter G., Feigel'man M.V., Geshkenbein V.B., Larkin A.I., Vinokur V.M. (1994) Rev Mod Phys 66:1125 and references therein
2. Civale L. (1992) In: Jin S. (Ed.) Processing and Properties of High-T_c Superconductors, vol I – Bulk Materials, World Sci., Singapore, 299
3. Krusin-Elbaum L., Thompson J.R., Wheeler R., Marwick A.D., Li C., Patel S., Shaw D.T., Lisowski P., Ullmann J. (1994) Appl Phys Lett 64:3331
4. Becchetti F.D., Jaenecke J., Lister P., Kwiatowski K., Karwowski H., Zhou S. (1983) Phys Rev C 28:276;
   Vaishnene L.A., Andronenko L.N., Kovshevny G.G., Kotov A.A., Solyakin G.E., Neubert W. (1981) Z Phys A 302:143
5. Thompson J.R., Christen D.K., Paranthaman M., Krusin-Elbaum L., Marwick A.D., Civale L., Wheeler R., Ossandon J.G., Lisowski P., Ullmann J. (1996) In: Klamut J., Veal B.V., Dabrowski B.M., Klamut P.W., Kazimierski M. (Eds.) Recent Developments in High Temperature Superconductivity, Lecture Notes in Physics vol 475, Springer, Berlin Heidelberg, 321–335

6. Thompson J.R., Krusin-Elbaum L., Kim Y.C., Christen D.K., Marwick A.D., Wheeler R., Li C., Patel S., Shaw D.T., Lisowski P., Ullmann J. (1995) IEEE Trans Appl Supercond 5:1876

7. Safar H., Cho J.H., Fleshler S., Maley M.P., Willis J.O., Coulter J.Y., Ullmann J.L., Lisowski P.W., Riley G.N., Rupich M.W., Thompson J.R., Krusin-Elbaum L. (1995) Appl Phys Lett 67:130

8. Thompson J.R., Krusin-Elbaum L., Christen D.K., Song K.J., Paranthaman M., Ullmann J.L., Wu J.Z., Ren Z.F., Wang J.H., Tkaczyk J.E., DeLuca J.A. (1997) Appl Phys Lett 71:536

9. Krusin-Elbaum L., Lopez D., Thompson J.R., Wheeler R., Ullmann J., Chu C.W., Lin Q.M. (1997) Nature 389:243

10. Li C., Patel S., Ye J., Narumi E., Shaw D.T. (1993) Appl Phys Lett 63:2558

11. Kim Y.C., Thompson J.R., Christen D.K., Paranthaman M., Specht E.D. (1995) Physica C 253:457

12. Gun S.H., Wu J.Z., Tawdry S.C., Eckert D.R. (1996) Appl Phys Lett 68:2565

13. Fietz W.A., Webb W.W. (1969) Phys Rev 178:657

14. Sun Y.R., Amm K.M., Schwartz J. (1995) IEEE Trans Appl Supercond 5:1870

15. Usami R., Itoh M., Fukuoka A., Wu X.J., Tanabe K. (1996) Physica C 269:193

16. Thompson J.R. (1998) In: Narlikar A.V. (Ed.) Studies of High Temperature Superconductors, vol 26, Nova Sci. Publ., Commack, NY, 113–131

17. Blatter G., Geshkenbein V.B., Larkin A.I. (1992) Phys Rev Lett 68:875

18. Krusin-Elbaum L., Blatter G., Thompson J.R., Petrov D.K., Wheeler R., Ullmann J., Chu C.W. (1998) Phys Rev Lett 81:3948

19. Thompson J.R., Ossandon J.G., Krusin-Elbaum L., Song K.J., Christen D.K., Ullmann J.L. (1999) Appl Phys Lett 74:3699

20. Morias Smith C., Caldeira A.O., Blatter G. (1996) Phys Rev B 54:R784; Morias Smith C., Caldeira A.O., Blatter G. (1996) Czech J Phys 46:1739

21. Maley M.P., Willis J.O., Lessure H., McHenry M.E. (1990) Phys Rev B 42:2639

22. Niederöst M., Suter A., Visani P., Mota A.C., Blatter G. (1996) Phys Rev B 53:9286

Bose–Einstein to BCS Crossover as a Model for High-T_c Cuprate Superconductors

Y. J. Uemura

Physics Department, Columbia University, New York, NY 10027, USA

Abstract. Crossover from Bose–Einstein (BE) to BCS condensation can be a guiding principle in understanding the evolution of high-T_c cuprate superconductors as a function of carrier doping. This picture is developed by combining two experimental results: (1) the "universal correlations" between T_c and n_s/m^* (superconducting carrier density / effective mass) found in μSR measurements of the magnetic field penetration depth λ and (2) the "pseudo gap" behavior observed in NMR, neutron scattering, dc- and optical conductivity, specific heat, and most–recently in angle–resolved photo–emission (ARPES) measurements. Here we provide a critical review of these experimental results and the relevant theoretical work in order to elucidate the essential features of this crossover picture and to discuss condensation mechanisms in the cuprates.

1 Introduction

Superconductivity and superfluidity are representative macroscopic quantum phenomena. Superfluidity of liquid ^{4}He is understood to be a result of Bose–Einstein (BE) condensation of He atoms with a macroscopic number of them occupying the zero–momentum state below T_c [1]. In this case the condensing bosons (^{4}He) exist even at temperatures much higher than T_c. Superconductivity in conventional metals (Al, Sn, etc.) is explained by the Bardeen–Cooper–Schrieffer (BCS) theory [2] which assumes: (1) formation of a Cooper pair (i.e. condensing boson) from 2 fermion charge carriers, and (2) condensation of Cooper pairs into zero–momentum state, with (1) and (2) occurring at the same temperature T_c. Although these two condensation phenomena share some fundamental aspects in common, they are apparently quite different, not only in whether the condensing bosons are formed at $T \gg T_c$ or at $T = T_c$, but also in spatial extent/overlap of these bosons. ^{4}He is very compact and only weakly overlapping, while thousands of Cooper pairs overlap with each other in conventional BCS superconductors.

A short coherence length ξ (or equivalently a high upper critical field H_{c2}) of high-T_c cuprate superconductors has been noted, following the discovery of the cuprates [3], as a major characteristic feature. The inferred small size (10–30 Å) of the superconducting pairs in real space has stimulated development of models of high-T_c superconductors based on Bose–Einstein (BE) condensation [4]. On the other hand, BE condensation has apparent difficulties as a theory for cuprates, at least in its simplest form for non–interacting local bosons: the condensation temperature T_B estimated from the doped hole density and the effective mass m^* becomes 4–5 times larger than the observed transition temperature T_c, and some results for specimens in the "optimally doped" region look more compatible with conventional BCS condensation. Since the highly successful BCS theory [2] has been widely regarded as a concept in opposition to BE condensation, there has been a tendency to develop theories adopting either of these two condensation mechanisms while discarding the other.

In this article, we show that a crossover from BE to BCS condensation with smooth interpolation provides a more successful account for high-T_c cuprate systems. This crossover picture is inferred from two experimental observations: (1) the "universal correlations" between T_c and n_s/m^* (superconducting carrier density / effective mass) found in muon spin relaxation (μSR) measurements of the magnetic field penetration depth λ, and (2) the "pseudo gap" behavior found in the underdoped region of cuprate systems by various experimental techniques. The accumulation of these experimental studies has turned the crossover phenomena from a subject of purely formalistic interest into an actual possibility. After reviewing these experimental results, we discuss the historical and theoretical background of the crossover picture, its relevance to phase fluctuations/coherence of the order parameter, the criterion which characterizes the "crossover region".

2   Penetration Depth and Universal Correlations

The magnetic field penetration depth λ is one of the most fundamental parameters of superconductivity. The London equation leads to a simple relation

$$\frac{1}{\lambda^2} = \frac{4\pi n_s e^2}{m^* c^2} , \tag{1}$$

which implies that the screening of an external field is due to a supercurrent, whose density is represented by n_s/m^*. It is important to recognize that the superconducting carrier density n_s here corresponds to the (effective) density of *all the* charge carriers, different from the population n_{mod} of $Q = 0$ Cooper pairs, defined as fermions in the modified density of states near the Fermi energy. In weak–coupling BCS superconductors, n_{mod} is a fraction of n_s: $n_{mod} \sim n_s \times \Delta/\varepsilon_F$, where Δ is the energy gap and ε_F the (effective) Fermi energy. To confirm the relationship between λ and n_s, one can substitute the known carrier densities and effective masses for simple metals, such as Al, and compare with observed values of λ. The situation realized in superconductors, with all the charge carriers participating in the superfluid at $T = 0$ but only a fraction of them ($\sim n_{mod}$) are in the actual $Q = 0$ pairs, is somewhat analogous to the case of superfluid ^{4}He in which 100% of the atoms form the superfluid but only about 10% of them are in the $Q = 0$ state [4], due to the overlap and interaction between bosons.

Muon spin relaxation (μSR) [5,6] is a powerful method for measuring both the absolute value and the temperature dependence of the penetration depth

λ of type-II superconductors. The muon spin relaxation rate σ is proportional to the width ΔH of the inhomogeneous internal magnetic field, and consequently to $1/\lambda^2$ below T_c [7,8]. Since high-T_c cuprate superconductors fall in the clean limit where the mean free path l is much longer than the coherence length ξ, a small correction factor $[1+\xi/l]^{-1}$, to be included in (1), becomes close to unity. This leads to the simple relation $\sigma \propto \lambda^{-2} \propto n_s/m^*$. For highly–anisotropic cuprate superconductors, the relaxation rate σ observed in ceramic specimens selectively reflects the in–plane penetration depth λ_{ab} [9] and the in–plane effective mass m^*_{ab}. Although σ is proportional to the 3–dimensional (3-d) carrier density n_s even in a quasi 2-d system, n_s can be converted into a 2-d carrier density n_{s2d} on the CuO_2 planes by multiplying n_s and c_{int}, the average distance between conducting planes.

Application of the μSR technique for measuring the penetration depth of the cuprates started [10–12] in early 1987. To study the doping dependence of T_c, we initially plotted T_c versus the low temperature relaxation rate $\sigma(T \to 0) \propto n_s/m^*$ for specimens of four different cuprate systems [13]. Subsequently, we accumulated results from many other specimens and demonstrated [14] the existence of the "universal" correlations between T_c and σ, followed by cuprates made with various different methods of doping, as well as having different crystal structures with single (214), double (123 and 2212) and triple (2223) CuO_2 planes. As shown in Fig. 1, T_c increases with initial doping of carriers following the linear relationship, whose slope is common to most of the cuprates, in the underdoped region. With further carrier doping T_c shows a saturation around the optimum doping region, followed by a suppression in the weakly overdoped region. These results were well reproduced by subsequent μSR results from various groups [15–18], while the saturation seen at $T_c = 90\,\mathrm{K}$ for the 123 systems has been shown to be caused by the effect of the CuO chain [18,19].

Since all the cuprate systems shown in Fig. 1 have an average interplanar distance $c_{int} \sim 6\,\text{Å}$, the horizontal axis of this figure can be regarded as proportional to n_{s2d}/m^*_{ab}. Note that this quantity is proportional to the 2-d Fermi energy $\varepsilon_F = (\hbar^2\pi)n_{2d}/m^*$ of a non–interacting electron gas. The linear relation between T_c and σ in the underdoped region is not compatible with the weak–coupling BCS model, in which T_c depends only weakly on carrier concentration through the density of states at the Fermi level. Instead, we suggested [14] that the linear relation implies $T_c \propto \varepsilon_F$, which is expected when the energy scale $\hbar\omega_B$ of exchange bosons is comparable to or larger than ε_F, i.e. when the pair–mediating interaction is non–retarded (we shall come back to this point later).

Subsequent μSR measurements in organic BEDT, doped C_{60}, Chevrel phase, $(Ba,K)BiO_3$ (BKBO) and heavy-fermion superconductors [20,21] yielded points in Fig. 1 lying rather close to the linear relationship found for the cuprates. These new "exotic" type-II superconductors have various features in common, such as short coherence length ξ, strongly–correlated electronic

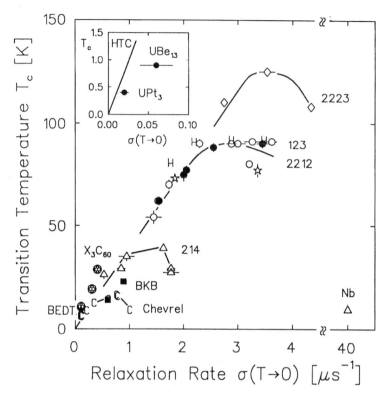

Fig. 1. A plot of the muon spin relaxation rate $\sigma(T \to 0) \propto n_s/m^*$ versus T_c. (*Open symbols* are from [14]; *closed circles* from [17], H symbols from [16]; BKB, BEDT, Chevrel, UPt$_3$ and UBe$_{13}$ from [20] and references therein; and X$_3$C$_{60}$ from [21])

structures, closeness to magnetic and/or structural phase instability, etc. The implications of Fig. 1 become clearer if we convert the horizontal axis into an effective Fermi temperature T_F, which can be directly done for 2-d systems as mentioned above. For 3-d systems, we can combine $\sigma \propto n_s/m^*$ with the Sommerfeld constant $\gamma \propto n_s^{1/3} m^*$ to obtain

$$k_B T_F = \frac{\hbar^2}{2} \frac{(3\pi^2)^{2/3} n_s^{2/3}}{m^*} \propto \sigma^{3/4} \gamma^{-1/4} .$$

The obtained value of T_F represents an average energy scale of superconducting carriers for translational motion (in a sense similar to the Drude spectral weight of optical conductivity), but not necessarily the actual Fermi temperature in its strict definition [22], especially in strongly–correlated electron systems having a complicated density of states. Note that, in principle, T_F can be obtained even for an ideal Bose gas in this procedure.

Figure 2 shows the resulting plot of T_c versus T_F [20]. We find that the cuprates and other exotic superconductors have a rather high T_c normalized by T_F, with $T_c/T_F \sim 0.01 - 0.1$, in contrast to the case of conventional BCS superconductors, with $T_c/T_F \ll 0.01$. The BE condensation temperature T_B for an ideal non–interacting Bose gas, calculated for a given density $n_s/2$ and mass $2m^*$ of the carriers, is shown by the broken line in Fig. 2. Compared to T_B, the transition temperatures of underdoped cuprates and BEDT systems are reduced only by a factor of $4 \sim 5$, suggesting closeness of these systems to the BE limit. Moreover, the points from cuprates lying parallel to T_B suggest that the universal linear relationship $T_c \propto n_{s2d}/m^*$ in the cuprates can now be viewed as $T_c \propto T_B$, i.e. T_c of the underdoped cuprates may be determined by BE condensation. This picture also gives a natural explanation for the "universal" behavior of T_c, since the BE condensation temperature is essentially determined by the density of bosons and the boson mass, regardless of the microscopic interaction / mechanisms for the pair (boson) formation.

Figure 2 provides an empirical way of classifying various superconductors in terms of energy scales between two limiting cases: BE condensation for a non–interacting Bose gas ($T_c = T_B \sim T_F/4$) and weak–coupling BCS condensation ($T_c/T_F \ll 0.01$). In [20], we stressed the importance of formulating theories which interpolate between these two limits to account for high-T_c superconductivity. The $4 - 5$ times reduction of T_c from T_B in cuprates is the feature discussed in the Introduction as a difficulty of the BE model. Such a reduction is, however, not unexpected in a system with overlapping carriers and strong interactions, since even the lambda transition of ^{4}He is reduced from $T_B = 3.2\,\mathrm{K}$ to $2.2\,\mathrm{K}$ due to the effect of finite particle size and inter–particle interaction. Note that about 2 to 6 pairs exist (i.e. overlap) in the coherence area ξ^2 on the conducting plane of cuprates and organic BEDT superconductors [23]. Pistolesi and Strinati [24] discussed an implication of Fig. 2 in a length–scale argument with respect to ξ.

Dimensionality is another important factor. Theoretically, BE condensation does not occur in purely 2-d systems. A small interplanar coupling will actually be sufficient to achieve BE condensation [25], but with a somewhat reduced T_c, in a way analogous to low–dimensional magnetic systems in which the ordering temperature is reduced from the energy scale of the mean–field exchange interaction. Consequently, a comparison of 2-d and 3-d superconductors is a non–trivial matter. We believe that the calculation of T_F used to obtain Fig. 2 is one of the least prejudiced ways in such a comparison. The impressive overlap of the points seen in Fig. 2 for 2-d systems (cuprates and BEDT) and some 3-d systems might be accidental, but may indicate that there is an intrinsic upper limit of T_c/T_F common to 2-d and 3-d superconductors.

3   Pseudo Gaps

Since 1989, the existence of a low-energy "spin gap" in underdoped materials has been noticed first in NMR [26–30] as the increase of $1/T_1T$ and the Knight

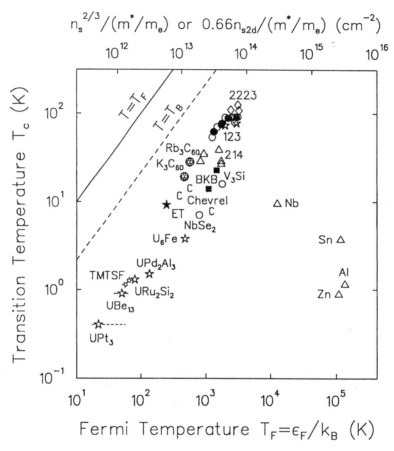

Fig. 2. A plot of the effective Fermi temperature T_F versus T_c in various superconductors (from [20,21]). The *broken line* shows the BE condensation temperature T_B
calculated for a non–interacting hard–core Bose gas with density $n_s/2$ and mass
$2m^*$. Fermi temperatures for Sn, Al and Zn are from known values

shift above T_c; and subsequently in neutron scattering [31,32] as a missing
inelastic scattering intensity below a characteristic energy scale $\sim kT^*$. With
increasing temperature, the effect of the spin gap gradually disappears around
the pseudo–gap temperature T^*. This feature indicates a formation of singlet
spin correlations below T^*, with a certain (gap) energy required to dissociate
the correlation. As shown in Fig. 3, T^* is significantly higher than T_c in
the underdoped region, but decreases with increasing doping, approaching
T_c near the "optimum doping" region. The pseudo gap behavior in NMR
is replaced by a standard Korringa behavior above T_c in the optimum and
overdoped regions.

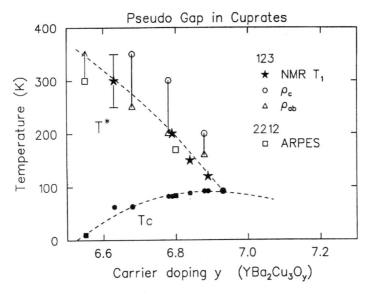

Fig. 3. The pseudo gap energy scale T^* obtained in the 123 and 2212 cuprate systems from various techniques. The NMR result for $y = 6.63$ is from Ref. [28], while the three other NMR results are for $Y(Ba_{1-x}La_x)_2Cu_3O_z$ from [30] plotted against the corresponding planar hole density of the 123 system. The NMR results represent the temperature at which $1/T_1T$ departs from a constant in the case of oxygen NMR and from the $1/T$ Curie behavior in the case of Cu NMR (for which the relaxation time T_1 is affected by the antiferromagnetic spin fluctuations). Symbols ρ_c and ρ_{ab} represent the temperature at which the c–axis and ab–plane resistivity, respectively, deviate from the linear-T behavior at higher temperatures in single crystals of 123 systems (from [34]). The ARPES results from [41] in underdoped Bi2212 systems are plotted against the corresponding doping level of the 123 systems

The next important observation comes from the dc–conductivity in single crystals. In 1991, Ito *et al.* [33] reported that the c–axis conductivity of effectively underdoped cuprate systems exhibits semiconducting/insulating behavior below a temperature that is well above T_c. This suggests that some "gap" also exists in the c–axis charge transport. In an optimally–doped 123 system, this "gap"–like behavior disappears, and the c–axis conductivity recovers a T–linear behavior above T_c, as demonstrated in Fig. 4. This feature has been confirmed in subsequent work with various single crystal specimens [34]. In 1993, Homes *et al.* [35] demonstrated the existence of this gap in c–axis conduction more directly by measurements of the ac– (or optical) conductivity.

By 1992, a corresponding anomaly was also found [36] in the ab–plane conductivity, as a reduction of resistivity and a departure from the T–linear behavior below T^*, as if the formation of the "gap" above T_c would reduce the scattering of charge carriers on the plane. These conductivity studies

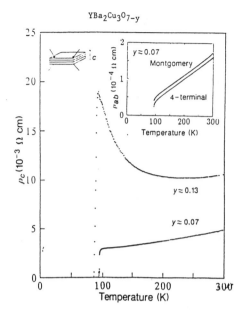

Fig. 4. The resistivity along the c-axis and ab-plane observed in fully oxygenated and moderately underdoped single crystals of $YBa_2Cu_3O_{7-y}$ (from [33])

demonstrated that the "gap" first noticed in spin phenomena now appears in transport phenomena. The energy scale T^* of this gap estimated by spin probes (NMR, susceptibility, neutrons) agrees rather well with that estimated by charge probes (dc- and ac-conductivity, and Hall effect), as discussed in [37,38] and as demonstrated in Fig. 3. This indicates that charge and spin gaps are manifestations of the same phenomenon. In specific heat measurements [39], a large amount of entropy release was found in the underdoped region above T_c. All these measurements indicate the formation of some singlet correlations with a binding energy (reminiscent of the pair formation), occurring in the normal state above T_c in the underdoped region. This phenomenon is now commonly referred to as a "pseudo gap".

Most recently, in 1996, angle-resolved photo-emission (ARPES) measurements of underdoped Bi2212 systems, performed independently by the Stanford [40] and Argonne [41] groups, have demonstrated that the pseudo gap can be observed in momentum space, with a profile close to that of the superconducting gap observed in an optimally-doped system. This feature strongly suggests that the pseudo gap in the underdoped region evolves smoothly from the superconducting gap in the optimally-doped and overdoped region.

4   BE-BCS Crossover Picture for Cuprates

Let us now consider a simple model in which an attractive interaction between carriers is mediated by an exchange boson having an energy scale $\hbar\omega_B$, and let us sketch the expected doping dependence by varying carrier concentration. In the dilute limit, pairs will form at a high temperature corresponding to the effective binding energy, and then these bosons will undergo BE condensation at a much lower temperature. The condensation temperature T_c will follow the behavior of T_B in the low–density region. In the high–density limit where many carriers are overlapping, we would expect the BCS–like behavior for a simple metal, with both pair formation and condensation occurring at the same temperature T_c. T_c in this region is expected to be rather low. If we assume that these two regions connect smoothly, we obtain the phase diagram shown in Fig. 5(a). The low–density side is characterized by the BE condensation of rather local and "pre–formed" pairs, while the pair–formation and condensation line should merge near the crossover region. The high–density side would connect to a typical BCS behavior in momentum space. The crossover will take place at a carrier concentration for which the effective Fermi energy kT_F becomes comparable to the exchange–boson energy scale $\hbar\omega_B$. In other words, the crossover region separates a non–retarded interaction in the low–density side from a retarded interaction in the high–density side.

This picture can be naturally adopted for the cuprate systems if we identify the pseudo gap as a signature of the "pre–formed" pairs in the normal state and recognize that the "universal correlations" in the penetration depth results represent $T_c \propto T_B$ in the underdoped region. The insulating behavior (above T_c) of the c–axis transport below T^* can be understood if one assumes that a carrier pair with charge 2e has a much smaller probability of tunnelling/hopping between conducting planes than does an unpaired fermion carrier with charge e. Since the pair formation/dissociation is an activation process, the c–axis conductivity below T^* will be determined primarily by the population of remaining unpaired carriers, following an Arrhenius behavior.

In several conferences since 1993 [42–44], the present author has proposed this BE-BCS crossover picture for the cuprates (Fig. 5(a)). The expected behavior of the energy gap in this picture is sketched in Fig. 5(b). The real gap in the BCS region connects to the pseudo gap in the low–density region and finally becomes the binding energy of a local pair in the dilute limit. Due to the gap formation (i.e. pairing) at T^*, T_B in the BE condensation region is not associated with a gap opening. Various properties, including the gap, should evolve smoothly in this crossover. Since most underdoped cuprate specimens lie rather close to the crossover region, we expect that the pairs will have substantially non–local features and that the pseudo gap will exhibit a momentum dependence similar to that of the real gap in the optimally–doped or overdoped regions, as was observed in photo–emission results [40,41].

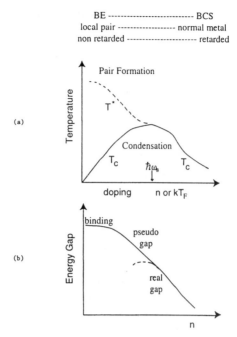

Fig. 5. The doping dependence of (**a**) the pair formation temperature T^*, and the condensation temperature T_c, and (**b**) the pseudo and real energy gaps, expected for a smooth crossover from Bose–Einstein to BCS condensation (from [43,44])

5   Theoretical Developments

Theoretically, there has been a stream of papers related to the concept of BE-BCS crossover. Although the condensation in BCS theory involves the BE condensation of Cooper pairs, occurring at the same temperature at which these pairs are formed, it has been widely considered that BCS and BE condensations are somewhat mutually exclusive concepts. Since BCS theory achieved remarkable success in explaining the superconductivity of simple metals and alloys, and since no superconducting system had been found to require a drastically non–BCS description until rather recently, the crossover from BCS to BE condensation was studied mainly for the superfluidity of ^{3}He and/or excitons until the mid 1980's. The only exception was a 1969 paper by Eagles [45], who considered a possibility of this crossover in superconducting doped $SrTiO_3$ systems. Pioneering works of Legett [46] and of Nozières and Shmitt–Rink [47] showed that the BE-BCS crossover should occur "smoothly", when the coupling strength is varied. The negative-U Hubbard model provided a lattice approach to the crossover problem, although here T_c does not follow the continuum results in the strong–attraction limit due to the increasing effective mass of moving carriers pairs to adjacent site: Micnas, Raninger and Robaszkiewicz [48] published a detailed review article, including these theories and a survey of existing superconductors possibly related to an effective local attraction. Most of the progress in the theory of BE-BCS crossover was reviewed in a 1993 conference article by Randeria [49]. Lee and collaborators [50], as well as Ranninger [51], worked on theories of coexisting bosons and fermions, while many other authors [52–60] contributed to theoretical descriptions related to the BE-BCS crossover.

Soon after the discovery [1] of cuprate systems, the connection of BE condensation to high-T_c superconductivity was suggested by Mott and other scientists [61], based on the short coherence length. Initially, such theories claimed that the entire doping region should be described by a local–pair BE condensation. A 1992 paper of Randeria et al. [59] was one of the first to connect the BE-BCS crossover concept to the spin gap in the underdoped cuprates observed in NMR studies. The connection of the penetration depth and c–axis dc–conductivity results to the crossover picture was subsequently made in [43–44]. From a somewhat different perspective, by considering universality–classes for critical behavior, Schneider and Keller [62] found that the results for cuprates are consistent with a 3d-XY model, thereby reinforcing the picture that the condensation of cuprate systems is deeply related to BE condensation.

Pictures similar to the BE-BCS crossover have also been developed by several authors who considered phase fluctuations of the order parameter. It has been known for a long time that T_c of superconductors corresponds to the point at which the phase of the order parameter achieves long range coherence [63]. Doniach and Inui, in their 1990 paper [64], noticed that superconductivity is taken over by an insulating (i.e. not metallic) state when $T_c \rightarrow 0$ with decreasing carrier concentration in the underdoped cuprates. They took this as a signature that pre–formed pairs exist above T_c in the underdoped region: localization of these pairs leads to the insulating state in the low doping limit. Although their paper was written at a time before the pseudo gap had been established by various experimental results, they interpreted susceptibility and NMR results in underdoped cuprates as a signature of singlet pair formation, i.e. the formation of the pre–formed pairs, and predicted a gapped response in optical conductivity. Using the Ginzburg–Landau formalism with a term representing phase fluctuations, they obtained a phase diagram which exhibits the pseudo–gap behavior.

Recently, Emery and Kivelson [65] published another paper regarding the phase fluctuation picture. In conventional BCS superconductors, the build–up of the amplitude of the order parameter (or equivalently the gap Δ) determines T_c, since T_c occurs at a temperature lower than T^θ where the phase fluctuations would destroy the coherence. However, if the mean–field effective attraction energy is higher than T^θ, then it would be the phase fluctuations which determine T_c, while the amplitude of the order parameter would remain finite above T_c. To estimate the maximum energy scale $T^\theta_{\max}$ for these phase fluctuations, Emery and Kivelson calculated the kinetic energy required for spatial phase modulation, leading to $T^\theta_{\max} \propto 1/\lambda^2 \times a$, where a is a length parameter.

In 2-dimensional systems, substitution of the interplanar distance c_{int} for a makes $T^\theta_{\max}$ essentially equivalent to the Kosterlitz–Thouless transition temperature T_{KT}, which is $1/8$ of the corresponding 2-d Fermi temperature T_{F} in the limit of strong coupling. In Figure 2, the line for T_{KT} for 2-d systems becomes parallel to that of T_{B} for 3-d systems with a factor of ~ 2 reduction. In 3-d systems, however, Emery and Kivelson adopted the coherence length ξ for a, which led to a very high value of $T^\theta_{\max} \gg T_{\text{B}}$ and also to an apparent difference between 2-d and 3-d systems. As already mentioned, the penetration depth λ is related to all of the charge carriers, while ξ is the size of a $Q = 0$ Cooper pair. The mixed use of parameters for these two different entities in [65] resulted in an overestimate of $T^\theta_{\max}$. Instead, if one uses the interparticle distance as a with λ^{-2}, one recovers $T^\theta_{\max} \sim T_{\text{B}}$ for 3-d systems as well. Then Table 1 of [65] reduces to Fig. 2 of the present article (i.e. Fig. 3 of [20]). Indeed, it is quite natural for T_{B}, defined for non–interacting ideal bosons, to serve as an upper bound for T_{c}, since every known effect of fermion overlapping, interboson interactions, and dimensionality would lower the condensation temperature from T_{B}. In this sense, there is no essential difference between the phase fluctuation picture and the BE-BCS crossover picture. The BE-BCS crossover picture, however, involves some important concepts and interesting consequences that are not directly visible in the phase fluctuation picture.

There exists a stream of thoughts rather different from the BE-BCS concept that leads to a similar phase diagram for the cuprates. In 1987, Anderson [67] suggested that the electronic states of the cuprates can be understood using the concept of a resonating valence bond (RVB), which was initially developed to account for geometrically–frustrated spin systems [68]. Doping a hole into the RVB environment would create spin (spinon) and charge (holon) quasiparticles, which may exist apart from each other to some extent, i.e. the spin charge separation. Subsequently, this concept was adopted in $t - J$ models and/or gauge field theories [69-70], resulting in a phase diagram similar to that of Fig. 5(a). There, the pseudo gap signals formation of the gap in the spin degrees of freedom, and thus the beginning of the spin–charge separation. The evolution of T_{c} in the underdoped region is explained as a BE condensation of holons. The tendency towards spin–charge separation is expected to diminish as the doping progresses (i.e. as the system goes away from half filling), leading to the merger of the spin gap and condensation temperatures near the optimum doping region.

A major difference between this RVB picture and the BE-BCS picture can be found in the normal state of underdoped cuprates above T_{c} but below T^*: in the former, carriers are charged with single e, while in the latter they are charged partly with e and partly with 2e. Spin–charge separation in 1-dimensional systems is readily predicted by theories, and experimental results now exist directly supporting this concept in a 1-d spin system [71]. Whether or not such a thing is actually happening in quasi 2-d cuprate systems is a fundamental question yet to be answered by experimental studies.

6   The Crossover Criterion

Cooper pairs, once formed at T_c, would immediately undergo a long awaited BE condensation even within the BCS theory. In this sense, T_c is always associated with a BE condensation in any doping region in Fig. 5(a). It is, however, only in the low density region, which we call the "BE region", that T_c is determined by the condensation temperature. Theoretically, a pseudo–gap and/or fluctuating order parameter corresponds to the idea that finite momentum pairs with $Q \neq 0$ exist above T_c in addition to the $Q = 0$ Cooper pairs [66]. In the low density limit, most of the local bosons have a finite momentum at temperatures between T^* and T_B. Therefore, this "finite Q pair" provides a very natural and smooth evolution connecting Cooper pairs and local bosons. The existence of the "pseudo gap" would help prevent the immediate decay of such a finite Q pair into fermions. In an analytical argument extending a BCS–like self–consistency treatment for the energy gap / pairing field, Tchernyshyov [66] have shown that the condensation temperature T_c of such a system with fluctuating pairs is given by $T_c \sim T_B$, as the pair–formation temperature departs from T_c with the carrier concentration decreasing from the high–density BCS side. For numerical attempts to obtain T_c in the crossover region, see a recent review of Micnas [53] and references therein. Thus, $T_c \propto T_B$ can be expected not only in the low–density local boson limit but also near the crossover region. This provides theoretical support for the phase diagram in Fig. 5(a).

Where should the crossover occur? Although we are much accustomed to the language of weak–coupling BCS theory, it is easier to consider this question from a strong coupling point of view. In the low–density limit, the binding energy (or the pseudo gap Δ) would be determined directly by the energy scale of the exchange interaction J (comparable to the mediating boson energy $\hbar\omega_B$), analogous to the way the gap is determined in Haldane or spin–Peierls systems. So, Δ may be between a tenth and a half of J. Doping carriers would lower Δ, i.e. the T^* line, again analogous to the doping dependence of a many–body gap in magnetic systems. The T_c line in the low–density limit would follow $T_B \sim T_F/4$, yet with reduced absolute values due to overlap/interaction. Then these two lines would merge ($T^* \sim T_c$) approximately when $\varepsilon_F \sim \hbar\omega_B$. This is a heuristic argument which supports the picture that the crossover occurs at the energy scale which separates retarded and non–retarded interactions.

In length–scale arguments, one can characterize this crossover region by a carrier density for which the interparticle distance ($\sim 1/k_F$) becomes comparable to the "size" of a fermion pair. In a strong–coupling superconductor, the potential energy due to the attractive interaction dominates over the kinetic energy due to the zero–point motion of fermions. We shall resort to the concept of "retardation length" $\xi_{ret} = v_F/\omega_B$ (v_F is the Fermi velocity), which represents the range of the potential and consequently the size of a pair in the strong coupling case. (In weak–coupling superconductors dominated by the kinetic energy, the coherence length $\xi \gg \xi_{ret}$, represents the "size" of a pair.) If ξ_{ret} is longer than the interparticle distance ($\sim 1/k_F$), two fermions would move apart, still feeling the attraction, just like Copper pairs in the BCS region. If opposite, two fermions would move "together" like a composite boson. Then the crossover between these two different kinds of particle motion should occur at $v_F/\omega_B \sim 1/k_F = \hbar/m^*v_F$, or equivalently at $1/\hbar\omega_B \sim 1/m^*v_F^2 \propto 1/\varepsilon_F$, leading to the same conclusion as given by the above–mentioned energy argument.

Once we assume that $\hbar\omega_B \sim \varepsilon_F$ defines the crossover criterion, we can estimate the energy scale $\hbar\omega_B$ of the mediating bosons from the measured values of ε_F in Fig. 2. Typically, high-T_c systems show the "optimal region" around $T_F \sim 2,000\,K$. This energy scale corresponds well with the energy of the mid–infrared reflection (MIR) observed in the optical conductivity of cuprate systems [72]. This suggests that the MIR manifests the pairing interaction, whereas the Drude part stands for translational motion. The $\sim 2\,000\,K$ energy scale also corresponds well with the antiferromagnetic exchange interaction $J \sim 1\,500\,K$ of adjacent spins on the CuO_2 planes. This gives support [73] to the viewpoint that the exchange of spin fluctuations is indeed the microscopic mechanism for pairing in cuprate systems.

7   Discussions and Summary

There are two distinct aspects to the understanding of superconductivity: (a) the microscopic pairing mechanism, including arguments regarding pair–mediating bosons (such as phonons or magnons or else), origin of attractive interaction between fermion carriers, and local symmetry of superconducting wave functions, etc.; and (b) the thermodynamic condensation mechanism including description of the phase transition. In this review, we focused on the latter aspect (b), and suggested that the actual situation in cuprate systems may well lie between the well known limits of BE and BCS condensation. Theoretically, the crossover region exists between strong–coupling (BE) and weak–coupling (BCS) limits. A complete formation of theories for this "intermediate coupling" region is yet to be established. The role played by dimensionality, i.e. the 2-d versus 3-d issue, has also to be explored.

From the naive picture shown in Fig. 5(a), one would expect that the parameter n_s/m^* would keep increasing with increasing doping in the high–density BCS region, since the carrier density would increase while the increasing overlap of carriers would widen the band width, decreasing m^*. For a retarded interaction, various physical parameters in the normal state would be directly convertible to superconducting parameters. In view of this, it was quite surprising that μSR results in the overdoped Tl2201 systems [74,75] exhibited an entirely opposite trend, namely a decreasing spectral density n_s/m^* with increasing carrier doping. This feature can be interpreted in terms of (a) microscopic phase separation between normal and superconducting states below T_c [74,43,44] and/or (b) strong pair–breaking scattering in the overdoped region [75]. This at least suggests that the high density side of the cuprate system is much more complicated than a simple BCS superconductor. Due to the limited space available here, we defer the overdoped region and the (Cu,Zn) substitution effect [73] elsewhere. We only note here that in all these cases, where superconductivity is suppressed by impurity/overcrowding effects, T_c more or less scales with n_s/m^*, reflecting a fundamental aspect of Bose condensation.

The most direct evidence for the crossover picture would be obtained if one observes the "2e" pairs between T_c and T^* in the underdoped region. There are some obstacles against this, such as: (1) the 2e pairs are expected to be short–lived; (2) most of transport measurements can not distinguish conduction of N particles charged with "e" from that of $N/2$ particles charged with "2e"; and (3) most of the cuprates have too "high" T_c, which prevents detailed quantum transport studies. Nevertheless, it would be invaluable to pursue measurements to observe "2e" charges.

In summary, we have shown that the experimental results regarding correlations between T_c and n_s/m^* and the pseudo gap behavior observed in underdoped cuprate systems lead to a picture of a crossover from Bose–Einstein to BCS condensation as a plausible framework for understanding high-T_c superconductors. As shown in this paper, various experimental and theoretical works have contributed to the development of this model. More than 70 years have past since the concept of BE condensation was proposed [76]. Thanks to recent experimental confirmation of a BE condensation in very–weakly–interacting Bose gas systems of laser–cooled Rb and Na atoms [77,78], we are now observing a "renaissance" of BE condensation. It would be very gratifying if the concept of a crossover in cuprate superconductors can enrich this renaissance.

Acknowledgement

The author would like to acknowledge stimulating discussions with O. Tchernyshyov on theoretical aspects of the crossover picture, and NEDO International Joint Research Grant (Japan) and NSF (USA) (DMR-95-10453, –10454, 98–02000) for financial support.

References

1. See, for example: Tilley D.R., Tilley J. (1990) Superfluidity and Superconductivity, Adam Hilger, Bristol

2. Bardeen J., Cooper L.N., Schrieffer J.R. (1957) Phys Rev 108:1175–1204
3. Bednorz J.G., Müller K.A. (1986) Z Phys B 64:189;
   Bednorz J.G., Müller K.A. (1988) Rev Mod Phys 60:585–600
4. See, for example, papers published in:
   Salje E.K.H., Alexandrov A.S., Liang W.Y. (Eds.) (1995) Polarons and
   Bipolarons in High-T_c Superconductors and Related Materials, Cambridge
   University Press, and references therein
5. For general aspects and historical development of the μSR technique, see
   proceedings of the seven previous international conferences on μSR, published
   in Hyperf Interact (1979) 6; (1981) 8; (1984) 17–19; (1986) 31; (1990) 63–65;
   (1994) 85–87; (1996)
6. For textbooks and/or review related to the μSR technique, see, for example:
   Schenck A. (1986) Muon Spin Rotation Spectroscopy, Adam Hilger, Bristol;
   Yamazaki T., Nakai K., Nagamine K. (Eds.) (1992) Perspectives of Meson
   Science, North Holland, Amsterdam;
   Karlsson E.B. (1995) Solid State Phenomena as Seen by Muons, Protons and
   Excited Nuclei, Oxford University Press
7. Redfield A.G. (1967) Phys Rev B 162:367–374
8. Pincus P. (1964) Phys Lett 13:21–22
9. Barford W., Gunn J.M.F. (1988) Physica C 156:515–522
10. Aeppli G., et al. (1987) Phys Rev B 35:7130–7132
11. Kossler W.J., et al. (1987) Phys Rev B 35:7133–7136
12. Gygax F., et al. (1987) Europhys Lett 4:437–439
13. Uemura Y.J., et al. (1988) Phys Rev B 38:909–912
14. Uemura Y.J., et al. (1989) Phys Rev Lett 62:2317–2320
15. Pümpin B., et al. (1990) Hyperf Interact 63:25–31
16. Glückler H. et al. (1990) Hyperf Interact 63:155–160
17. Seaman C.L. et al. (1990) Phys Rev B 42:6801–6804
18. Tallon J.L. et al. (1995) Phys Rev Lett 74:1008–1011
19. Basov D.N. et al. (1995) Phys Rev Lett 74:598–601
20. Uemura Y.J. et al. (1991) Phys Rev Lett 66:2665–2668
21. Uemura Y.J. et al. (1991) Nature 352:605–607;
    Uemura Y.J. et al. (1994) Physica C 235–240:2501–2502
22. Uemura Y.J., Luke G.M. (1993) Physica B 186–188:223–228;
    Uemura Y.J., Le L.P., Luke G.M. (1993) Synthet Met 55–57:2845–2852
23. Uemura Y.J. et al. (1990) In: Kresin V.Z., Little W.A. (Eds.) Organic
    Superconductivity, Plenum, New York, 23–29;
    Le L.P. et al. (1992) Phys Rev Lett 68:1923–1926
24. Pistolesi E., Strinati G.C. (1994) Phys Rev B 49:6356–6359
25. Friedberg R., Lee T.D., Ren H.C. (1990) Phys Rev B 42:4122–4134;
    Friedberg R., Lee T.D., Ren H.C. (1991) Phys Lett A 152:423–429
26. Yasuoka H., Imai T., Shimizu T. (1989) In: Fukuyama H. (Ed.) Spin
    Correlations and Superconductivity, Springer, Berlin, 254–261;
    Yasuoka H. (1997) Hyperf Interact 105:27-34 and references therein
27. Warren W.W. et al. (1989) Phys Rev Lett 62:1193–1196;
    Walstedt R.E. et al. (1990) Phys Rev B 41:9574–9576
28. Takigawa M. et al. (1989) Physica C 162–164:853–856;
    Takigawa M. et al. (1991) Phys Rev B 43:247–257
29. Alloul H., Ohno T., Mendels P. (1989) Phys Rev Lett 63:1700–1703

30. Matsuura M. et al. (1995) J Phys Soc Jpn 64:721–724
31. Rossat-Mignod J. et al. (1991) Physica C 185–189:86–92;
    Rossat-Mignod J. et al. (1992) Phys Scr T45:74
32. Tranquada J.M. et al. (1992) Phys Rev B 46:5561–5575
33. Ito T. et al. (1991) Nature 350:596–598
34. Nakamura Y., Uchida S. (1993) Phys Rev B 47:8369–8372;
    Takenaka K. et al. (1994) Phys Rev B 50:6534–6537
35. Homes C.C. et al. (1993) Phys Rev Lett 71:1645–1648
36. Takagi H. et al. (1992) Phys Rev Lett 69:2975–2978;
    Ito T., Takenaka K., Uchida S. (1993) Phys Rev Lett 70:3995–3998
37. Batlogg B. et al. (1994) Physica C 235–240:130–133
38. Nakano T. et al. (1994) Phys Rev B 49:16 000–16 008
39. Loram J.W. et al. (1993) Phys Rev Lett 71:1740–1743
40. Marshall D.S. et al. (1996) Phys Rev Lett 76:4841–4844;
    Marshall D.S. et al. (1996) Science 273:325–329
41. Ding H. et al. (1996) Nature 382:51–54
42. Uemura Y.J. (1993) At: Euroconference on Superconductivity in Fullerenes,
    Oxides and Organic Materials, Pisa, Italy, January 1993, not published
43. Uemura Y.J. (1995) In: Salje E.K.H., Alexandrov A.S., Liang W.Y. (Eds.)
    Proc. of the Workshop on "Polarons and Bipolarons in High-T_c
    Superconductors and Related Materials", Cambridge, UK, April 1994.
    Cambridge University Press, 453–460;
    Uemura Y.J. (1995) In: Feng S., Ren H.C. (Eds.) High-T_c Superconductivity
    and the C_{60} Family, CCAST Symposium on. . . , Beijing, May 1994. Gordon
    and Breach, 113–142
44. Uemura Y.J. (1997) Physica C 282–287:194–197 [1]
45. Eagles D.M. (1969) Phys Rev 186:456–463
46. Leggett A.J. (1980) In: Pekalski A., Przystawa R. (Eds.) Modern Trends in
    the Theory of Condensed Matter, Springer, Berlin
47. Nozières P., Schmitt-Rink S. (1985) J Low Temp Phys 59:195–211
48. Micnas R., Ranninger J., Robaszkiewicz S. (1990) Rev Mod Phys 62:113–171
49. Randeria M. (1995) In: Griffin A., Snoke D.W., Stringari S. (Eds.) Bose
    Einstein Condensation, Cambridge University Press, 355–391
50. Lee T.D. (1994) Physica C 235–240:186–188;
    Friedberg R., Lee T.D., Ren H.C. (1994) Phys Rev B 50:10 190–10 217
51. Ranninger J. (1994) Physica C 235–240:277–280
52. Pistolesi F., Strinati G.C. (1996) Phys Rev B 53:15 168–15 192; and references
    therein
53. Micnas R., Robaszkiewicz S., Kostyrko T. (1995) Phys Rev B 52:6863–6879;
    Micnas R., Kostyrko T. (1996) In: Klamut J. et al. (Eds.) Recent
    Developments in High Temperature Superconductivity, Lecture Notes in
    Physics vol. 475, Springer, Berlin Heidelberg, 221–242
54. Pietronero L., Strässler S. (1992) Europhys Lett 18:627–633
55. Enz C.P. (1989) Helv Phys Acta 62:122–138;
    Enz C.P., Galasiewicz Z.M. (1988) Solid State Commun 66:49–50;
    Enz C.P., Galasiewicz Z.M. (1993) Physica C 214:239–246

---

[1] There is a typographic error in p. 195, right column, 10 lines from the bottom: "underdoped cuprates with retarded interaction" should read "underdoped cuprates
with non–retarded interaction"

56. van der Marel D., Mooij J.E. (1992) Phys Rev B 45:9940–9950
57. Tokumitsu A., Miyake K., Yamada K. (1993) Phys Rev B 47:11 988–12 003
58. Drechsler M., Zwerger W. (1992) Ann Physik 1:15–23
59. Randeria M. et al. (1992) Phys Rev Lett 69:2001–2004
60. Sá de Melo C.A.R., Randeria M., Engelbrecht J.R. (1993) Phys Rev Lett 71:3202–3205
61. Mott N.F. (1990) Adv Phys 39:55–81
62. Schneider T., Keller H.J. (1992) Phys Rev Lett 69:3374–3377
63. See, for example: Nozières P. (1995) In: Griffin A., Snoke D.W., Stringari S. (Eds.) Bose Einstein Condensation, Cambridge University Press, 15–31
64. Doniach S., Inui M. (1990) Phys Rev B 41:6668–6678
65. Emery V.J., Kivelson S.A. (1995) Nature 374:434–437
66. Tchernyshyov O. (1997) Phys Rev B 56:3372
67. Anderson P.W. (1987) Science 235:1196–1198
68. Anderson P.W. (1973) Mater Res Bull 8:153;
    Fazekas P., Anderson P.W. (1970) Phil Mag 30:432–440
69. Nagaosa N., Lee P.A. (1990) Phys Rev Lett 64:2450-2453;
    Lee P.A., Nagaosa N. (1992) Phys Rev B 46:5621–5639
70. Suzumura Y., Hasegawa Y., Fukuyama H. (1988) J Phys Soc Jpn 57:2768;
    Tanamoto T., Kuboki K., Fukuyama H. (1991) J Phys Soc Jpn 60:3072–3092
71. Kim C. et al. (1996) Phys Rev Lett 77:4054–4057
72. See, for example: Timusk T., Tanner R.B. (1989) In: Ginsberg G.M. (Ed.) Physical Properties of High–Temperature Superconductors I, World Sci., Singapore, 339–407;
    Tanner R.B., Timusk T. (1992) In: Ginsberg G.M. (Ed.) Physical Properties of High–Temperature Superconductors III, World Sci., Singapore, 363–469, and references therein.
    See also: Uchida S. (1995) In: Feng S., Ren H.C. (Eds.) Proc. CCAST Symposium on High-T_c Superconductivity and the C_{60} Family, Beijing, May 1994, Gordon and Breach, London, 199–249
73. Uemura Y.J. (1996) In: Batlogg B. et al. (Eds.) Proc. 10th Anniversary Workshop of High-T_c Superconductors, Houston, March 1996, World Sci., Singapore, 68–71;
    Uemura Y.J. (1997) Hyperf Interact 105:35–46
74. Uemura Y.J. et al. (1993) Nature 364:605
75. Niedermayer Ch. et al. (1993) Phys Rev Lett 71:1764
76. Bose S.N. (1924) Z. Phys. 26:178;
    Einstein A. (1924) Sitzber Kgl Preuss Akad Wiss 261;
    Einstein A. (1925) Sitzber Kgl Preuss Akad Wiss 3
77. Jin D.S. et al. (1996) Phys Rev Lett 77:420–423
78. Mewes M.-O. et al. (1996) Phys Rev Lett 77:416–419

Strong Correlations and Electron–Phonon Interaction in Superconductors

Karol I. Wysokiński

Institute of Physics, M. Curie-Skłodowska University, ul. Radziszewskiego 10A, PL–20-031 Lublin, Poland (E-mail: karol@tytan.umcs.lublin.pl)

1   Introduction

In the last two decades or so, a number of new materials have been discovered which do not comply to the standard theoretical description of their normal and broken symmetry, mainly superconducting, phases. Among them, a unique role is played by the high temperature superconductors (HTS) [1], with their unusual doping dependence of the normal and superconducting properties. Other materials, such as heavy fermion systems and their alloys [2], alkali metal fullerides [3,4], carbon nanotubes [5], borocarbide superconductors [6], organic superconductors [7], etc., also show interesting features. In some of the mentioned systems, clear non–Fermi liquid behavior in the normal state and strong departures from BCS–Eliashberg [8,9] theory predictions in the superconducting state have been observed. The important issue in this context has been the symmetry of the order parameter, the origin of the pseudogap in the underdoped materials, and the operating mechanism of superconductivity.

While there seems to arise a consensus that the symmetry of the order parameter is of $d_{x^2-y^2}$ type [10,11], the origin of the pseudogap and the superconducting mechanism in HTS remains still under debate. Superconducting alkali–metal–doped fullerides seem to be less controversial with most of the data pointing at the electron–phonon driven superconductivity, while in borocarbides the interesting competition between magnetism and superconductivity has been observed.

The complexity of the behavior of new superconductors and apparent non–applicability of the standard theory has also contributed to new work aimed at deeper understanding of the conventional superconductivity with electron–phonon interaction (EPI) serving as a driving force of electron attraction. Study of the competition between strong Coulomb correlations and EPI has shown *inter alia* a possibility of anisotropic superconductivity [12] with the d wave type order parameter.

The main purpose of this work is to review some results obtained for models which assume the evolution of superconducting phases as arising from the competition between strong electron–electron and electron–phonon interactions. The Coulomb correlations are known to play crucial role in establishing the properties of various systems [13,14,15], and in particular, strong correlation related features have been found in normal and superconducting state of all above mentioned materials. A short discussion of the electron–phonon interaction in the normal state of correlated systems can be found in [16].

In the next section, we shall briefly describe the experiments showing signatures of EPI. The early theoretical work, based mainly on the BCS–Eliashberg type approach, which takes into account the presence of singularities in the electron spectrum and anharmonic electron–phonon (EP) couplings is described in Sect. 3. The competition and mutual influence of the electron correlations and EPI in the framework of Migdal–Eliashberg (ME) theory is discussed in Sect. 4. The recent attempts to go beyond ME approach are reviewed in Sect. 5. We finish with conclusions and remarks Sect. 6.

2   Experimental Signatures of EPI

Here we review experiments which have been interpreted as a clear indication that phonons and electron–phonon interactions participate in the phenomenon of superconductivity. There are two ways in which this participation occurs. First, it is EPI which drives the system to be superconducting and this leads to the major changes in phonon–related characteristics when one crosses the superconducting transition temperature. Second, the superconductivity is connected with quite different interactions, but the changes of electron spectrum indirectly influence the phonons and EPI, which is always present, particularly at temperatures T around T_c.

The most convincing arguments in favor of phonon mechanism, coming from measurements of isotope effect and tunneling spectra, have been gathered for $Ba_{1-x}K_xBiO_3$, $BaPb_{1-x}Bi_xO_3$ as well as K_3C_{60}. In HTS, there have also been found Raman and infrared–active modes [17] with strongly temperature dependent linewidth. Small, but nonzero, isotope shift exponent α has been interpreted [18,19] as a strong indication of the electron–phonon coupling. Similar conclusions have been inferred from tunneling and photoemission [20], neutron [21,22], specific heat [23], thermal conductivity [24,25] and other experiments [26].

We should, however, express a word of caution that in some cases the unambiguous interpretation of experiments as to the role of EPI is difficult to achieve. For a well studied example, see section 3.1. The critical discussion of the lattice properties and electron lattice interaction can be found in [27].

3   BCS–Eliashberg Theory
   with "Real Life" Complications

From the very beginning of the HTS era, there appeared papers in the literature which suggested that the EP interaction is responsible for the phenomenon. From the well-known work on maximum attainable supercon-ducting transition temperature resulting from electron–phonon interaction in classic superconductors [28,29] it was clear that this mechanism had to be supplemented by various real "metal" complications in order to account for T_c around 100 K.

The location of the Fermi level, E_F, close to the Van Hove singularity in the density of states (DOS) was one of the earliest complications taken into account [30,31]. This later evolved into what is known as the Van Hove scen-ario [32]. Strong repulsive, short range electron–electron interactions, usually described by the Hubbard model, the vicinity of the system to metal–insulator transition, the phase separation and the anharmonicities in the phonon spec-tra, and EP interaction were all considered to play a role in setting the high superconducting transition temperature and to be responsible for other un-usual properties of HTS.

In spite of the great diversity of the superconducting systems, the belief is that they fall into a single universal class. Indeed, some universal relations and correlations between different properties have been found. The best known are those connected with the dependence of the superconducting transition temperature T_c in HTS on the carrier concentration n [33], and the so called Uemura plots which relate T_c to the penetration depth λ [34]. The position of the superconductor on this plot has always been used as a "measure" of the "exoticity" of the material [35].

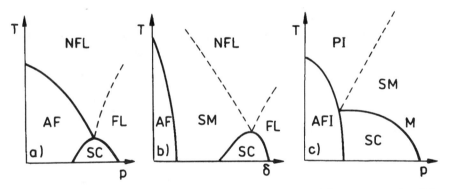

Fig. 1. The schematic phase diagrams of heavy fermions (**a**), copper oxides (**b**), and organic materials (**c**). Adapted from [36] and [37]; p denotes the pressure, δ – hole concentration. Various phases are denoted by: AF(I) – antiferromagnetic (insulating), PM(I) – paramagnetic (insulating), M(FL) – metallic (Fermi-Liquid), NFL – non–Fermi–Liquid, SM – strange metal

The similarities between organic [36] and high temperature superconduct-ing oxides, as well as heavy fermions and HTS [37], have also been discussed by direct comparison of their phase diagrams on the "temperature–pressure" and "temperature–doping" planes. In Fig. 1, we show the schematic phase diagrams of heavy fermions (a), high temperature copper oxides (b) and organic materials (c). Up to the details connected with slopes of various lines, there is close similarity between these seemingly different systems. For some values of the parameters, they show magnetic, superconducting and normal metal (i.e. Fermi–Liquid (FL)) behavior. In all cases, there is also a phase, or at least a region on the phase diagram, where the systems show unusual features of non–Fermi–Liquid (NFL) type. This "phase" is often referred to as 'peculiar', 'unconventional', 'strange' etc. At present, there is no clear definition of these 'exotic' phases.

3.1   Isotope Effect — What Does It Measure ?

One of the early arguments against the electron–phonon mechanism in HTS was based on the absence of isotope effect. This is very important parameter. The discovery of the isotope effect has played a crucial role in identifying the electron–phonon interaction as a superconductivity mechanism operating in classic superconductors. It is also believed to be an important factor in the studies of HTS [18] and other exotic superconductors.

As already discussed, a number of experiments have shown that electron–phonon interaction, though not necessarily solely responsible for the superconductivity in these materials, does play an important role and should not be abandoned completely. With regard to the isotope effect, the situation has been very puzzling. The early experiments on the oxygen isotope effect [38] have shown that the value of isotope exponent α ($T_c \sim M^{-\alpha}$) strongly depends on the technological details. It may take on small positive, zero or small negative values for HTSs, and moreover depends on the doping level δ, vanishing for optimally doped samples. On the other hand, this coefficient is quite large in alkali metal doped fullerides [39]. The situation seems to be even more complicated for the copper isotope effect which has also been measured [40] and has fairly large coefficient, taking on values around $1/3$ in the underdoped regime.

In the framework of BCS–Eliashberg theory one writes the superconducting T_c as

$$T_c^{\text{BCS}} = 1.13\,\bar{\omega}\,\exp\left[-\frac{1}{\lambda - \mu^\star}\right],\tag{1}$$

where $\bar{\omega}$ is the characteristic phonon frequency (often replaced by the Debye one ω_D), λ – electron–phonon coupling and $\mu^\star$ – Coulomb pseudopotential which describes effective repulsion renormalized from its bare value μ to

$$\mu^\star = \frac{\mu}{1 + \mu \ln E_F/\bar{\omega}},$$

due to retardation effects; E_F is the Fermi level. In this theory, the isotope coefficient is given by (assuming $\bar{\omega} \approx \omega_D \propto M^{-1/2}$; M is the isotope mass)

$$\alpha = \frac{1}{2}\left\{1 - \left(\frac{\mu^\star}{\lambda - \mu^\star}\right)^2\right\}\tag{2}$$

and the values of α are found to be smaller than $1/2$, with small α correlated to small T_c (cf. (1) and (2)).

The new superconductors we are talking about are complicated chemical systems with a large number of atoms per unit cell. In such a case, one defines partial isotope coefficients α_r for replacing atom r by the isotope. The total α is then defined as $\alpha_{\text{total}} = \sum_1^N \alpha_r$. Thus, replacing one of the atoms by corresponding isotope leads to α_{total} smaller than $1/2$.

It has recently been found that the value of the coefficient α is strongly influenced by such factors as magnetic impurities, proximity effect, non–adiabaticity [41], and a presence of the VHS or strong energy dependence of the electronic density of states (DOS) in the close vicinity of the Fermi level [42]. The solution of the Eliashberg equation with energy dependent DOS has shown the possibility for obtaining values of $\alpha < 1/2$ when there exists a peak in DOS and values of the order 1 or larger when the DOS has a dip [42].

The general conclusion is that the presence or absence of isotope effect can teach us a lot about the system (e.g. about the magnetic scattering, proximity and nonadiabaticity), but its presence or absence cannot be used as an argument in favor or against the electron–phonon mechanism [41].

3.2 Electron–Lattice Coupling in HTS

Weber [43] was among the first to propose the EP interaction to be responsible for superconductivity in $La_{2-x}Sr_xCuO_4$. His realistic calculations of lattice dynamics and the Eliashberg function $\alpha^2 F(\omega)$ lead him to conclude that it is strong coupling of some phonon modes to electrons ($\lambda \approx 2 - 2.5$) and light oxygen mass what allows for T_c values ~ 30–40 K.

The necessity of the fully self–consistent description of the normal and superconducting states in the framework of the Eliashberg approach has been stressed in [44]. These authors argued that strong electron–phonon interactions may lead to marginal Fermi liquid like behavior and that it gives a natural explanation of anomalous normal state and superconducting behavior in HTS.

The departure of the measured dimensionless thermodynamic ratios in superconductors (like the ratio of the gap to T_c and other) from universal BCS values can be most easily explained in the strong coupling Eliashberg theory, particularly if two or more interactions contribute to the Eliashberg function $\alpha^2 F(\omega)$. The case of electron–phonon and exciton interactions has been studied in [45].

The solution of Eliashberg equations for a system with a Van Hove singularity (VHS) in the electron spectrum lead Mahan [31] to conclude that VHS can contribute to an increase of T_c but only in the circumstances that the Fermi level is within 10% of the singular energy. On the other hand, this author has found that angular variation of pairing parameters can increase T_c

up to 50%. Similar calculations have recently been performed [46] with application to the tunneling conductance of HTS. It has been found that both high T_c values and d–wave symmetry of the order parameters can be obtained in reasonable agreement with the data on optimally doped $Bi_2Ca_2SrCu_2O_{8+\delta}$. They used the Eliashberg function, which consists of isotropic and anisotropic parts, and were able to fit the value of the gap, the critical temperature, and the shape of the density of states.

Some of the early applications of the Eliashberg theory to HTS have been reviewed and critically analyzed in [47] and [48].

3.3   Anharmonicity in the Phonon Spectrum and Electron–Phonon Coupling

Standard Eliashberg formalism used to derive values of λ assumes that lattice response is harmonic. The anharmonic effects, which in HTS are mainly connected with the motion of apex oxygen ions, modify the original picture. In the work [49] aimed at the description of 90 K superconducting $YBa_2Cu_3O_{7-\delta}$, it has been proposed that some phonon modes, namely those connected with motion of apex oxygens may be strongly anharmonic. The calculations [49] show that if the apex oxygen moves in a double–well potential, both the coupling λ (> 5) and averaged frequency $\langle\omega^2\rangle^{1/2}$ are subjected to large, temperature–dependent change. The resulting superconducting transition temperature, in the strong coupling limit $\lambda \gg 1$ calculated from the formula [50]

$$T_c = 0.18\left(\lambda\langle\omega^2\rangle\right)^{1/2},\tag{3}$$

has been found to take on values of the order of 100 K.

Both $YBa_2Cu_3O_{7-\delta}$ and La_2CuO_4 were argued [51] to be unstable mechanically, which can mean a strong enhancement of electron–lattice coupling, especially if the anharmonicity is connected with double–well potentials. In the general case, the parameters λ can become temperature dependent. The thorough analysis of the experiments [38] has lead Müller [18] to conclude that the motion of apex oxygen is highly anharmonic most probably due to double well potential along c axes.

The effect of anharmonicity on the electron spectrum, charge ordering and superconducting properties in exotic, mainly high temperature superconductors, has been studied by a number of groups [52,53,54].

Strong anharmonicity of the lattice potential can be induced by correlations. For the near half–filled Hubbard–Holstein model, such an effect has been demonstrated in [55]. This is due to polaronic self–localisation of carriers and can exist even in systems with relatively weak, bare electron–phonon coupling.

4   Electron Correlations and EPI–Eliashberg Theory

One of the theoretical arguments against the electron–phonon interaction and in favor of Coulomb mechanism in HTS is the apparent $d_{x^2-y^2}$ symmetry of the order parameter. EPI, in metals, is only weakly anisotropic due to its locality in real space and strong screening and thus leads to s–wave gap. In many exotic superconductors and most notably in oxides, the screening is weak and the anisotropy of EPI much stronger. The explicit solutions of Eliashberg equations [56] with different strength of EPI at different regions of the Fermi surface, has led to solutions of s– and d–wave symmetry being nearly degenerate in energy. They could also account for high values of T_c, suppression of the isotope effect, and large ratio $2\Delta/T_c$.

We shall see later in this Section that short coherence length and strong on–site Coulomb repulsion severely modify the original Eliashberg equation. As a result, the symmetry of the resulting order parameter depends on the doping level. The d–wave symmetry is found for hole–doped materials and the s–wave symmetry – for electron–doped materials [12,57].

As mentioned, the short range Coulomb correlations are strong in most exotic superconductors including organic ones [58]. At the same time, the superconducting coherence length ξ is short. All of this means that correlations cannot be neglected. Correlations cause the half–filled system to be Mott–Hubbard insulator. Superconductivity appears at a finite doping level, as seen in Fig. 1(b).

There exists in literature a number of papers which study the superconductivity in the presence of both EPI and correlations [12,57,59–63]. Both covalent (i.e. resulting from modulation of the hopping amplitude) EP coupling and ionic one (i.e. on–site) have been studied in the framework of three band or effective single band models.

Before we continue, let us remark that competition between interactions can lead to unexpected results. Transition temperature T_c, calculated for a model with electron–phonon coupling and "correlated hopping" [64] terms, was increased or decreased by EPI depending on the parameters [65]. In this particular case, such behavior was connected with momentum dependence of the order parameter.

4.1   Eliashberg Equations in Correlated Systems

To study the interplay of strong short range Coulomb correlations and electron–phonon coupling, we use the single band model Hamiltonian, however, with both on–site and intersite electron–lattice coupling [57]. In electron representation this reads

$$\mathcal{H} = \sum_{ij\sigma}(t_{ij} - \mu\delta_{ij})\,\tilde{c}^{+}_{i\sigma}\tilde{c}_{j\sigma} + \sum_{ijs\sigma}T^{\prime\alpha}_{ijs}\,u^{\alpha}_{s}\,\tilde{c}^{+}_{i\sigma}\tilde{c}_{j\sigma} + U\sum_{i}\tilde{n}_{i\uparrow}\tilde{n}_{i\downarrow} + \mathcal{H}_{\text{ph}}\ . \quad (4)$$

Here, $\tilde{c}_{i\sigma}^{+}(\tilde{c}_{i\sigma})$ is the creation (annihilation) operator for a spin σ electron at a site i of the lattice, u_i^{α} is the α-th component of the displacement vector of the ion, $\tilde{n}_{i\sigma} = \tilde{c}_{i\sigma}^{+}\tilde{c}_{i\sigma}$, μ denotes the chemical potential, t_{ij} is the hopping integral assumed to take on nonzero value $-t$ for i,j being nearest neighbor sites, U is the Hubbard on–site repulsion of carriers. $\mathcal{H}_{\mathrm{ph}}$ denotes the Hamiltonian of the lattice. We assume the validity of the harmonic approximation. The electron–lattice interaction described by the second term in (4) has two components

$$T_{ijs}^{\prime\alpha} = T_{ijs}^{\alpha} + V_{js}^{\alpha}\,\delta_{ij}\,,$$

the first of which, T_{ijs}^{α}, has been derived from modulation of the hopping integral t_{ij} in the deformed lattice. It is related to the derivative of t_{ij} taken at equilibrium position of an ion. The second part of the interaction, i.e. the term V_{js}^{α}, is connected with fluctuations of the crystal field. Due to the ionic character of high temperature superconductors, this term is expected to be more important and has to be taken into account even if the equilibrium value of crystal field $\varepsilon_i = t_{ii}$ is zero. The systematic derivation and the discussion of the electron–ion interaction in the context of superconducting oxides can be found in [66].

To proceed, we shall assume the validity of the Migdal theorem [67] and use the simple version of the slave boson technique to deal with strong correlations U in Hamiltonian (4). For large U values, the states corresponding to doubly occupied sites are pushed to high energies. The limit in which we are interested here, (i.e. $U = \infty$), they are not important at all and we expect the sites to be singly occupied or empty. Thus in general, the condition quantifying this limit is $n_i = \sum_{\sigma} \tilde{c}_{i\sigma}^{+}\tilde{c}_{i\sigma} \leq 1$.

The idea of slave bosons [68] is to represent the physical electron in the $U = \infty$ limit by the fictitious fermion, described by the operators $c_{i\sigma}^{+}$ ($c_{i\sigma}$) and an auxiliary boson b_i^{+} (b_i). The condition of no double occupancy of a site can now be expressed in formally exact form

$$Q_i = \sum_{\sigma} c_{i\sigma}^{+}c_{i\sigma} + b_i^{+}b_i = 1\,. \tag{5}$$

To keep track of the constraint, one usually introduces a set of Lagrange multipliers Λ_i and adds to the Hamiltonian a term $\sum_i \Lambda_i(Q_i - 1)$, neglecting at the same time the term proportional to U. The intersite terms in Hamiltonian (4) contain the additional operators b_i, b_j^{+} while the on–site terms do not. This clearly distinguishes the influence of strong correlations on the covalent and ionic EPI.

To illustrate the modifications of the Eliashberg theory, we shall write the equations obtained in the mean field approach to slave bosons. The details of the derivation can be found in [12] and [57]. This last reference also contains results beyond mean field approximation.

It has been shown that mean field description can be obtained by assuming average values $\langle b_i \rangle = \langle b_j^{+} \rangle = r$ and $\Lambda_i = \Lambda$ at each site. Two parameters, r

and Λ entering the mean field Hamiltonian can be chosen so as to make minimal the ground state energy $E_{\mathrm{GS}} = \langle \mathcal{H} \rangle$. One gets [68]

$$r^2 = 1 - n \, ,$$

$$-\Lambda = \frac{1}{N} \sum_{ij\sigma} t_{ij} \langle c_{i\sigma}^+ c_{j\sigma} \rangle + \frac{1}{N} \sum_{ijs\alpha\sigma} T_{ijs}^{\prime\alpha} \langle u_s^\alpha c_{i\sigma}^+ c_{j\sigma} \rangle \, .$$

Here, $n = \frac{1}{N} \sum_{i\sigma} \langle c_{i\sigma}^+ c_\sigma \rangle$ denotes the concentration of electrons in the band ($n < 1$). There are two modifications of the spectrum of electrons encountered on the mean field level. First is the band narrowing described by r^2 and its shift described by Λ. The spectrum of noninteracting fermions in the mean field is given by $(r^2 \epsilon_k - \mu + \Lambda)$ instead of $(\epsilon_k - \mu)$ of original electrons (at $U = 0$). For the half–filled band $n = 1$, the system is localized ($r = 0$). Here, ϵ_k is the Fourier transform of t_{ij}.

To properly describe the superconducting state in the system at hand, one has to work in site representation. The important point is that in the considered $U = \infty$ limit the double occupation of a given site is strictly forbidden. This means, *inter alia*, that correlation functions $\langle c_{i\uparrow} c_{j\downarrow} \rangle$ describing superconducting pairs vanish exactly for $i = j$, i.e. the on–site pairing is forbidden. On the other hand, the correlations of the type $\langle c_{i\sigma}^+ c_{i\sigma} \rangle$ measure the average number of carriers at site i, and are allowed to enter into the formula. This important fact has first been noted by Zieliński and coworkers [12] and leads, as we shall see, to severe changes in the form of Eliashberg equations.

The matrix self–energy $\hat{\Sigma}_k(\omega)$ is written as usual (with $\hat{\tau}_i$ denoting Pauli matrices) [57]

$$\hat{\Sigma}_k(\omega) = \omega[1 - Z_k(\omega)] \hat{\tau}_0 + \phi_k(\omega)\hat{\tau}_1 + \chi_k(\omega)\hat{\tau}_3 \tag{6}$$

and calculated up to a second order in EPI. We get

$$\omega[1 - Z_k(\omega)] = \frac{1}{2} \int d\omega_1 \int d\omega_2 \frac{\tanh(\beta\omega_1/2) + \coth(\beta\omega_2/2)}{\omega - \omega_1 - \omega_2} \frac{1}{N} \sum_{k'} K_{kk'}^{\mathrm{N}}(\omega_2) \cdot$$
$$(-\frac{1}{\pi}) \, \mathrm{Im} \frac{\omega_1 Z_{k'}(\omega_1)}{A_{k'}(\omega_1)} \; ; \tag{7a}$$

$$\phi_k(\omega) = \frac{1}{2} \int d\omega_1 \int d\omega_2 \frac{\tanh(\beta\omega_1/2) + \coth(\beta\omega_2/2)}{\omega - \omega_1 - \omega_2} \frac{1}{N} \sum_{k'} K_{kk'}^{\mathrm{S}}(\omega_2) \cdot$$
$$(-\frac{1}{\pi}) \mathrm{Im} \frac{-\phi_{k'}(\omega_1)}{A_{k'}(\omega_1)} \; ; \tag{7b}$$

$$\chi_k(\omega) = \frac{1}{2} \int d\omega_1 \int d\omega_2 \frac{\tanh(\beta\omega_1/2) + \coth(\beta\omega_2/2)}{\omega - \omega_1 - \omega_2} \frac{1}{N} \sum_{k'} K_{kk'}^{\mathrm{N}}(\omega_2) \cdot$$
$$(-\frac{1}{\pi}) \mathrm{Im} \frac{r^2 \epsilon_{k'} - \mu + \Lambda + \chi_{k'}(\omega_1)}{A_{k'}(\omega_1)} \, , \tag{7c}$$

where we denoted $A_{\boldsymbol{k}'}(\omega_1) = [\omega_1 Z_{\boldsymbol{k}'}(\omega_1)]^2 - [\phi_{\boldsymbol{k}'}(\omega_1)]^2 - [r^2\epsilon_{\boldsymbol{k}'} - \mu + \Lambda + \chi_{\boldsymbol{k}'}(\omega_1)]^2$
and

$$K^{\mathrm{N}}_{\boldsymbol{k}\boldsymbol{k}'}(\omega_2) = \sum_\nu \left|M^\nu_{\boldsymbol{k},\boldsymbol{k}-\boldsymbol{k}'}\right|^2 \left(-\frac{1}{\pi}\right) \mathrm{Im} D_{\boldsymbol{k}-\boldsymbol{k}'\nu}(\omega_2 + \mathrm{i}0), \qquad (8)$$

$$K^{\mathrm{S}}_{\boldsymbol{k},\boldsymbol{k}'}(\omega_2) = \frac{1}{N}\sum_{\boldsymbol{q}\nu} \left|M^\nu_{\boldsymbol{k},\boldsymbol{q}}\right|^2 \left(-\frac{1}{\pi}\right) \mathrm{Im} D_{\boldsymbol{q},\nu}(\omega_2 + \mathrm{i}0)\,\gamma(\boldsymbol{k}-\boldsymbol{q}-\boldsymbol{k}'). \qquad (9)$$

Here, $D_{\boldsymbol{k}\nu}(\omega)$ is the phonon propagator and electron–phonon matrix element $M^\nu_{\boldsymbol{k},\boldsymbol{q}}$ is given by

$$\left|M^\nu_{\boldsymbol{k},\boldsymbol{q}}\right|^2 = \left|g_\nu(\boldsymbol{q})\right|^2 \left|\sum_\alpha e^\alpha_\nu(\boldsymbol{q})\left[r^2 V_c(\tilde{v}^\alpha_{\boldsymbol{k}-\boldsymbol{q}} - \tilde{v}^\alpha_{\boldsymbol{k}}) + V_i V^\alpha_{\boldsymbol{q}}\right]\right|^2, \qquad (10)$$

with $g_\nu(\boldsymbol{q}) = (\hbar/2\omega_\nu(\boldsymbol{q})\cdot M)^{1/2}$, $\tilde{v}^\alpha_{\boldsymbol{k}} = (1/\hbar)(\partial\epsilon_{\boldsymbol{k}}/\partial k^\alpha)$. Here, $e^\alpha_\nu(\boldsymbol{q})$ is the α-th component of the phonon polarization operator, while $V^\alpha_{\boldsymbol{q}}$ is the Fourier transform of the ionic part of electron–phonon interaction, $\omega_\nu(\boldsymbol{q})$ denotes the phonon dispersion and M – the ionic mass. V_c (V_i) denotes the strength of the covalent (ionic) part of electron–phonon interaction. Note the different renormalization by r^2 of the covalent and ionic parts of EPI. The parameter $\gamma(\boldsymbol{k})$ entering (9) is of geometrical origin. It takes care of short superconducting coherence length and is given by the sum of phase factors $\exp[\mathrm{i}\boldsymbol{k}\boldsymbol{r}_i]$ over nearest neighbor sites $\boldsymbol{r}_i$ in a given lattice.

4.2   Consequences of Generalized Eliashberg Equations

There are a few differences between above and usual Eliashberg equations for superconductors. The most important, beside the renormalization of the spectrum, is the presence of two different types of kernels $K^{\mathrm{N}}_{\boldsymbol{k}\boldsymbol{k}'}$ and $K^{\mathrm{S}}_{\boldsymbol{k}\boldsymbol{k}'}$ determining normal ($Z_{\boldsymbol{k}'}, \chi_{\boldsymbol{k}'}$) and anomalous ($\phi_{\boldsymbol{k}'}$) parts of the self–energy, respectively.

The equation for $\chi_{\boldsymbol{k}}(\omega)$ is usually neglected. In the theory valid for strongly correlated superconductors with short coherence length, this has to be taken into account as it leads to nontrivial modification of the relation between chemical potential μ and carrier concentration n.

The complete solution of the above generalized equations is complicated. We shall discuss here only most important consequences. First, one notes that the equation for $Z_{\boldsymbol{k}}(\omega)$ has precisely the structure as in standard theory [48]. This means that the wave function renormalization is mainly of s–wave type. By this, we mean that in the decomposition of $Z_{\boldsymbol{k}}$ into Fermi surface harmonics (for a two dimensional square lattice these are $1, \cos k_x + \cos k_y, \cos k_x - \cos k_y$, etc., for s, extended s (denoted by $s^\star$), d, ... symmetry) the largest contribution will come from the spherically symmetric contribution. The same is not true for the order parameter, and in fact the largest component is of s–wave type for small electron concentration in the band, whereas it is a d wave type for small hole concentration in the system. This is illustrated in Fig. 2, where we plot the corresponding superconducting electron–phonon coupling constants λ_s and λ_d.

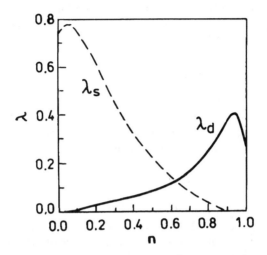

Fig. 2. The carrier concentration dependence of the s and d wave components of the coupling constants calculated from the kernel entering equation (7b) for the order parameter

Two different kernels $K^N_{k,k'}$ and $K^S_{k,k'}$ of Eliashberg equations imply the existence of two different Eliashberg functions $\alpha^2 F_I(\omega)$, $I =$ N,S for normal (N) and superconducting (S) properties defined by

$$\alpha^2 F_I(\omega) = N(E_F) \langle \langle K^I_{k,k'}(\omega) \rangle_{k'_F} \rangle_{k_F} , \qquad (11)$$

where the sign $\langle \langle \ldots \rangle_{k_F} \rangle_{k'_F}$ denotes the Fermi surface average and $N(E_F)$ is the density of states at the Fermi energy. One also defines two different electron–phonon coupling constants $\lambda^I = 2 \int d\omega / \omega \cdot \alpha^2 F_I(\omega)$.

There exists [69] still another Eliashberg function which in the present case can be defined as

$$\alpha^2 F_{tr}(\omega) = N(E_F) \left\langle \left\langle K^N_{k,k'}(\omega) \frac{[v(k) - v(k')]^2}{2 \langle v(k)^2 \rangle_{k_F}} \right\rangle_{k'_F} \right\rangle_{k_F} \qquad (12)$$

and corresponding coupling λ^{tr} which describes transport (tr) properties of the system. Calculations using the kernel K^N (8) show that the couplings λ^{tr} and λ^N are of the same order [70]. Vertex corrections strongly reduce λ^{tr} (see below and in Fig. 3).

It is the peculiarity of the correlated system that we have to define two different sets of electron–phonon couplings λ^I_i, $i = s, s^\star, d$; one for normal

$I = $ N and another for superconducting $I = $ S part of the generalized Eliashberg equations. In principle, all these parameters can be deduced from experiments [69]: λ^{tr} from high temperature transport measurements, and λ^{N} from specific heat. The superconducting coupling λ^{S} is expected to enter the formula for T_c. To see this, assume that of normal state parameters λ_i^{N} only the λ_s^{N} component is nonzero. The superconducting transition temperature into the phase with i–th symmetry can be written approximately in McMillan–like form as [71]

$$T_c^i = 1.13\,\bar{\omega}\exp\left[-\frac{1+\lambda_s^{\text{N}}}{\lambda_i^{\text{S}}}\right].\tag{13}$$

The stability of the obtained superconducting phases with respect to the changes of the electron spectrum has been checked [72] by performing the calculations for the band structure given by

$$\varepsilon(\mathbf{k}) = -2t(\cos k_x a + \cos k_y a + (2t'/t)\cos k_x a \cos k_y a).\tag{14}$$

Interestingly enough, for $t'/t \leq 0$ the d–wave phase remains stable for small hole doping. In the opposite limit of $t'/t > 0$, the s–wave coupling constant is larger than the d–wave one for all values of carrier concentration n. It is worth noting, in this context, that the minus sign is proper for the spectrum of HTS which also possesses d–wave order parameter.

4.3   Vertex Corrections Due to Electron–Electron Interaction

The theory presented above has neglected the vertex corrections – both those resulting from electron–phonon as well as those due to electron–electron interactions. Some aspects related to the electron–phonon vertex corrections will be discussed in the next section. Here, we shall concentrate on the electron–electron interactions. These vertex corrections have been considered e.g. in [61] and [60].

Correlations induce strong dependence of the vertex corrections on $\mathbf{v}_{\text{F}}\mathbf{q}/\omega$ ($\mathbf{v}_{\text{F}}$ is the Fermi velocity, $\mathbf{q},\omega$ stand for momentum and frequency transfer). For a three–band Hubbard model, they have been shown to suppress quasiparticle coupling in the $\mathbf{v}_{\text{F}}\mathbf{q}/\omega \gg 1$ limit. In the opposite limit, the effective EPI has been increased. EPI was found to induce a phase separation instability. Near the phase separation instability, the system develops superconducting instabilities in both s–wave and d–wave channels.

Momentum and frequency dependence of vertex corrections complicate analysis of the experiments which probe different regions of the $(\mathbf{q},\omega)$–space. The detailed calculations for a single band Hubbard model supplemented with local electron–phonon interaction has revealed much larger reduction of the transport Eliashberg function $\alpha^2 F_{\text{tr}}(\omega)$ than $\alpha^2 F(\omega)$. This is connected with the factor $[\mathbf{v}(\mathbf{k}) - \mathbf{v}(\mathbf{k}')]^2$ which favors large momentum transfers in $\alpha^2 F_{\text{tr}}(\omega)$

(cf. (12)). The function $\alpha^2 F(\omega)$ is also reduced by vertex corrections but only by a factor of 3–4 [60].

The momentum dependence of the vertex $\gamma(k = k_F, q)$ on q, along the diagonal $q = (q, q)$ of the two–dimensional Brillouin zone calculated in [60] by means of Hubbard X operator technique and $1/N$ expansion, is plotted in Fig. 3. Note the strong doping dependence of the vertex.

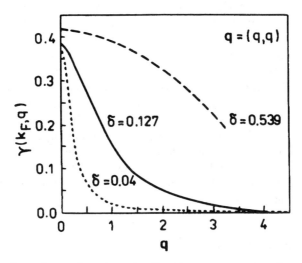

Fig. 3. The dependence of static vertex function on momentum transfer calculated in [60] for few dopings. Note strong momentum and doping dependence

The semi–quantitative analysis of phonon mediated superconductivity in a correlated system is given in [73] with special emphasis on HTS. This author also argues, *inter alia*, that the pseudogap behavior observed in underdoped cuprates can be explained by invoking phonons and EPI in the strong coupling limit ($\lambda > 2$) and concludes that the anisotropy of the pseudogap is a direct indication of the anisotropy of the coupling.

The EPI, as a mechanism of superconductivity in correlated HTS, has been recently shown to lead to universal dependence of T_c on the doping level δ [74].

5   The Role of EP Vertex Corrections – Beyond Migdal–Eliashberg Theory

Here, we want to comment on the issue of vertex corrections due to electron–phonon interaction. They have been argued long ago by Migdal [67] to be unimportant for metals. The higher order diagrams have been found to give contributions of the relative order ω_{ph}/E_F, where ω_{ph} is the characteristic phonon frequency and E_F – the Fermi energy. In wide band metals, this parameter is of the order of 10^{-3} or smaller and neglecting vertex corrections is justified. In the materials, we are talking about the bands which are narrowed by correlation effects (particularly near half–filling), the ratio ω_{ph}/E_F is of the order of 0.1 or larger, and the vertex corrections become important. Moreover, the reduced dimensionality of the system makes vertex corrections the more important as the relevant parameter is no longer ω_{ph}/E_F but $\omega_{ph}/E_F \log(E_F/\omega_{ph})$ [75]. Again, the largest corrections appear in restricted frequency regions and make them important even for relatively small values of ω_{ph}/E_F.

The same, which has been said in connection with the normal state [75], remains valid for the superconducting state. Here, one expects the renormalization of coupling constant(s) λ^I, transition temperature T_c, penetration depth, isotope coefficient etc. A number of researchers have recently addressed questions related to vertex corrections [76,77].

An important conclusion, in the present context, has been obtained in [77]. The vertex corrections practically do not influence the transition temperature of the s–wave superconducting phase. On the other hand, the d–wave phase superconducting transition temperature is substantially diminished by vertex corrections.

6   Remarks and Conclusions

Shortly after the discovery of high T_c superconducting (HTS) oxides with transition temperatures T_c around 100 K, *much* exceeding the most optimistic estimation of T_c resulting from the electron–phonon interaction, a hot discussion started about the applicability of BCS–Eliashberg theory and the electron–phonon (EP) mechanism. I hope that I have convinced the reader that the issue is still far from being solved, and that it is interesting and puzzling enough to be studied both experimentally and theoretically.

On the theoretical part, better controlled approximation schemes and numerically exact results for larger systems are very desirable. We have seen that in most of exotic superconductors and, in particular, in high T_c oxides, the delicate balance of various interactions exists. In various respects they seem to be "marginal," or falling into the crossover region. The coherence length put them neither in the limit of BCS pairing nor Bose–Einstein condensation. They fall neither in the adiabatic nor strong anti–adiabatic limit. All this causes the well–established theories to break down. The relevant expansion parameters are all "of order 1".

As already noted, this shows a need for more complete understanding of the origin of classic superconductivity and models used to describe it. The simplest such model is that introduced by Holstein [78] and studied recently in various nontrivial limits [79]. In order to take strong correlations into account, the Holstein model is supplemented by the Hubbard U term [80] describing on–site repulsion of electrons. The resulting Holstein–Hubbard model has also recently been intensively studied [81].

The experimentally observed signatures of EPI are often weak, not because the interaction itself is weak, but as discussed above, because of interference with other interactions. This makes proper analysis of experiments and comparison with competing theories difficult. For an example, see the interesting exchange of comments on the interpretation of oxygen isotope effect on the effective mass of charge carriers in $YBa_2Cu_3O_{6.94}$ [82].

Useful information about the electron–(exchanged) boson(s) interactions is contained in the Eliashberg function $\alpha^2 F(\omega)$. Various thermodynamic and transport coefficients depend on it. The important issue is how to extract the piece of $\alpha^2 F(\omega)$ due to EPI and distinguish it from contributions possibly due to other interactions, e.g. spin–fluctuation.

Can one unambiguously identify the contribution from EPI to tunneling and optical conductivity? In particular, it would be desirable to know how to extract the information on phonons, spin fluctuations, or other interaction from optical conductivity in normal and superconducting state. Schachinger and Carbotte [83] have recently asked that question. To answer it, they used a model in which pairing proceeds through exchange of antiferromagnetic spin fluctuations supplemented with EPI and solved the resulting Eliashberg equations. They have shown that the presence of phonons shows up qualitatively both in frequency and temperature dependence of conductivity, but the comparison with existing experimental data cannot confirm or rule out large electron–phonon contribution. They used the isotropic electron–phonon contribution to the spectral function $\alpha^2 F(\omega)$ and the anisotropic one due to spin–fluctuation exchange. These authors did not consider the possibility of two different kernels and thus two different Eliashberg functions considered in the present work.

We have to emphasize that even an effect such classical as the isotope effect, when measured in exotic systems shows up previously unknown and unexpected faces. This forces further theoretical work to be done.

We have been interested here in the EPI induced superconductivity in strongly correlated systems and have seen that the traditional Eliashberg description needs modification. Due to correlations and short coherence length, the equations become highly anisotropic and in particular lead to the d–wave superconducting order parameter.

All of this complicates the identification of the mechanism of superconductivity and the role of EPI in it. At the same time, we have seen that much experimental data could be explained by invoking EPI. Thus, this interaction remains a serious candidate of mechanism of superconductivity at least in strongly correlated systems.

Acknowledgements

This work has been supported by KBN through grant No. 2 P03B 031 11.

References

1. Bednorz J.G., Müller K.A. (1986) Z Phys B 64:189;
   Bednorz J.G., Müller K.A. (1988) Rev Mod Phys 60:585
2. Andraka B., Stewart G.R. (1993) Phys Rev B 47:3208;
   Bernal O.O. et al. (1995) Phys Rev Lett 75:2023;
   Maple B. et al. (1995) J Low Temp Phys 99:223
3. Hebbard A.F. et al. (1991) Nature 350:600
4. Knupfer M., Fink J. (1997) Phys Rev Lett 79:2714
5. Iijima S. (1991) Nature 354:56
6. Nagarajan R. et al. (1994) Phys Rev Lett 72:274
7. Wiliams J.M. et al. (1991) Science 252:1501;
   Jerome D. (1991) Science 252:1509
8. Bardeen J., Cooper L.N., Schrieffer J.R. (1957) Phys Rev 108:1175
9. Eliashberg G.M. (1960) Zh Eksp Teor Fiz 38:966 [Engl. in:
   Sov Phys–JETP 11:696]
10. Pines D. (1996) In: Klamut J. et al. (Eds.) Recent Developments in High
    Temperature Superconductivity, Lecture Notes in Physics vol 475, Springer,
    Berlin Heidelberg, 201–220
11. Annett J.F., Goldenfeld N., Leggett A.J. (1996) In: Ginsberg D.M. (Ed.)
    Physical Properties of High Temperature Superconductors, Vol 5, World Sci.,
    Singapore
12. Zieliński J., Matlak M., Entel P. (1992) Phys Lett A 165:285;
    Mierzejewski M., Zieliński J. (1995) Phys Rev B 52:3079;
    Mierzejewski M., Zieliński J., Entel P. (1994) Physica C 235–240:2143;
    Zieliński J. et al. (1995) J Supercond 8:135
13. Fulde P. (1993) Electron Correlations in Molecules and Solids, Springer Series
    in Solid–State Sciences, Springer, Berlin Heidelberg;
    Fulde P. (1999) Solids with Weak and Strong Electron Correlations, to be
    published
14. von Lohneysen H. (1998) Phil Trans Roy Soc A 356:139
15. Raychhuri A.K. (1995) Adv Phys 44:21
16. Wysokiński K.I. (1996) In: Klamut J. et al. (Eds.) Recent Developments in
    High Temperature Superconductivity, Lecture Notes in Physics, vol 475,
    Springer, Berlin Heidelberg, 243–252
17. Chrzanowski J., Gygax S., Irvin J.C., Hardy W.N. (1988)
    Solid State Commun 65:139;
    Cooper S.L., Klein M.V., Pazol B.G., Rice J.P., Ginsberg D.M. (1988)
    Phys Rev B 37:5817
18. Muller K. Alex (1990) Z Phys B 80:193
19. Bornemann H.J., Morris D.E. (1991) Phys Rev B 44:5322;
    Nickel J.H., Morris D.E., Ager J.W., III (1993) Phys Rev Lett 70:81
20. Hinks D.G. et al. (1989) Physica C 162–164:1405;
    Zasadzinski J.F. et al. (1989) Physica C 162–164:1053;
    Namatame H. et al. (1994) Phys Rev B 50:13 674
21. Mook H.A., Mostoller M., Harvey J.A., Hill N.W., Chakoumakos B.C.,
    Sales B.C. (1990) Phys Rev Lett 65:2712
22. Arai M., Yamada K., Hidaka Y., Itoh S., Bowden Z.A., Taylor A.D.,
    Endoh Y. (1992) Phys Rev Lett 69:359

23. Reeves M.E., Ditmars D.A., Wolf S.A., Verah T.A., Kresin V.Z. (1993) Phys Rev B 47:6065
24. Cohn J.L., Wolf S.A., Verah T.A. (1992) Phys Rev B 45:511
25. Jeżowski A., Klamut J., Dabrowski B. (1995) Phys Rev B 52:R7030
26. Tsuda N., Shimada D., Miyakawa N. (1991) Physica C 185–189:1903–1904;
    Vedeneev S.I., Stepanov V.A. (1989) JETP Lett 49:588;
    Aoki R., Murakami H. (1996) J Low Temp Phys 105:1231
27. Ranninger J. (1991) Z Phys B 84:167
28. see e.g.:
    Varma C.M. (1982) In: Buckel W., Weber W. (Eds.) Superconductivity in d–
    and f–Band Metals, KfK Karlsruhe, 603
29. For a review see e.g.:
    Allen P.B., Mitrović B. (1982) In: Ehrenreich H., Seitz F., Turnbull D. (Eds.)
    Solid State Physics, Academic, New York, 1;
    see also: Ginzburg V.L., Kirzhnits D.A. (Eds.) (1977) Problems of High
    Temperature Superconductivity, Nauka, Moscow (in Russian)
30. Labbé J., Bock J. (1987) Europhys Lett 3:1225
31. Mahan G.D. (1993) Phys Rev B 48:16 557
32. For an excellent review of the Van Hove scenario see:
    Markiewicz R. (1997) J Chem Phys Solids 58:1173
33. Schneider T., Keller H. (1992) Phys Rev Lett 69:3374;
    Zhang H., Sato H. (1993) Phys Rev Lett 70:1697;
    Rao C.V.N., Trodahl H.J., Tallon J.L. (1994) Physica C 225:45
34. Uemura Y.J. et al. (1989) Phys Rev Lett 62:2317;
    Uemura Y.J. et al. (1991) Phys Rev Lett 66:2665
35. Brow B. (1998) Phys Rept 296:1.
    This author uses the term 'exotic superconductors' which was much earlier
    used by:
    Gorkov L.P. (1985) Phys Scr 32:6;
    in reference to the heavy fermion superconductors
36. McKenzie R.H. (1997) Science 278:820
37. Varma C. (1998) Phys World 11(Oct):22
38. Battlog B. et al. (1997) Phys Rev Lett 58:2333;
    Battlog B. et al. (1987) Phys Rev Lett 59:912;
    Bornemann H.J., Morris D.E., Liu H.B. (1991) Physica C 182:132;
    Crawford M.K. et al. (1990) Phys Rev B 41:8933–8936;
    Franck J.P. et al. (1991) Phys Rev B 44:5318
39. Ebbesen T.W. et al. (1992) Nature (Lond) 355:620;
    Zakhidov A.A. et al. (1992) Phys Lett A 164:355;
    Ramirez A.P. et al. (1992) Phys Rev Lett 68:1058;
    Crespi V.H., Cohen M.L. (1995) Phys Rev B 52:3619
40. Franck J.P. (1996) Phys Scr T 66:220;
    Zhao G.M. et al. (1996) Phys Rev B 54:14 956
41. Kresin V.Z., Wolf S.A. (1994) Phys Rev B 49:3652;
    Kresin V.Z. et al. (1997) Phys Rev B 56:107
42. Yokoya Y. (1997) Phys Rev B 56:6107
43. Weber A. (1987) Phys Rev Lett 58:1371
44. Ginzburg V.L., Maximov E.G. (1994) Physica C 235–240:193
45. Marsiglio F., Carbotte J.P. (1987) Phys Rev B 36:3937

46. Ummarino G.A., Gonnelli R.S. cond-mat/9809262 (preprint)
47. Maximov E.G., Savrasov D.Yu., Savrasov S.Yu. (1997) Usp Fiz Nauk 167:353 [In Russian]
48. Carbotte J.P. (1995) Rev Mod Phys 62:1027
49. Hardy J.R., Flocken J.W. (1988) Phys Rev Lett 60:2191
50. Bulaevski L.N. et al. (1989) Mod Phys Lett B 3:101;
    Golubov A. (1988) Physica C 156:286;
    Marsiglio F., Carbotte J.P. (1987) Solid State Commun 63:419
51. Cohen R.E. et al. (1988) Phys Rev Lett 60:817
52. Drechsler S.L., Plakida N.M. (1987) phys stat sol (b) 144:K113;
    Galbaatar T. et al. (1991) Physica C 176:496
53. Cyrot M. et al. (1994) Phys Rev Lett 72:1388
54. Stasyuk I.V., Shvaika A.M., Schachinger E. (1993) Physica C 213:57
55. Zhong J., Schüttler H.B. (1992) Phys Rev Lett 69:1600
56. Santi G. et al. (1995) J Supercond 8:405
57. Wysokiński K.I. (1996) Phys Rev B 54:3553
58. Girlando A. et al. (1993) Mol Cryst Liquid Cryst 254:145
59. Kim J.H., Levin K., Wentzovitch R., Auerbach A. (1991) Phys Rev B 44:5148;
    Kim Ju H., Tesanović Zlatko (1993) Phys Rev Lett 71:4218
60. Kulić M.L., Zeyher R. (1994) Phys Rev B 49:4398;
    Zeyher R., Kulić M.L. (1996) Phys Rev B 53:2850
61. Grilli M., Castellani C. (1994) Phys Rev B 50:16 880
62. Zimanyi G.T., Kivelson S.A., Luther A. (1988) Phys Rev Lett 60:2089
63. Plakida N.M., Hayn R. (1994) Z Phys B 93:313;
    Minh-Tien C., Plakida N.M. (1992) Mod Phys Lett B 6:1309
64. Micnas R., Ranninger J., Robaszkiewicz St. (1990) Rev Mod Phys 62:113
65. Marsiglio F. (1989) Physica C 160:305
66. Petru Z.K., Plakida N.M. (1994) In: Paszkiewicz T., Rapcewicz K. (Eds.) Die Kunst of Phonons, Plenum Press, New York
67. Migdal A.B. (1958) Zh Eksp Teor Fiz 34:1438 [Engl. in: Sov Phys – JETP 7:996]
68. Barnes S.E. (1976) J Phys F 6:1375;
    Coleman P. (1984) Phys Rev B 29:3035;
    Newns D.M., Read N. (1987) Adv Phys 36:799
69. Grimvall G. (1981) The Electron–Phonon Interaction in Metals, North Holland, Amsterdam
70. Wysokiński K.I. (1997) Acta Phys Pol A 91:329
71. McMillan W.L. (1968) Phys Rev 167:331
72. Wysokiński K.I. (1997) Physica C 282–287:1701
73. Varelogiannis G. (1998) Phys Rev B 57:13 743
74. Mierzejewski M., Zieliński J., Entel P. (1997) Solid State Commun 104:253
75. Kostur V.N., Mitrović B. (1993) Phys Rev B 48 16 388
76. Pietronero L. (1992) Europhys Lett 17:365;
    Pietronero L., Strässler S. (1992) Europhys Lett 18:627;
    Pietronero L. et al. (1995) J Low Temp Phys 99:535;
    Ikeda M.A., Ogasawara A., Sugihara M. (1992) Phys Lett A 170:319;
    Zheng H., Avignon M., Bennemann K.H. (1994) Phys Rev B 49:9763;
    Cai J., Lei L., Xie L.M. (1989) Phys Rev B 39:11 618;
    Hu S., Overhauser A.W. (1988) Phys Rev B 37:8618

77. Mierzejewski A., Zieliński J. (1998) J Supercond 11:331
78. Holstein T. (1959) Ann Phys (NY) 8:325
79. Ranninger J., Thibblin U. (1992) Phys Rev B 45:7730;
    Marsiglio F. (1995) Physica C 244:21;
    Freericks J.K., Jarrell M. (1994) Phys Rev B 50:6939;
    Alexandrov A.S., Kabanov V.V., Ray D.K. (1994) Phys Rev B 49:9915
80. Hubbard J. (1963) Proc Roy Soc A (Lond) 276:238
81. Sakai T., Poilblanc D., Scalapino D.J. (1997) Phys Rev B 55:8445;
    Trapper U. et al. (1994) Z Phys B 93:465;
    Alder B.J., Runge K.J., Scaletter R.T. (1997) Phys Rev Lett 79:3022;
    Hotta T., Takada Y. (1997) Phys Rev B 56:13 916;
    Feshke H. et al. (1995) Phys Rev B 51:16 582;
    Freericks J.K. (1993) Phys Rev B 48:3881
82. Kresin V.Z., Wolf S.A. (1996) Phys Rev B 54:15 543;
    Zhao G.–M., Morris D.E. (1996) Phys Rev B 54:15 545
83. Schachinger E., Carbotte J.P. (1999) Phonons in High-T_c Superconductors, to
    be published

Penetration Depth in Pure
and Zinc–Substituted $La_{2-x}Sr_xCuO_4$

A. J. Zaleski[1] and J. Klamut[1,2]

[1] Institute of Low Temperature and Structure Research,
   Polish Ac. Sci., P.O. Box 1410, PL-50-950 Wrocław, Poland
[2] International Laboratory of High Magnetic Fields
   and Low Temperatures, 95 Gajowicka Str.,
   PL-53-529 Wrocław, Poland

Abstract. In the paper we present measurements of the in–plane $\lambda_{ab}(T)$ ($\boldsymbol{H} \parallel \boldsymbol{c}$) and out–of–plane $\lambda_\perp(T)$ ($\boldsymbol{H} \perp \boldsymbol{c}$) penetration depths in $La_{2-x}Sr_xCu_{1-y}Zn_yO_4$ for $x = 0.08, 0.1, 0.125, 0.15$, and 0.2, and for $y = 0, 0.005, 0.01$, and 0.02. The penetration depth was obtained from ac susceptibility measurements of powdered samples, immersed in wax and magnetically oriented in a static magnetic field of 10 T. For unsubstituted, underdoped samples ($x < 0.15$) penetration depth varies linearly with temperature for low temperature region. For the samples from the overdoped region ($x > 0.15$) the measured points can be fitted by the exponential function of temperature. Our results support the view that for underdoped samples we are dealing with Bose–Einstein condensation while for overdoped ones the superconductivity is BCS–like. Extrapolated to $T = 0$, penetration depth values may be described by the quadratic function of strontium concentration similarly as the $T_c(x)$ dependence. For zinc–doped, underdoped $La_{1.85}Sr_{0.15}Cu_{1-y}Zn_yO_4$ the temperature dependences of penetration depths can be described by power laws, but with exponents n varying linearly with substituent content. These exponents n increase at a rate of about 2.5 per at% of zinc substitution. We found that the penetration–depth anisotropy is dependent on substituent content in $La_{1.85}Sr_{0.15}Cu_{1-y}Zn_yO_4$, decreasing to a minimum at $x \simeq 0.015$ and increasing for higher substitutions and can be described by a quadratic function. Our results strongly suggest that both the effective mass and the density of charge carriers must be taken into account in theories describing high–temperature superconductivity.

1 Introduction

They are three key issues for high temperature superconductivity in the cuprate superconductors: the physical characteristics and physical origin of the anomalous spin and charge behavior in the normal state; the mechanism for high temperature superconductivity; and the superconducting pairing state. Identifying the pairing state symmetry may help to narrow the choice of potential mechanisms. For conventional, low temperature superconductors phonon mechanism is responsible for the existence of the Cooper pairs. According to the BCS theory [1] in this case gap function is isotropic and the pairing is s-wave type. It is already well established that also for high temperature superconductors (HTSC) Cooper pairs are responsible for carrying the charge [2]. But up to now there is still some controversy if the symmetry of the gap function is isotropic [3–6] or anisotropic [7–8]. In some experiments also mixed $s + d$ type symmetry was observed [9]. For a d-wave pairing state the superconducting gap vanishes in some places on the Fermi surface, i.e. it exhibits nodes. Zeroing of the gap results in the quasiparticle excitations existence even in the limit of zero temperature. This, in turn, should lead to the power–law dependence of different thermodynamic functions [10].

One such function, for which temperature dependence may reflect the symmetry of the order parameter is magnetic penetration depth. It was studied employing different methods for different families of high temperature superconductors [11–14].

Many different techniques have been employed to obtain the penetration depth, depending upon whether the superconducting material was in the form of single crystal, thin film, or bulk ceramics. For single–crystalline samples, surface impedance [15], microwave inductance [16], infrared reflectivity or transmission [17], magnetic torque [18], and muon–spin–relaxation are the main techniques [19] that have been used. For thin films, the mutual–inductance method [20] has been the primary technique employed, but also infrared [21] and transmission–line techniques [22] have been used. For ceramic or powdered (usually magnetically aligned [23]) samples, the penetration depth has been derived from magnetization measurements in both low fields [24] and high fields, where the magnetization is reversible [25]. The penetration depth also has been determined from tunneling [26], magneto–optical [27], scanning Hall probe [28], and ESR [29] measurements.

The most reproducible determinations of the penetration depth seem to have been obtained from positive muon–spin–rotation measurements. This technique yields only the penetration depth λ_{ab} within the superconducting CuO_2 planes, which are common to all high-T_c superconductors. But the most distinctive difference between the low–temperature superconducting materials and the high–temperature superconducting (HTS) cuprates is the magnitude of the anisotropy of superconducting properties in the latter. These cuprates may be considered as stacks of superconducting CuO_2 planes separated by bridging blocks, which act as charge reservoirs for the planes. The different properties of these materials are associated with different strengths of interlayer Josephson coupling. However, since HTS properties also depend upon the quality of the blocking layers and the number of CuO_2 planes in a unit cell, it is important to examine also the penetration depth perpendicular to the CuO_2 planes. This can be done effectively using magnetically aligned superconducting powders [23].

In the weak–coupling limit, the square of the penetration depth is proportional to the effective mass of the superconducting carriers and inversely proportional to their density [30]:

$$\lambda_\parallel^2 = c^2 m^* / 4\pi n_s e^2 (1 + \xi/l), \tag{1}$$

where n_s is the concentration of superconducting carriers and m^* – the effective mass, ξ – the coherence length, and l – the mean free path. So the measurements of the penetration depth within the ab–plane and perpendicular to it may enable to find penetration depth tensor. As it can be seen from the above equation, this tensor is proportional to the density of the superconducting carriers and inversely proportional to their effective masses. The latter value reflects in some way interaction between the carriers itself and also the crystal lattice.

Although lanthanum–based 214–type family may be treated as a model one of HTSC, there are only few papers dealing with penetration depth studies; mainly for the in–plane component λ_{ab} [31–33].

We are aware of the only paper in which the study of penetration depth tensor for $La_{2-x}Sr_xCuO_4$ was presented [34]. The authors applied surface impedance method for single crystals with typical dimensions equal to $1 \times 1 \times 0.2\,mm^3$ with large faces parallel or perpendicular to the $a - b$ planes. They found that the anisotropy of the penetration depth decreases with the increase of strontium content from about 25 for strontium content equal to $x = 0.09$ to about 10 for $x = 0.19$. The magnitude of $\lambda_c(0)$ was in agreement with the Josephson–coupled layer model and the temperature dependence of λ_c was different from the local–limit BCS formula.

Because of limited resolution of their apparatus the authors were not able to present the penetration depth behavior for low temperatures, so they could not distinguish whether this dependence may be described by exponential or power–law. This was one of the reasons why we decided to present our results of penetration depth anisotropy measurements of $La_{2-x}Sr_xCuO_4$ obtained by ac–susceptibility method for the magnetically aligned powders.

Substitution of copper in superconducting planes with magnetic or non-magnetic ions provides an additional way of distinguishing between different symmetries of the pairing state. The spin fluctuation model [35] predicts that nonmagnetic zinc will be a stronger pair breaker than magnetic nickel when it substitutes for copper in the CuO_2 plane. The different pair–breaking properties of Zn and Ni also should be also revealed in London penetration–depth studies. It was another reason for the study how Cu–site substitution with isovalent magnetic or nonmagnetic ions influences the temperature dependence of the penetration depth in ceramic, magnetically oriented $La_{1.85}Sr_{0.15}Cu_{1-x}M_xO_4$ ($M = $ Ni or Zn).

2   Experimental

Polycrystalline $La_{1.85}Sr_{0.15}Cu_{1-y}Zn_yO_4$ samples were prepared using the standard solid–state reaction method, with appropriate amounts of La_2O_3, $SrCO_3$, CuO, ZnO, and NiO taken as starting materials. After two cycles of grinding and calcining at $920\,^\circ C$ and $945\,^\circ C$, the powder was ground, pressed into pellets, and fired at $1020\,^\circ C$ (24 h), $1050\,^\circ C$ (24 h), and $1100\,^\circ C$ (48 h).

The samples then were annealed at $920\,^{\circ}$C ($12\,$h), cooled to $445\,^{\circ}$C, annealed at this temperature for $48\,$h, and slowly cooled to room temperature [36].

X-ray data showed that all our samples were single–phase with the K_2NiF_4–type structure. Ceramic samples were reground in an agate mill for about $100\,$min. Powdered samples were magnetically aligned in molten Okerin wax in a static magnetic field of $10\,$T. The wax was cooled through the melting point while in this field. Parallelepiped blocks of dimensions $2 \times 2 \times 10\,$mm^3, with their c–axes parallel and perpendicular to the longest dimension, were cut from the composite using a saw. Grain size distributions were determined from SEM photographs of different parts of these blocks.

AC measurements were carried out using a commercial Lake Shore ac sus-ceptometer with an amplitude of $0.1\,$mT and a frequency of $111.1\,$Hz. From SEM photographs it was seen that the individual grains were well separated. However, since the densities of the wax and powders were different, some sedimentation was unavoidable. We therefore performed additional tests by measuring the linearity of the susceptometer output voltage as a function of the ac–field amplitude and frequency. The output voltage was linear within the limits of $\pm 0.1\%$, indicating that there were no weak links between the in-dividual grains.

The temperature dependence of the penetration depth was determined by a method used earlier by Porch *et al.* [37] from the measured ac signal, volume of superconductor, and measured grain–size distribution using the formulas:

$$\frac{\chi_i}{\chi_0} = \frac{2}{3} \frac{\nu_{AC}}{V_s N - \nu_{AC} f(\frac{1}{3} - D)} \, , \tag{2}$$

$$\frac{\chi_i}{\chi_0} = \frac{\int \left(1 - \frac{3\lambda}{r} \coth \frac{r}{\lambda} + \frac{3\lambda^2}{r^2}\right) r^3 g(r)\,\mathrm{d}r}{\int r^3 g(r)\,\mathrm{d}r} \, , \tag{3}$$

where χ_i is the measured susceptibility in the ab plane or perpendicular to it, χ_0 – the susceptibility of a perfectly diamagnetic spherical grain, ν_{AC} – the measured ac signal (in μV after subtraction of empty holder signal), V_s – the volume of superconductor (in mm^3), N – the calibration factor for ac apparatus (in μV/mm^3 for perfectly diamagnetic superconductor with $D = 0$), f – the volume of superconductor divided by the total volume of the composite, D the demagnetizing factor of a grain, r – the radius of a grain, and $g(r)$ – the measured grain–size distribution function.

The above method yields the in–plane penetration depth λ_{ab} when the ac field is oriented perpendicular to the CuO_2 planes, since the induced currents all flow parallel to these planes. We cannot distinguish between the penetra-tion depths λ_a and λ_b for currents along the a and b direction, respectively. When the ac field is oriented parallel to the CuO_2 planes, currents are induced both parallel and perpendicular to the planes, and thus the method yields an effective penetration depth which is a complicated mean value of the in–plane penetration depth λ_{ab} and the out–of–plane, c–axis penetration depth λ_c.

If the values of the anisotropic penetration depth are to be obtained from susceptibility measurement on magnetically aligned powders, first great care has to be taken that as much as possible of the sample volume consists of properly aligned grains. To achieve this goal, grains of the powder should be single domain. Ball milling rather than grinding was therefore employed. Our powders consisted predominantly of nearly spherically shaped grains. However, grains with different shapes, ranging from short needles to small plates, were also present (an SEM photograph of $La_{1.85}Sr_{0.15}CuO_4$ grains oriented perpendicular to the magnetic field is presented in Fig. 1 as an example). Since a deviation from spherical grain shape influences mainly the magnitude of the penetration depth (our estimated error is below 5%) but has very little effect upon the temperature dependence of λ, we decided to use the demagnetization factor of a sphere for our calculations.

Fig. 1. SEM (scanning electron microscope) photograph of $La_{1-85}Sr_{0.15}CuO_4$ powder in Okerin wax, oriented in magnetic field. White marker is 0.1 mm in length. The cross–section area of the grains was used for the defining their radius

We tested our procedure for penetration–depth evaluation by measuring seven small tin spheres. We chose tin rather than lead, as it was much easier to prevent the tin surface from oxidizing, and therefore the spheres' diameters were more precisely defined. The measured penetration depth extrapolated to zero temperature was 54 nm, which is in excellent agreement with the values obtained by Parr [38].

3   Results and Discussion

The quality of the powder alignment may be easily inferred from the X–ray diffraction patterns. They are depicted in Fig. 2a,b (one of them in logarithmic scale for clarity) for most of the measured samples. As it is seen, $00l$ reflections are present almost exclusively, their intensity being about 40 times higher than that for non–oriented samples. Traces of other than $00l$ reflections result from the volume of non–aligned material. To evaluate the percentage of non–aligned grains we have applied the procedure described in [39] and found that in all our samples the amount of non–oriented material was below 15%.

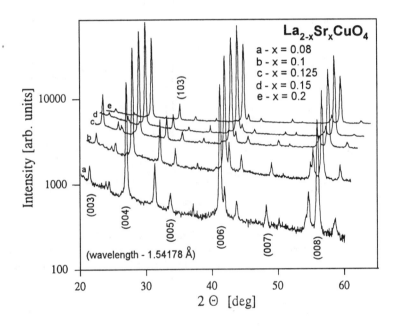

Fig. 2.a X–ray diffraction patterns for aligned powders of $La_{2-x}Sr_xCuO_4$ for $y = 0$ (in logarithmic scale; for clarity, curves are shifted by increments of one degree along the x direction and one arbitrary unit along the y direction)

Unit–cell parameters were used to calculate the density of superconducting material. These values, together with the results of density measurements for powders immersed in wax, were used to evaluate the volume of superconducting material in the measured samples. The unit–cell parameters for the unsubstituted and the Ni– and Zn–substituted samples are shown in Table 1. The results for the unsubstituted sample are close to the values presented in other papers [19,32,41]. Although the results for the Ni–substituted samples differ significantly from those presented in [42], they are similar to those cited

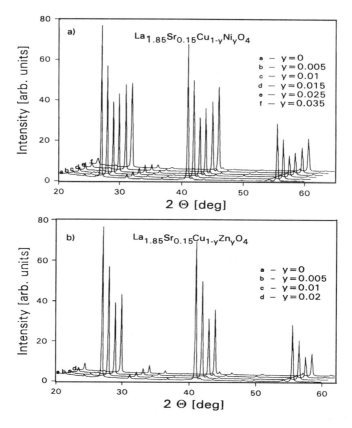

Fig. 2. b X-ray diffraction patterns for aligned powders of $La_{2-x}Sr_xCu_{1-y}M_yO_4$ for $x = 0.15$ (in logarithmic scale; for clarity, curves are shifted by increments of one degree along the x direction and one arbitrary unit along the y direction)

by other authors [43]. Results for the Zn–substituted samples are similar to those published elsewhere [44].

The distribution of the grains was obtained from SEM photographs of different parts of the samples. Obtained histograms were fitted with the same function for all samples $\left(a\sin^2(bx)\exp[cx+d]\right)$. One of histograms, together with distribution function used, is presented in Fig. 3 as an example. The diameter of the grains varied between 0.5 to 10 μm with the maximum for about 1 μm for all our samples. Although from the SM pictures it was seen that there are some grains with the shape of platelets and needles, the demagnetizing factor for sphere was used for calculation. Such a choice should influence only the penetration depth values – not their temperature dependence or anisotropy.

Table 1. Critical temperature, lattice constants, unit cell volume and penetration depth of compounds studied

Compound	T_c (K)	a, b (nm)	c (nm)	V (nm^3)	$\lambda_{ab}(0)$ (nm)	$\lambda_\perp(0)$ (nm)
$La_{1.92}Sr_{0.08}CuO_4$	26.5	0.37728	1.31634	0.187373	327	431
$La_{1.9}Sr_{0.1}CuO_4$	36.9	0.37769	1.32021	0.188331	304	449
$La_{1.875}Sr_{0.125}CuO_4$	34.8	0.37832	1.32321	0.189383	563	693
$La_{1.85}Sr_{0.15}CuO_4$	39.5	0.37949	1.32143	0.190302	234	547
$La_{1.8}Sr_{0.2}CuO_4$	33.9	0.37696	1.32241	0.187914	252	497
$La_{1.85}Sr_{0.15}Cu_{0.995}Ni_{0.005}O_4$	36.1	0.37804	1.3227	0.189033	264	521
$La_{1.85}Sr_{0.15}Cu_{0.99}Ni_{0.01}O_4$	33.6	0.37816	1.3229	0.189181	351	453
$La_{1.85}Sr_{0.15}Cu_{0.985}Ni_{0.015}O_4$	31.7	0.37758	1.3240	0.188760	431	436
$La_{1.85}Sr_{0.15}Cu_{0.975}Ni_{0.025}O_4$	26.8	0.37925	1.3218	0.190110	376	508
$La_{1.85}Sr_{0.15}Cu_{0.965}Ni_{0.035}O_4$	23.8	0.37839	1.3213	0.189187	210	752
$La_{1.85}Sr_{0.15}Cu_{0.995}Zn_{0.005}O_4$	31.9	0.37983	1.3223	0.190776	602	932
$La_{1.85}Sr_{0.15}Cu_{0.99}Zn_{0.01}O_4$	27.9	0.37951	1.3219	0.190384	739	968
$La_{1.85}Sr_{0.15}Cu_{0.98}Zn_{0.02}O_4$	17.5	0.37879	1.3251	0.190129	664	962

We should add that the log–normal plot usually used to approximate the shape of powder grain size distribution gave worse approximation than the function used by us (especially for the small grain size values).

Temperature dependence of penetration depth derived from ac susceptibility measurements for the samples from the underdoped and optimally doped regions (except from the samples with $x = 0.125$ composition) is presented in Fig. 4a. This dependence is linear below about $0.5 T_c$ for both λ_{ab} and $\lambda_\perp$ for all samples as is expected for the material with the nodes on the Fermi surface. Such linear behavior was already observed in LaSrCuO by muon spin rotation technique [31]. It was also observed for other high temperature superconductors [4,12].

For the samples with strontium content equal to $x = 0.125$, linear behavior is not observed down to the lowest temperature used, i.e. about 4 K. Low temperature behavior of both in–plane and out–of–plane penetration depth can be satisfactorily fitted by a polynomial of the third order. It is depicted in Fig. 4b.

Also for the samples from the overdoped region linear temperature dependence of penetration depth is not observed. The in–plane component of penetration depth can be satisfactorily described by a fifth order polynomial and the out–of–plane component – by a seventh order one (Fig. 4c). Nevertheless, it should be admitted that fitting with polynomials of such high

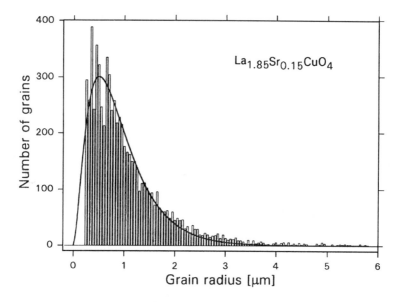

Fig. 3. An example of histogram of the grain size distribution for sample of $La_{1.85}Sr_{0.15}CuO_4$, measured from SEM photographs of sections perpendicular to the orienting magnetic field. Solid curve is distribution function which parameters were taken for penetration–depth evaluation

orders differs only marginally from the exponential behavior. Therefore, the samples from the overdoped region may represent BCS–type behavior.

So whereas for under– and optimally–doped samples the existence of d–wave pairing existence is supported by our measurements, BCS–type behavior is observed rather for the samples from the overdoped region. Such behavior agrees with expectations resulting from the Bose–Einstein to BCS crossover picture suggested by Uemura [45] (see also paper by Y. J. Uemura presented in this volume, p. 193).

We have obtained extrapolated to zero temperature penetration depth values. Its dependence on strontium concentration is presented in Fig. 5. The obtained values for $\lambda_{ab}(0)$ are similar to those presented by Locquet *et al.* [32]. There are few features of these dependences which are not obvious.

For underdoped samples the in–plane penetration depth decreases with the increase of the carrier number, but for overdoped region it starts to increase again. Such an effect was first reported by Locquet *et al.* [32] and showed that in the overdoped state, despite the increase of the number of charge carriers, the density of superconducting carriers may decrease and/or their effective mass may increase. Such an increase of the effective mass of the carriers was postulated by us [51] to explain the results of our measurements of penetration depth of zinc– and nickel–substituted $La_{1.85}Sr_{0.15}CuO_4$.

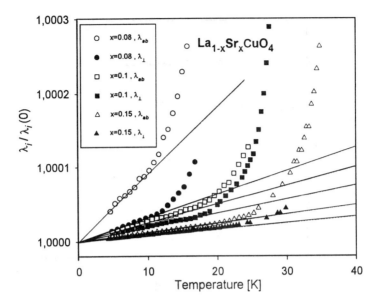

Fig. 4.a Temperature dependence of the penetration depth of $La_{2-x}Sr_xCuO_4$ for $x = 0.08$, 0.1, and 0.15 (reduced values)

The out–of–plane penetration depth increases with strontium content in the underdoped range and slightly decreases in the overdoped one.

For the samples with composition $La_{1.875}Sr_{0.125}CuO_4$, both in–plane and out–of–plane penetration depths are much greater than that for other compositions reflecting the fact that just for this composition the critical temperature is suppressed. It is probably connected both with the low temperature structural phase transition and with static order of the stripes correlations of holes and spins, discovered by Tranquada *et al.* [46]. In the case of existence of regular pattern of normal and superconducting stripes, it is obvious that the effective penetration depth increased. The fact that λ_{ab} increased more strongly than $\lambda_\perp$ supports our view that the part of evaluated $\lambda_\perp$, which is connected with the screening currents flowing within the CuO planes, is rather meaningful.

Extrapolated to zero temperature penetration depth on strontium doping dependences are pretty well described by a quadratic function (of course with the exception of $x = 0.125$).

The dependence of critical temperature on strontium concentration is presented in Fig. 6 together with the same dependence of the anisotropy of the penetration depth defined as $\lambda_\perp(0)/\lambda_{ab}(0)$. It is seen that both critical temperature and penetration depth anisotropy dependences follow similar trends. It is well established that transition temperature T_c is proportional to the density of superconducting carriers and inversely proportional to their

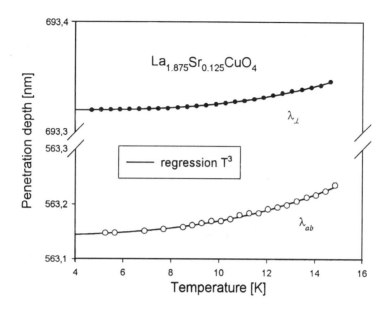

Fig. 4. b Temperature dependence of the penetration depth of $La_{2-x}Sr_xCuO_4$ for $x = 0.125$ (reduced values). Also polynomial fits to the measured points are presented

effective mass n_s/m^* in the underdoped region [11,47] (so–called Uemura plot). For overdoped material this proportionality breaks up, what may imply that apart of increase of density of carriers with doping there is also increase of their effective masses in this region. We should remind that the Uemura plot was constructed from the results obtained by positive muon rotation technique, which probes only superconducting ab–planes, so is responsible for λ_{ab} only. The increase of the effective mass of carriers within the ab–planes is equivalent to the decrease of the penetration depth anisotropy (if defined as we did, i.e. $\lambda_\perp(0)/\lambda_{ab}(0)$). As for overdoping region the critical temperature also diminishes, it is not surprising that both T_c and $\lambda_\perp(0)/\lambda_{ab}(0)$ have similar doping dependence.

The anisotropy of the penetration depth is highest for optimally doped samples and decreases for under– and overdoped branches of the dependence. It is consistent with the crossover between Bose–Einstein to BCS picture proposed by Uemura [45]. The increase of the anisotropy with doping (there is no Bose–Einstein condensation (BEC) in purely 2D systems) decreases the temperature of BEC and drives it close to that one resulting from the BCS theory just for the optimally doped samples.

For the strontium composition equal to $\frac{1}{8}$, the anisotropy is smallest of all the samples. Apparently for $La_{1.875}Sr_{0.125}CuO_4$ the deviation from the

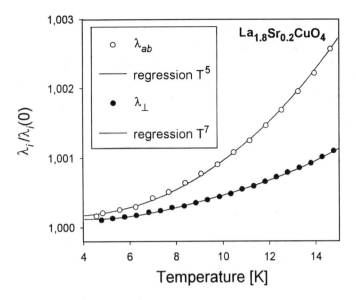

Fig. 4.c   Temperature dependence of the penetration depth of $La_{2-x}Sr_xCuO_4$ for $x = 0.2$ (reduced values). Also polynomial fits to the measured points are presented

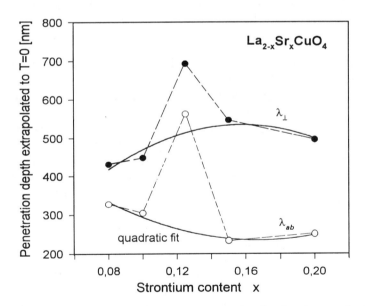

Fig. 5. Extrapolated to zero temperature penetration depth values as the function of strontium content (*solid lines* are quadratic fit to the measured points – except for $x = 0.125$ – and are guides for an eye only)

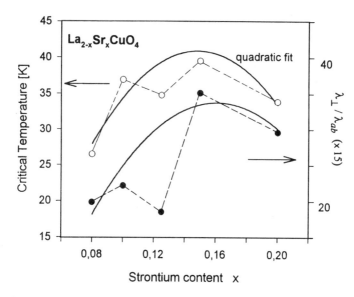

Fig. 6. Critical temperature (*left scale*) and anisotropy of the penetration depth $\lambda_\perp/\lambda_{ab}$ (multiplied by the factor of 15 – *right scale*) dependence on strontium content (*solid lines* are quadratic fit to the measured points except for $x = 0.125$ and are guides for an eye only)

optimal composition is connected with the increasing level of isotropisation of the penetration depth in the material.

The first results of the penetration depth anisotropy measurements for $La_{2-x}Sr_xCuO_4$ were presented by Shibauchi et al. [34]. They have employed surface impedance method for measurements of small single crystals with typical dimensions $1 \times 1 \times 0.2\,mm^3$, having large faces parallel and perpendicular to the ab planes (cut from bigger crystals). Critical temperatures of their samples were lower than previously published for crystals used by them [48] and known from other papers [33]. They have not found the traces of critical temperature suppression for their sample with composition $x = 0.12$, which was usually found by other authors [10,33,49].

The values of in–plane penetration depths obtained by Shibauchi et al. [34] are very similar to the discussed in our paper. They have found that the temperature dependence of λ_{ab} is roughly explained by the local, clean limit BCS theory, while in contrast λ_c behavior is consistent with Josephson–coupled layer model. The anisotropy measured by them was the decreasing function of the strontium content, and changes from about 25 for strontium content equal to $x = 0.09$ to about 10 for $x = 0.19$.

Our results for the out–of–plane penetration depth $\lambda_\perp(0)$ are lower than obtained by Shibauchi et al. – $\lambda_c(0)$. We are not convinced that the contributions of λ_{ab} and λ_c in measured out–of–plane penetration depth are equal, as suggested by Shibauchi *et al.* [34]. This assumption may be one of the sources of discrepancies between our results. The other source may be the Schibauchi's single crystals enhanced demagnetization factor which would be very difficult to assess, especially for the case when larger faces of the samples were parallel to the magnetic field direction.

The anisotropy of resistivity measured for the samples used in Shibauchi's study by Kimura *et al.* [48] was much higher then the penetration depth ones and equal to about 4000 for strontium content $x = 0.06$ and about 160 for $x = 0.266$. So also results obtained by Shibauchi *et al.* for penetration depth anisotropy are unexpectedly low. We think that the reason of it is similar both in their and our case. The difference between us may be caused by the different kind of samples used and different method employed to study penetration depth behavior.

A much lower value of the anisotropy of penetration depth in $La_{1.85}Sr_{0.15}CuO_4$ was also obtained by Uchida *et al.* [33]; i.e. about 2–5 times. But the authors obtain these results by comparing out–of–plane penetration depth derived from optical reflectivity spectra with the in–plane values measured by positive muon relaxation rate by Uemura *et al.* [50]. In their case anisotropy was also decreasing function of doping.

Although the values of our $\lambda_\perp$ and Shibauchi's λ_c were different, their temperature and composition dependence should be rather similar. But we definitely did not observe the BCS–type behavior of the temperature dependence of the in–plane penetration depth, not only for optimally doped material but for all underdoped one. The anisotropy of penetration depth measured by us is also not a decreasing function of strontium content as presented by Shibauchi *et al.* [34], but has shape of critical temperature on strontium concentration dependence. This behavior was in fact one of the reasons for carrying out of the present study and presenting their results.

From the results presented above and from that of our previous paper [51] we can state that in underdoped state the penetration depth is connected with the changes of superconducting charge carriers density, while for overdoped region also interaction between the charge carriers, i.e. their effective masses should be taken into account.

In conventional superconductors magnetic ions such as nickel are strong pair–breakers, while nonmagnetic ions such as zinc have a minor effect on T_c. Prior to discussing the influence of substitution into copper site it is important to remember that the localized spins are present on the Cu^{+2} ions and exhibit dynamic antiferromagnet intralayer order throughout the metallic and insulating composition regimes [52].

The dependences of the critical temperature T_c (defined as the onset of the magnetic transition) upon the Zn and Ni concentrations are presented in Fig. 7 and Table 1. The critical temperature decreases linearly in both cases but with different slope, namely $-4.1\,K/at\%$ for Ni and $-15\,K/at\%$ for Zn. Similar to the behavior in $YBa_2Cu_3O_{7-\delta}$, the effect of zinc substitution on

T_c is about three times stronger than in the case of nickel [53]. This behavior is different from what is observed in conventional superconductors. In our view, this clearly supports the idea that substitution by nonmagnetic Zn is nevertheless connected with the induction of a static magnetic moment on the Zn–substituted site [54]. A substituted nonmagnetic zinc atom removes a Cu^{2+} spin and causes its closest Cu neighbors to have partially noncompensated magnetic moments. Nickel, carrying a magnetic moment after substituting copper, introduces only a partial, incremental moment, which may be maximally equal to the difference between Ni and Cu. Thus, substitution of copper by magnetic nickel effectively disturbs the CuO_2 plane less than substitution by nonmagnetic zinc. Thanks to these noncompensated Cu moments around the Zn ion, the area around a zinc impurity also may be excluded from superconductivity, leading to phase separation as in the "Swiss cheese" model [30]. It is important to bear in mind that since both Zn and Ni have a very similar effect on the normal–state resistivity, their normal–carrier scattering potential is evidently quite similar [55].

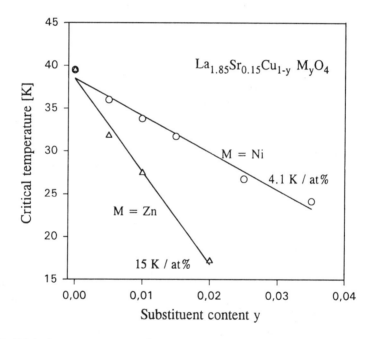

Fig. 7. Critical temperature *vs* substituent concentration. *Solid lines* are linear fits to the measured points

Substitution of copper by nickel or zinc changes the exponent n of the power law describing the low–temperature behavior of the penetration depth $[\lambda_i(T) - \lambda_i(0)]/\lambda_i(0) = AT^n$. This is depicted in Fig. 8a and b, where the normalized penetration depths are plotted below 20 K on logarithmic scales for nickel (Fig. 8a) and zinc (Fig. 8b) substitutions. For nickel, the exponent n changes from one for the unsubstituted sample (Fig. 4a) to about four for a Ni concentration of 3.5 at%. For zinc, these changes are stronger; n varies from one for the unsubstituted sample (Fig. 4a) to about five for a Zn concentration of 2 at%. The exponent n varies nearly linearly with Ni or Zn concentration, with a slope of about 1 per at% of nickel and about 2.5 per at% of zinc.

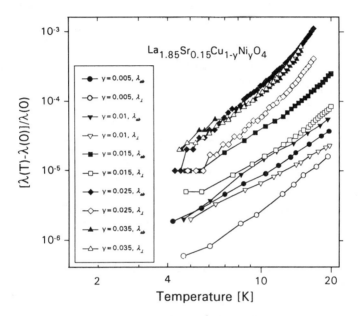

Fig. 8.a Double logarithmic plots showing the normalized penetration depths λ_{ab} and $\lambda_\perp$ vs temperature for $La_{1.85}Sr_{0.15}Cu_{1-y}Ni_yO_4$

Values of the penetration depths $\lambda_{ab}(0)$ and $\lambda_\perp(0)$ for magnetic fields parallel and perpendicular, respectively, to the c–axis, extrapolated to zero temperature, are summarized in Table 1 for our samples of $La_{1.85}Sr_{0.15}CuO_4$ substituted with Ni and Zn. The difference between the zinc– and nickel–substitution influence on the superconducting properties of $La_{1.85}Sr_{0.15}CuO_4$ is seen more clearly in Fig. 9a and b, where the penetration depths $\lambda_{ab}(0)$ and $\lambda_\perp(0)$ for the two substituents are compared.

Substitution of nickel first increases the in–plane penetration depth λ_{ab} and then, for concentrations above 1.5 at%, decreases it. In contrast, $\lambda_\perp$ first decreases and above 1.5 at.% of Ni, increases again. It appears that T_c decreases in substituted $La_{1.85}Sr_{0.15}CuO_4$ because the substituents act as impurities, which shorten the mean free path of the carriers. This mean–free–path shortening (see (1)) leads to the initial increase of the penetration

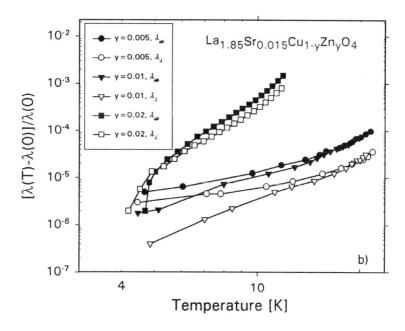

Fig. 8.b Double logarithmic plots showing the normalized penetration depths λ_{ab} and $\lambda_{\perp}$ vs temperature for $La_{1.85}Sr_{0.15}Cu_{1-y}Zn_yO_4$

depth λ_{ab}. The subsequent decrease of λ_{ab} can be explained by the decoupling of CuO_2 planes.

It has been shown [56] that, for tetragonal non–chain superconductors with a gap order parameter of $d_{x^2-y^2}$ symmetry, the c-axis hopping integral is a function of the in–plane momentum. This hopping integral is vanishingly small along the c-axis in the vicinity of gap nodes in clean systems. In the case of induced disorder, if the impurity scattering is anisotropic, impurity–assisted hopping [57] might be more important than coherent hopping, and a new conduction channel can be opened, which has a direct contribution to the c-axis superfluid density [56]. In our opinion, this mechanism can explain the observed initial decrease for nickel–substituted $La_{1.85}Sr_{0.15}CuO_4$.

It was also stated in [56] that in the case of a superconductor with d-wave order–parameter symmetry, the disorder connected with substitution has opposite effects on the superfluid response in the ab–plane and along the c-axis. Thus, changing the impurity concentration should have opposite effects upon the in–plane penetration depth λ_{ab} and the out–of–plane penetration depth λ_c (and hence $\lambda_{\perp}$). This is just what is observed for nickel substitution. The initial increase of λ_{ab} with Ni addition evidently is connected with the decrease of $\lambda_{\perp}$; and for large Ni content y, the decrease of λ_{ab} is accompanied by an increase of $\lambda_{\perp}$. It seems that nickel influences the superconducting properties of $La_{1.85}Sr_{0.15}CuO_4$ by acting as a simple impurity scatterer.

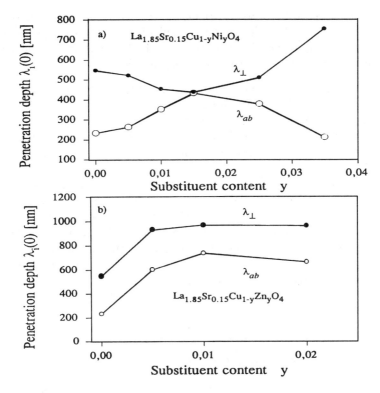

Fig. 9. Penetration depth, extrapolated to zero temperature, vs substituent content for (**a**) Zn– and (**b**) Ni–substituted $La_{1.85}Sr_{0.15}CuO_4$

The character of changes in λ_{ab} is similar for both substituents, although zinc additions have a much stronger effect upon the penetration depth (see Fig. 9b). It is known [58] that in overdoped material, superconductivity can be strongly suppressed by electron–electron scattering, which may be interpreted as a change of the effective mass of the electrons. Zinc doping, shifting the material to the underdoped region, also causes a decrease of electron interaction, reduction of the effective mass and some decrease of the in–plane penetration depth.

The main qualitative difference between the influence of both substituents on superconductivity in $La_{1.85}Sr_{0.15}CuO_4$ is seen in the behavior of $\lambda_\perp$ ($\boldsymbol{H} \perp \boldsymbol{c}$) (Fig. 9b). Substitution of zinc increases the effective penetration depth very strongly, even for low doping levels.

From the paper of Nachumi $et\ al.$ [30] it is seen that $La_{1.85}Sr_{0.15}CuO_4$ may be treated as overdoped and that increasing the zinc concentration shifts the compound towards the universal $T_c(n_s)$ curve, i.e. towards underdoped region (the "Swiss cheese" effect). It also proves that, in spite of the fact that zinc substitution is isovalent, there are still changes of the effective superconducting carrier concentration. So $\lambda_\perp(y)$ behavior is easily understood. Normal micro–regions created in the vicinity of substituents decrease the coupling between the CuO_2 planes, increase the out–of–plane penetration depth λ_c, and hence increase the effective penetration depth $\lambda_\perp$.

It is interesting to compare the changes of the penetration–depth aniso-tropy produced by Ni and Zn substitution. These are plotted in Fig. 10 as $(\lambda_\perp - \lambda_{ab})/\lambda_{ab}$ vs Ni or Zn concentration. Quadratic fits to the measured points are presented for Ni (dashed curve) and Zn (solid curve). What is sur-prising is that both fitting curves are almost the same within experimental error. The penetration–depth anisotropy first decreases, almost vanishes for about 1.5% of substitution, and then increases. The magnitude of the penct-ration depth in zinc–substituted samples is almost twice as high as in nickel–substituted ones. The morphology of both sets of samples is the same, and the mean grain radius is also similar. It appears that both impurities change the anisotropy of the penetration depth almost identically, independent of their magnetic properties. This is a surprising fact, especially if one compares the very different behaviors of $\lambda_\perp$ shown in Fig. 9.

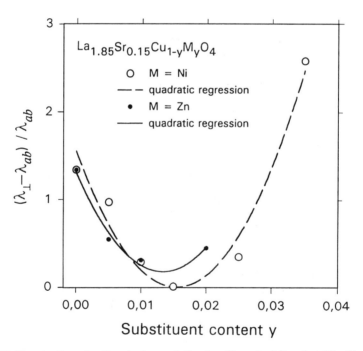

Fig. 10. Penetration–depth anisotropy defined as $(\lambda_\perp - \lambda_{ab})/\lambda_{ab}$ (see Fig. 9), vs sub-stituent concentration in Ni– and Zn–substituted $La_{1.85}Sr_{0.15}CuO_4$. *Dashed curve:* quadratic fit to measured points (*open circles*) for nickel substitutions; *solid curve:* quadratic fit to measured points (*closed circles*) for Zn substitutions

The existence of a minimum in the penetration–depth anisotropy *vs* temperature can be inferred from theoretical papers dealing with c–axis properties of cuprates [56,57]. It may appear in high-T_c materials, in which the order parameter possesses either s-wave or d-wave symmetry. For s-wave–type materials, the minimum should be rather flat and appear above $0.5T_c$, whereas for superconductors with order parameters of d-wave symmetry, the minimum should be more pronounced, appearing below $0.5T_c$. Thus, the existence of the minimum of anisotropy on substituent content is not surprising. What is unexpected is the fact that, for substituents whose influence on superconductivity is so very different, the minimum in the penetration–depth anisotropy occurs at the same concentration in $La_{1.85}Sr_{0.15}Cu_{0.985}M_{0.015}O_4$.

It is seen that the penetration depth anisotropy of zinc– and nickel–substituted $La_{1.85}Sr_{0.15}CuO_4$ is not connected with the value of the critical temperature. Since we have used isovalent substitutions of copper, the density of charge carriers should be the same for all samples studied. However, the density of carriers taking part in superconductivity may vary if the mean free path is changed by impurities or if strong magnetic (or other) pair–breaking exists.

4   Conclusions

In summary, we have evaluated penetration depth in $La_{2-x}Sr_xCuO_4$ for $x = 0.08$, 0.1, 0.125, 0.15, and 0.2 from AC susceptibility measurements of magnetically aligned single–crystalline powders. We found that temperature dependence of both in–plane and out–of–plane penetration depth is the linear function for the samples from underdoped region. The exceptions are the samples with $x = 0.125$. Their penetration depth on temperature dependence may be described by the polynomial of third order. For overdoped samples exponential behavior of both $\lambda_{ab}(T)$ and $\lambda_\perp(T)$ is observed. Such behavior was expected from Uemura model [45] of crossover from Bose–Einstein condensation to BCS–type mechanism of superconductivity. Extrapolated to zero temperature penetration depth may be described by a simple quadratic function of strontium content, except for the concentration $x = 0.125$. Samples with this composition behave as if made from different material. The anisotropy of the penetration depth follows the critical temperature on strontium concentration dependence.

The influence of isovalent substitution of Zn and Ni for Cu upon the penetration depth in magnetically oriented, ceramic powders of $La_{1.85}Sr_{0.15}CuO_4$ was studied for a broader range of concentrations than has been reported in the literature.

Similar to the behavior in $YBa_2Cu_3O_7$, substitution of Zn for Cu in $La_{1.85}Sr_{0.15}CuO_4$ reduces the critical temperature more strongly (by a factor of about three) than does substitution of Ni. Such behavior may be explained in the framework of the "Swiss cheese" model [30] in the case of zinc substitution and simple impurity scattering in the case of nickel.

The temperature dependence of the penetration depth is linear for low temperatures in unsubstituted material. For nickel– and zinc–substituted $La_{1.85}Sr_{0.15}CuO_4$, however, the temperature dependence is still described by a power law (T^n), but with an exponent n different from unity. The dependence of n upon the substituent concentration is stronger for zinc than for nickel, by more than a factor of two.

The in–plane penetration depth λ_{ab} ($\boldsymbol{H} \parallel \boldsymbol{c}$), extrapolated to zero temperature, has a similar concentration dependence for both Ni and Zn, although the underlying physics of this behavior may be quite different. A qualitative difference was found for the influence of Ni and Zn on the effective penetration depth $\lambda_\perp$ ($\boldsymbol{H} \perp \boldsymbol{c}$). The behavior of the nickel–substituted samples can be explained by the influence of impurity scattering. To account for the influence of zinc on the penetration depth, however, the idea of an effective mass change in under– and over–doped compounds has to be employed.

We have found that nearly the same quadratic law describes the penetration–depth anisotropy vs substituent concentration for both Zn and Ni. From our study, it appears that magnetic nickel acts simply as an impurity, decreasing the mean free path of carriers within the CuO_2 planes and increasing impurity–assisted hopping between the planes [56]. Our results support the "Swiss cheese" model [30], in which some area around each Zn impurity is excluded from superconductivity. This might be connected with the ability of zinc atoms to effectively suppress spin fluctuations [59]. To explain the properties of zinc– and nickel–substituted $La_{1.85}Sr_{0.15}CuO_4$, we found it necessary to treat this superconductor as having d-wave pairing.

Our observation of a minimum in the penetration–depth anisotropy vs substituent content suggests that a complete theory of high–temperature superconductivity must account not only for the impurity concentration and the kind of impurities but also for the anisotropy of the effective mass of the carriers. Since different physical mechanisms have different influences on the effective mass components (m_{ab} parallel and m_c perpendicular to the CuO_2 planes), they lead not only to different temperature dependences predicted in [57] but also to different dependencies on the impurity concentration. This, evidently, is the reason for a minimum of the penetration–depth anisotropy vs temperature [57] and vs impurity concentration (we found this to occur for $x_{\min} \approx 1.5\%$). According to our measurements, the effective masses m_{ab} and m_c have different impurity concentration dependencies. We believe that taking into changes of both the effective mass and the density of excitations will permit a more effective interpretation of the phenomena connected with high–temperature superconductivity.

Our results also show that coupling between the CuO_2 planes plays an important role in the high–temperature superconductors.

Acknowledgements

This work was sponsored under Grant KBN No 2 PO3 B05411 (partially).

References

1. Bardeen J., Cooper L.N., Schriffer J.R. (1957) Phys Rev 108:1175
2. Hoevers H.F.C., van Bentum P.J.M., van der Leemput L.E.C.,
   van Kempen H., Schellingerhout A.J.G., van der Marel D. (1988)
   Physica C 152:105
3. Fiory A.T., Hebard A.F., Mankiewich P.M., Howard R.E. (1988)
   Phys Rev Lett 61:1419
4. Hardy W.N., Bonn D.A., Morgan D.C., Liang R., Zhang K. (1993)
   Phys Rev Lett 70:3999
5. Klein N., Tellmann N., Schultz H., Urban K., Wolf S.A., Kresin V.Z. (1993)
   Phys Rev Lett 71:3355
6. Anlage S., Langley B.W., Deutscher G., Halbritter J., Beasley M.R. (1991)
   Phys Rev B 44:9764
7. Hammel P.C., Takigawa M., Heffner R.H., Fisk Z., Ott K.C. (1989)
   Phys Rev Lett 63:1992
8. Ma Z., Taber R.C., Lombardo L.W., Kapitulnik A., Beasley M.R.,
   Merchant P., Eom C.B., Hou S.Y., Phillips J.M. (1993) Phys Rev Lett 71:781
9. Mesot J., Böttger G., Mutka H., Furrer A. (1998) Europhys Lett 44:498
10. Momono N., Ido M. (1996) Physica C 264:311
11. Zhao X., Sun X., Fan X., Wu W., Li X–G., Guo S., Zhao Z. (1998)
    Physica C 307:265
12. Panagopoulos C., Cooper J.R., Xiang T., Peacock G.B., Gameson I.,
    Edwards P.P. (1997) Phys Rev Lett 79:2320
13. Mendels P., Alloul H., Brewer J.H., Morris G.D., Duty T.L., Johnston S.,
    Ansaldo E.J., Collin G., Marucco J.F., Nidermayer C., Noakes D.R.,
    Stronach C.E. (1994) Phys Rev B 49:10 035
14. Basov D.N., Liang R., Bonn D.A., Hardy W.N., Dabrowski B., Quijada M.,
    Tanner D.B., Rice J.P., Ginsberg D.M., Timusk T. (1995)
    Phys Rev Lett 74:598
15. Maeda A., Shibauchi T., Kondo N., Uchinokura K., Kobayashi M. (1992)
    Phys Rev B 46:14234
16. Zhang K., Bonn D.A., Kamal S., Liang R., Baar D.J., Hardy W.N., Basov D.,
    Timusk T. (1994) Phys Rev Lett 73:2484
17. Basov D.N., Timusk T., Dabrowski B., Jorgensen J.D. (1994)
    Phys Rev B 50:3511
18. Janossy B., Prost D., Pekker S., Fruchter L. (1991) Physica C 181:51
19. Kossler W.J., Kempton J.R., Yu X.H., Schone H.E., Uemura Y.J.,
    Moodenbaugh A.R., Suenaga M., Stronach C.E. (1987) Phys Rev B 35:7133
20. Fiory A.T., Hebard A.F., Mankevich P.M., Howard R.E. (1988)
    Appl Phys Lett 52:2165
21. de Vaulchier L.A., Vieren J.P., El Azrak A., Guldner Y., Bontemps N.,
    Guilloux–Viry M., LePaven–Thivet C., Perrin A. (1995) Phys Rev B 52:564
22. Pond J.M., Carrol K.R., Horwitz J.S., Chrisey D.B., Osovsky M.S.,
    Cestone V.C. (1991) Appl Phys Lett 59:3033

23. Farrell D.E., Chandrasekhar B.S., DeGuire M.R., Fang M.M., Kogan V.G., Clem J.R., Finnemore D.K. (1987) Phys Rev B 36:4025
24. Cooper J.R., Chu C.T., Zhou L.W., Dunn B., Gröfner G. (1988) Phys Rev B 37:638
25. Sok J., Xu M., Chen W., Suh B.J., Gohng J., Finnemore D.K., Kramer M.J., Schwartzkopf L.A., Dabrowski B. (1995) Phys Rev B 51:6035
26. Sun A.G., Gajewski D.A., Maple M.B., Dynes R.C. (1994) Phys Rev Lett 72:2267
27. Moser N., Koblischka M.R., Kronmueller H., Gegenheimer B., Theuss H. (1989) Physica C 159:117
28. Oral A., Bending S.J., Humphreys R.G., Heinini M. (1997) Supercond Sci Technol 10:17
29. Puri M., Kevan L. (1992) Physica C 197:53
30. Nachumi B., Keren A., Kojima K., Larkin M., Luke G.M., Merrin J., Tchernyshov O., Uemura Y.J., Ichikawa N., Goto M., Uchida S. (1996) Phys Rev Lett 77:5421
31. Luke G.M., Fudamoto Y., Kojima K.M., Larkin M., Merrin J., Nachumi B., Uemura Y.J., Sonier J.E., Ito T., Oka K., de Andrade M., Maple M.B., Uchida S. (1997) Physica C 282–287:1465
32. Locquet J.-P., Jaccard Y., Cretton A., Williams E.J., Arrouy F., Machler E., Schneider T., Fisher Ø., Martinoli P. (1995) IBM Research Report RZ 2720, July 3rd,
33. Uchida S., Tamasaku K., Tajima S. (1996) Phys Rev B 53:14558
34. Shibauchi T., Kitano H., Uchinokura K., Maeda A., Kimura T., Kishio K. (1994) Phys Rev Lett 72:2263
35. Pines D. (1994) In: Bedell K., Wang Z., Meltzer D.E., Valatzky A., Abrahams E. (Eds.) Strongly Correlated Electronic Materials: Los Alamos Symposium 1993, Addison–Wesley, Reading, MA
36. Olejniczak J., Zaleski A.J., Ciszek M. (1994) Mod Phys Lett B 8:185
37. Porch A., Cooper J.R., Zheng D.N., Waldram J.R., Campbell A.M., Freeman P.A. (1993) Physica C 214:350
38. Parr H. (1975) Phys Rev B 12:4886
39. Imbert P., Jehanno G., Garciu C., Hodges J.A., Bahout-Moullem M. (1992) Physica C 190:316
40. Radelli P.G., Hinks D.G., Wagner J.L., Dabrowski B., Vandervoort K.G., Viswanathan H.K., Jorgensen J.D. (1994) Phys Rev B 49:4163
41. Chang J.-J., Scalapino D.J. (1989) Phys Rev B 40:4299
42. Tarascon J.M., Greene L.H., Bagley B.G., McKinnon W.R., Barboux P., Hull G.W. (1987) In: Wolf S.A., Kresin V.Z. (Eds.) Novel Superconductivity, Plenum, New York, NY, 705
43. Fujishita H., Sato M. (1989) Solid State Commun 72:529
44. Mirza K.A., Loram J.W., Cooper J.R. (1997) Physica C 282–287:1411
45. Uemura Y.J. (1997) Physica C 282–287:194
46. Tranquada J.M., Sternlieb B.J., Axe J.D., Nakamura Y., Uchida S. (1995) Nature 375:561
47. Uemura Y.J., Keren A., Le L.P., Luke G.M., Wu W.D., Kubo Y., Manako T., Shimakawa Y., Subramanian M., Cobb J.L., Markert J.T. (1993) Nature 364:605
48. Kimura T., Kishio K., Kobayashi T., Nakayama Y., Motohira N., Kitazawa K., Yamafuji K. (1992) Physica C 192:247

49. Müller A., Zhao G., Conder K., Keller H. (1998) J Phys Cond Matt 10:L291
50. Uemura Y.J., Luke G.M., Sternlieb B.J., Brewer J.H., Carolan J.F.,
    Hardy W.N., Kadono R., Kempton J.R., Kiefl R.F., Kreitzman S.R.,
    Mulhern P., Riseman T.M., Williams D.L., Yang B.X., Uchida S., Takagi H.,
    Gopalakrishnan J., Sleight A.W., Subramanian M.A., Chien C.L.,
    Cieplak M.Z., Xiao Gang, Lee V.Y., Statt B.W., Stronach C.E., Kossler W.J.,
    Yu X.H. (1989) Phys Rev Lett 62:2317
51. Zaleski A.J., Klamut J. (1999) Phys Rev B 59:14 023
52. Johnston D.C. (1989) Phys Rev Lett 62:957
53. Bonn D.A., Kamal S., Zhang K., Liang R., Baar D.J., Klein E., Hardy W.N.
    (1994) Phys Rev B 50:4051
54. Xiao G., Cieplak M.Z., Xiao J.Q., Chien C.L. (1990) Phys Rev B 42:8752
55. Sumner M.J., Kim J.–T., Lemberger T.R. (1993) Phys Rev B 47:12 248
56. Xiang T., Wheatley J.M. (1996) Phys Rev Lett 77:4632
57. Radtke R.J., Kostur V.N., Levin K. (1996) Phys Rev B 53:R522
58. Leggett A.J. (1994) Physica B 199–200:291
59. Stolbov S.V. (1997) J Phys Cond Matt 9:4691

Structural Sensitivity of Transport Properties of the High-T_c Cuprates

J.–S. Zhou and J. B. Goodenough

Texas Materials Institute, ETC 9.102, University of Texas at Austin, TX 78712–1063, USA

Abstract. Experiments demonstrating the structural sensitivity of the transport properties of the high-T_c copper oxides are reviewed. The superconductive phase is an intermediate phase appearing between an antiferromagnetic parent phase and a metallic phase. Transitions between localized and itinerant electronic behavior are first–order, and the normal state of the superconductive phase exhibits unusual transport properties resulting from strong electron–lattice interactions; there is no evidence that the transport properties are influenced by changes in the spin system. A model in which itinerant vibronic states become increasingly stabilized with decreasing temperature is proposed.

1 Introduction

High-T_c superconductivity in the cuprates is found in a distinguishable thermodynamic phase located between a phase exhibiting localized–electron behavior and one exhibiting Fermi–liquid itinerant–electron behavior. The transition from localized to itinerant electronic behavior is first–order, so we may anticipate that strong electron–lattice interactions play a critical role in stabilizing the superconductive phase and in imparting to the normal state of this phase its unusual transport properties. Here we review briefly the sensitivity to structural changes of the normal–state properties of the super-conductive phase in (a) ceramic samples of $La_{2-x}Ba_xCuO_4$ (b) single–crystal YBa_2CuO_8, and (c) single–crystal films of $La_{1.85}Sr_{0.15}CuO_4$.

2 Background

A hallmark of the superconductive state has been the opening of an energy gap at the Fermi energy within a partially filled itinerant–electron band, and the copper–oxide superconductors exhibit this property. On the other hand, the transport properties of the electrons in the normal state are not well–described with the conventional Boltzmann approximation.

The superconductive cuprates all have layered structures with strongly anisotropic transport properties, which mandates single–crystal measurements for any definitive test of the conventional Boltzmann description of the resistivity. Figure 1 shows the temperature dependence of the basal–plane resistivity $\rho(T)$ for a defect–free, single–crystal $La_{1.85}Sr_{0.15}CuO_4$ film grown by molecular–beam epitaxy on a $LaSrAlO_4$ substrate [1]. The $La_{2-x}Sr_xCuO_4$ system has no charge reservoir; all the mobile charge carriers reside in the basal–plane CuO_2 sheets, which are parallel to plane of the film. From 100 K to room temperature, the curve is perfectly linear, and this linear temperature dependence has been shown [2] to extend above room temperature to resistivity values that exceed the theoretical upper limit corresponding to one Cu–O–Cu distance for the mean–free path between scattering events. This behavior contrasts with the high–temperature breakdown of the Boltzmann approximation in Nb_3Sn and Nb_3Sb, insert of Fig. 1; the $\rho(T)$ curves for these A15 alloys bend over on approaching this upper limit and never cross it [3]. Moreover, extension of the $\rho(T)$ curve of Fig. 1 to low temperatures does not follow the Bloch–Grüneisen function, but extrapolates to zero near $T = 0\,K$. These data indicate that the conventional scattering mechanism of the Boltzmann approximation is not applicable to the high-T_c cuprates.

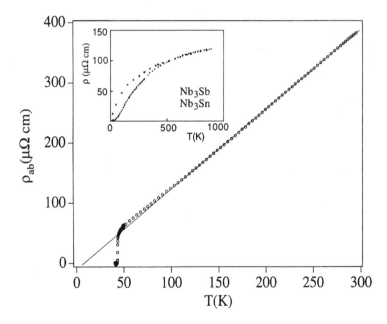

Fig. 1. Basal–plane resistivity $\rho_{ab}(T)$ for a single–crystal film of $La_{1.85}Sr_{0.15}CuO_4$ on a $LaSrAlO_4$ substrate. **Inset:** $\rho(T)$ data for Nb_3Sn and Nb_3Sb, after Ref. [3]

Although the thermoelectric power of polycrystalline samples fails to portray the anisotropic character of the electronic transport, the measurement is reliable and provides additional evidence of an unusual electronic behavior in the superconductive cuprates. Figure 2 shows a typical temperature dependence of the thermoelectric power $\alpha(T)$ for an underdoped ($0 < x < 0.1$) and an optimally doped ($x = 0.15$) polycrystalline sample of $La_{2-x}Sr_xCuO_4$.

In this system, only the CuO_2 sheets contribute to $\alpha(T)$. In the underdoped compositional range, a temperature–independent $\alpha(T)$ is found above about 240 K, Fig. 2(a), and this behavior extends to well above room temperature. A temperature–independent $\alpha(T)$ is characteristic of polaronic behavior in which the statistical term dominates the transport term. However, the magnitude of $\alpha(300\,\mathrm{K})$ is reduced from the value calculated for small polarons; it is necessary to use a fractional site occupancy $c = xN/N'$ where $N'/N \approx 5$, dashed line of Fig. 3 [4]. These data indicate that the nonadiabatic polarons are not small, but contain about 5 copper centers. Calculation [5] has shown that this unusual phenomenon could be due to a cooperative pseudo–Jahn–Teller deformation of the copper sites.

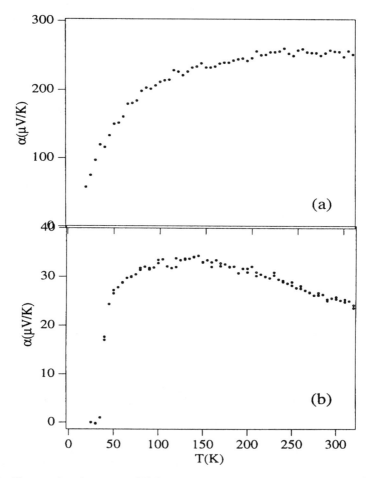

Fig. 2. Thermoelectric power $\alpha(T)$ for polycrystalline $La_{2-x}Sr_xCuO_4$ with (a) $x = 0.05$ and (b) $x = 0.15$

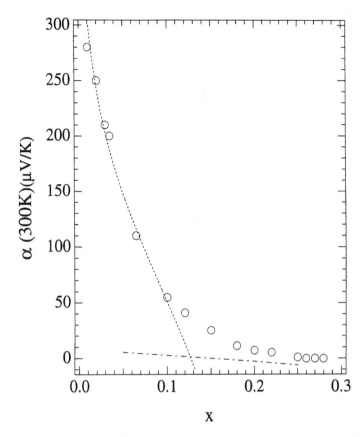

Fig. 3. Room–temperature thermoelectric power for the system $La_{2-x}Sr_xCuO_4$. *Dashed lines* are theoretical, see text

Curves similar to 2(a) are found for underdoped $La_2CuO_{4+\delta}$ samples in which a spinodal segregation into the parent antiferromagnetic and the superconductive phases occurs below room temperature as a result of a segregation into regions poor and rich in the interstitial excess oxygen [6,7]. Comparison of the $\alpha(T)$ data for $La_{2-x}Sr_xCuO_4$ and $La_2CuO_{4+\delta}$ indicates that a similar spinodal segregation occurs in underdoped $La_{2-x}Sr_xCuO_4$, $0 < x < 0.1$; but in this case, phase segregation must be accomplished by cooperative oxygen displacements since the atoms on regular crystallographic positions are not mobile at and below room temperature [8]. These cooperative oxygen displacements may allow for mobile phase boundaries, which would account for the remarkable isotope effect on the EPR signal that is reported at this conference [9] to occur in the normal state in the doping range $0 < x < 0.10$.

The $\alpha(T)$ curve for the optimally doped sample, Fig. 2(b), also has a nearly temperature–independent $\alpha(T)$ above room temperature (not shown) [8]. However, Fig. 3 shows that the magnitude of $\alpha(300\,\mathrm{K})$ deviates from both the polaronic curve applicable in the range $0 < x < 0.10$ and the Fermi–liquid curve that is approached in the overdoped samples with $x < 0.25$. Moreover, an enhancement below room temperature has a broad maximum at about $140\,\mathrm{K}$, which is too high for a conventional phonon drag. This unusual enhancement is a characteristic feature only of the normal state of superconductive sheets and chains in the copper oxides [10]. Since the superconductive phase appears to be thermodynamically distinguishable [8], the anomalous enhancement is not to be associated with a phase segregation, but rather with some unidentified electron–lattice interaction.

In summary, the transport properties of the system $La_{2-x}Sr_xCuO_4$ reflect the electronic behavior of the CuO_2 sheets without interference from a charge reservoir. They indicate the formation of intermediate–size, non-adiabatic polarons at higher temperatures and, in the range $0 < x < 0.10$, a spinodal, dynamic phase segregation at lower temperatures, into the antiferromagnetic parent phase and a thermodynamically distinguishable superconductive phase. The normal state of the superconductive phase shows unique $\rho(T)$ and $\alpha(T)$ behaviors that appear to reflect unusual electron–lattice interactions. Consistent with this interpretation is a remarkable sensitivity of these transport properties to changes in structure, as we now illustrate with three specific examples.

2.1   Polycrystalline $La_{2-x}Ba_xCuO_4$

The phase diagram for the system $La_{2-x}Ba_xCuO_4$, Fig. 4, shows a non-metallic low–temperature–tetragonal (LTT) phase below $60\,\mathrm{K}$ in a narrow compositional range $0.11 < x < 0.15$. The $La_{2-x}Sr_xCuO_4$ system has only the high–temperature–tetragonal (HTT) and low–temperature–orthorhombic (LTO) phases. In the LTO phase, the CuO_6 octahedra rotate cooperatively about the [110] axis and in the LTT phase they rotate in alternate CuO_2 sheets about [100] and [010] axes, respectively. At atmospheric pressure, superconductivity is suppressed in the range $x = 0.12 - 0.13$. Suppression of the superconductivity at $x \approx 1/8$ [11] has been correlated [12] with the appearance below T_N of an antiferromagnetic spin–density wave (SDW) and a charge–density wave (CDW) of half the wavelength of the SDW. The CDW is in the form of hole–rich and hole–poor stripes oriented in alternate CuO_2 sheets along [100] and [010] axes. The CDW is commensurate with the lattice at $x = 0.12$, and the cooperative rotations of the CuO_6 octahedra are in a direction that pins the stripes. However, superconductivity reappears under hydrostatic pressure [13], Fig. 5(b). Pressure straightens the $(180° - \varphi)$ Cu–O–Cu bond angle in the CuO_2 sheets and consequently depins the stripes.

Figure 5 shows the resistance $R(T)$ and the thermoelectric power $\alpha(T)$ under different hydrostatic pressures for $x = 0.11$ and $x = 0.12$. The $x = 0.10$ sample remains in the LTO phase to lowest temperatures. Figure 5(a) shows

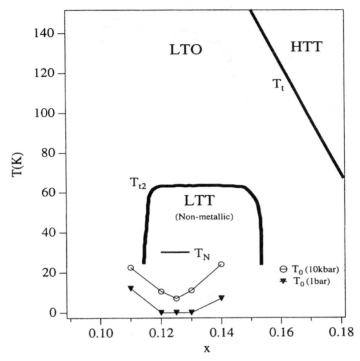

Fig. 4. Phase diagram for $La_{2-x}Ba_xCuO_4$: HTT = high–temperature tetragonal, LTO = low–temperature orthorhombic, LTT = low–temperature tetragonal

that pressure increases T_c and also $\alpha(300\,K)$ in this underdoped composition, which suggests that pressure stabilizes the trapping of polarons from the hole–poor phase into the superconductive phase to increase the volume of the hole–rich phase. On the other hand, the $x = 0.12$ sample undergoes an LTO/LTT phase transition on cooling below 60 K under atmospheric pressure. The onset of this transition correlates with the minimum in the $R(T)$ curve and a drop in $\alpha(T)$ to negative values. As can be seen in Fig. 5(b), pressure restores superconductivity within the LTT phase. Similarly, in the manganese–oxide perovskites $La_{1-x}Sr_xMnO_3$ with $x \approx 1/8$, pressure suppresses charge and orbital ordering into a static CDW [14]. By analogy, we presume that pressure also suppresses stabilization of the static CDW in $La_{2-x}Ba_xCuO_4$ with $x \approx 1/8$. Stabilization of a CDW demonstrates the presence of a strong electron–lattice coupling, and it follows that identification of superconductivity with a depinning of the stripes indicates that a dominant feature of the superconductive phase is strong coupling of the conduction electrons to dynamic, cooperative oxygen displacements as in a traveling CDW.

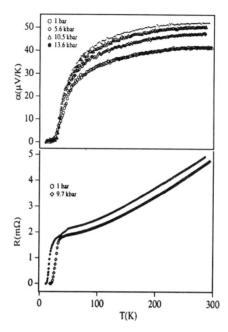

Fig. 5. (a)   The resistance $R(T)$ and thermoelectric power $\alpha(T)$ under different hydrostatic pressures for $La_{1.89}Ba_{0.11}CuO_4$

2.2   Single–Crystal $YBa_2Cu_4O_8$

The structure of $YBa_2Cu_4O_8$, illustrated in Fig. 6, contains pairs of edge–shared chains of Cu atoms in square–coplanar oxygen coordination between BaO–CuO_2–Y–CuO_2–BaO layers with Cu atoms in square–pyramidal oxygen coordination [15]. The a–axis resistivity $\rho_a(T)$ and thermoelectric power $\alpha_a(T)$ represent transport properties of the CuO_2 sheets; the b–axis parameters $\rho_b(T)$ and $\alpha_b(T)$ are a measure of the sum of chain and sheet contributions. As–grown $YBa_2Cu_4O_8$ crystals are naturally detwinned and stoichiometric with 0.25 holes in the Cu(III)/Cu(II) couple. Distribution of the holes between the chain and sheet Cu atoms leaves the CuO_2 sheets underdoped. Our single crystal of $YBa_2Cu_4O_8$ was thin in the c–axis direction, so we report c–axis resistance $R_c(T)$ rather than resistivity $\rho_c(T)$. Three issues motivated our study of this crystal [16]. We wished (1) to check whether a weak anomaly in $\rho_a(T)$ reflects the opening of a spin gap, as has been speculated in the literature; such an association would support the contention that the high-T_c phenomenon is driven by magnetic–exchange interactions rather than electron–lattice interactions; (2) to determine the relative contributions of the CuO_2 sheets and the double chains to the normal–state transport properties; and (3) to confirm electron transfer from the sheets to the chains under hydrostatic pressure as has been assumed to account for the large increase in T_c with pressure.

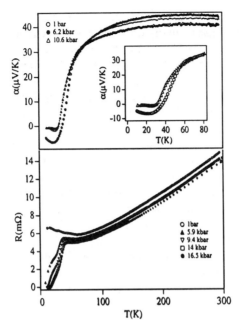

Fig. 5. (b)   The resistance $R(T)$ **and thermoelectric power** $\alpha(T)$ under different hydrostatic pressures for $La_{1.88}Ba_{0.12}CuO_4$

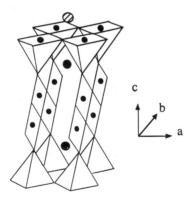

Fig. 6. Schematic of the $YBa_2Cu_4O_8$ structure. Small full circles are Cu, larger full circles Ba, and hatched circle – Y. Oxygen atoms occupy positions at intersections of lines

Figure 7 shows $\rho(T)$ and $\alpha(T)$ curves for the a, b, and c–axes of YBa$_2$Cu$_4$O$_8$ under different hydrostatic pressures. A remarkable anisotropy of the transport properties is evident. Whereas $\rho_b(T)$ is typical of a metal, $\rho_a(T)$ undergoes a change of slope near 160 K. It is obvious that the chain conductivity is much higher than that of the sheets, which makes the chain contribution dominant along the b–axis to render $\rho_b(T) \ll \rho_a(T)$. As a result, the contribution to $\alpha_b(T)$ from the chains dominates the contribution to $\alpha_b(T)$ from the sheets. The magnitude of $\alpha_a(300\,\mathrm{K})$ is a measure of the concentration of mobile holes in the CuO$_2$ sheets, and a $d\alpha_a(300\,\mathrm{K})/\,dP < 0$ confirms that pressure induces electron transfer from the underdoped sheets to the chains to give the large $dT_c/\,dP \approx 0.57\,\mathrm{K/kbar}$ that is observed. In the interval $T_c < T < 300\,\mathrm{K}$, $\alpha(T)$ is dominated by the unusual enhancement term $\delta\alpha(T)$ having a maximum magnitude near 140 K. It is evident that both the positive $\delta\alpha_a(T)$ from the CuO$_2$ sheets and the negative $\delta\alpha_b(T)$, which reflects the dominant chain contribution, are amplified as T_c increases with pressure. This observation adds to the considerable evidence that this $\delta\alpha(T)$ enhancement is associated with the high-T_c phenomenon; the $\delta\alpha(T)$ indicates, therefore, that the chains as well as the CuO$_2$ sheets become superconductive below T_c.

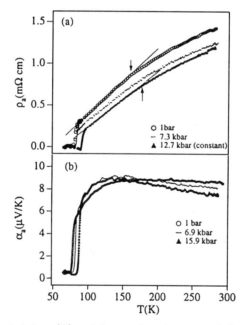

Fig. 7.(a)   The resistivity $\rho(T)$ and thermoelectric power $\alpha(T)$ curves under different hydrostatic pressures for single–crystal YBa$_2$Cu$_4$O$_8$ measured along the a–axis

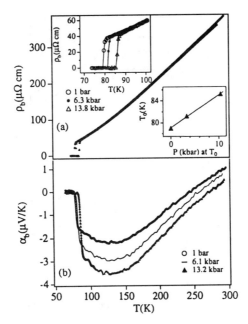

Fig. 7.(b) The resistivity $\rho(T)$ and thermoelectric power $\alpha(T)$ curves under different hydrostatic pressures for single–crystal YBa$_2$Cu$_4$O$_8$ measured along the b–axis. Insets show $\rho_b(T)$ and zero–resistance temperature T_0 for 1 bar, 6.3 kbar, and 13.8 kbar

The spin–gap temperature T^*, identified by NMR [17], decreases strongly with increasing hole concentration in the CuO$_2$ sheets. Therefore we can test whether the anomaly in $\rho_a(T)$ reflects an opening of the spin gap by changing the oxidation state of the CuO$_2$ sheets. Significantly, the anomaly in $\rho_a(T)$ remains independent of pressure whereas pressure oxidizes the CuO$_2$ sheets, increasing the hole concentration. We [18] have also reduced the CuO$_2$ sheets by doping a YBa$_2$Cu$_4$O$_8$ crystal with Ni. Increases in $\rho_b(T)$ and $\alpha_b(T)$ indicate that the Ni dopant enters preferentially the chain sites as low–spin Ni(III). An increase in $\alpha_b(300\,\mathrm{K})$ confirms that the CuO$_2$ sheets are reduced, and T_c is correspondingly lowered. Nevertheless, the anomaly in $\rho_a(T)$ remains near 160 K; it is not changed by either reduction or oxidation of the CuO$_2$ sheets. Moreover, $\alpha(T)$ shows no anomaly at $T \approx 160\,\mathrm{K}$ as should appear if the opening of a spin gap opened a pseudo gap at the Fermi energy. Therefore, we conclude that the anomaly in $\rho_a(T)$ is not associated with the opening of the spin gap. On the other hand, Fig. 7(a) shows, on cooling, an onset at 160 K of a transition in $\rho_c(T)$ from a semiconductive to a metallic temperature dependence; the decrease in $\rho_a(T)$ on cooling below 160 K is accompanied by a decrease in $\rho_c(T)$. Pair–density–function (PDF) analysis of pulsed neutron data [19] has revealed the onset of microdomain fluctuations in the chains below 160 K; these fluctuations may reflect a coupling between fluctuating

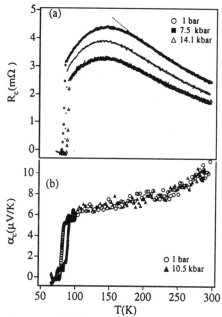

Fig. 7. (c) The resistance $R_c(T)$ and thermoelectric power $\alpha(T)$ curves under different hydrostatic pressures for single–crystal $YBa_2Cu_4O_8$ measured along the c–axis

CDW segments in the chains and sheets as they become more ordered at lower temperatures.

2.3   Single–Crystal $La_{2-x}Sr_xCuO_4$ Films

A broad solid–solution range $x = 0 - 0.3$ in the system $La_{2-x}Sr_xCuO_4$ allows doping from the antiferromagnetic parent phase to the non–superconductive overdoped phase. The LTO–HTT transition drops to lowest temperatures near $x = 0.22$ where the overdoped phase coexists with the superconductive phase [20]. High pressure stabilizes the HTT phase relative to the LTO phase, and T_c in the LTO phase has been found to increase with pressure until the Cu–O–Cu bond angle in the CuO_2 sheets becomes 180° in the tetragonal phase. For a given x, T_c reaches its maximum value, independent of hydrostatic pressure, in the HTT phase [20,21]. Moreover, substitution of Ca for Sr increases the bending of the $(180° - \varphi)$ Cu–O–Cu bond angle, and T_c decreases [22]. These observations demonstrate that T_c varies sensitively with the bending angle φ.

A defect–free $La_{1.85}Sr_{0.15}CuO_4$ single–crystal film was deposited layer–by–layer by MBE on a single–crystal $LaSrAlO_4$ substrate; the c–axis was normal to the film [1]. The bond–length mismatch across the substrate–film interface places the CuO_2 sheets of the $La_{1.85}Sr_{0.15}CuO_4$ film under a biaxial compressive stress, but the film remains orthorhombic. The T_c of the film increases under an applied hydrostatic pressure P up to 3.3 kbar; at higher pressures, T_c is pressure–independent, Fig. 8. The maximum $T_c = 43.8$ K is about 2 K higher than that obtained [23] with a ceramic specimen, which not only confirms that the tetragonal phase with strong Cu–O–Cu bonds gives the optimal T_c, but also indicates that the biaxial component of the pressure raises this optimum value. This latter conclusion is supported by a recent report [24] of a greatly enhanced T_c in an underdoped $La_{1.9}Sr_{0.1}CuO_4$ film made with the MBE technique.

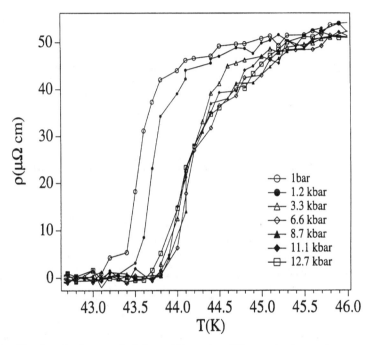

Fig. 8. The basal–plane resistivity $\rho(T)$ under different hydrostatic pressures of a $La_{1.85}Sr_{0.15}CuO_4$ single–crystal film on $LaSrAlO_4$

Figure 9 shows $\alpha(T)$ for the $La_{1.85}Sr_{0.15}CuO_4$ film for different hydrostatic pressures. Here also the low–temperature enhancement $\delta\alpha(T)$ increases with T_c, approaching saturation where T_c becomes temperature–independent, Fig. 10.

From quite general considerations, an increase in α reflects an increase in the asymmetry of the density of electron energies below and above the Fermi energy. Elegant angle–resolved photoemission data [25,26] on other copper–oxide compounds have demonstrated that, on cooling, there is a massive transfer of spectral weight from the (π,π) to the $(\pi,0)$ direction in the CuO_2 sheets; this spectral–weight transfer produces an extraordinary flattening of

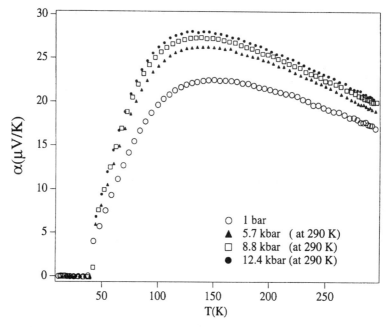

Fig. 9. Thermoelectric power $\alpha(T)$ under different hydrostatic pressures of the film of Fig. 8

the $\varepsilon(\boldsymbol{k})$ vs. $\boldsymbol{k}$ dispersion in the direction of the Cu–O–Cu bond axes. We therefore attribute the unusual $\delta\alpha(T)$ enhancement, which is peculiar to the high-T_c phenomenon, to this transfer of spectral weight. Since mobile CDW stripes are also associated with the high-T_c phenomenon, we have suggested [27] that the superconductive phase contains itinerant vibronic states in which itinerant–electron states of wave–vector $\boldsymbol{k}$ are stabilized by admixing with a phonon of wave–vector $\boldsymbol{q}$ moving along a Cu–O–Cu bond axis perpendicular to the mobile stripes. The resulting electron energies $\varepsilon(\boldsymbol{k}) \sim (\boldsymbol{k} \cdot \boldsymbol{q})^2$ would give rise to an increasing transfer of spectral weight from the (π, π) to the $(\pi, 0)$ direction as the CDW segments become more ordered with decreasing temperature.

3   Conclusion

High-T_c superconductivity in the copper oxides occurs in a thermodynamically distinguishable phase occurring as an intermediate phase within a first–order transition from localized to itinerant electronic behavior. The unusual transport properties of the normal state are sensitive to structural changes indicative of a strong electron–lattice coupling. At high temperatures, the normal state may be described as a strongly interacting gas of intermediate–size polarons; but on cooling below 300 K, the polarons condense into mobile

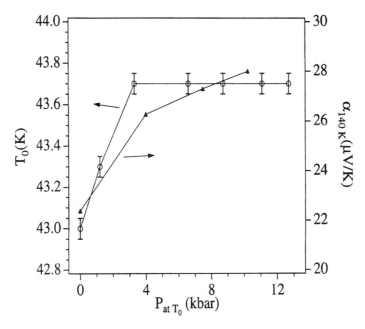

Fig. 10. Variation with pressure of the critical temperature for zero resistance, T_0, and the maximum thermoelectric power, $\alpha(T_{max})$, for the film of Fig. 8

CDW stripes that become increasingly ordered on cooling from 300 K to T_c. Pinning of commensurate stripes by the LTT distortion in $La_{2-x}Ba_xCuO_4$ near $x = 1/8$ introduces an activated electronic conduction and suppresses superconductivity. Superconductivity appears to be associated with mobile stripes and a strong coupling of itinerant electrons to phonons of the moving CDW. In this model, T_c is higher the more stable the ordering of the stripes, and ordering of mobile CDW stripes appears to be stabilized by straightening of the (180°) Cu–O–Cu bond and by shortening of the Cu–O bond length in the CuO_2 sheets relative to the c–axis Cu–O bond length under a biaxial stress.

In $YBa_2Cu_4O_8$, the transport properties vary smoothly through T^* unaffected by the opening of a spin gap. Moreover, the chains have a higher conductivity than the sheets and also become superconductive below T_c. A reduction in $\rho_a(T)$ below 160 K correlates with a reduction in $\rho_c(T)$ and the appearance of microdomain fluctuations in the chains. This correlation suggests the lowering of $\rho_a(T)$ and $\rho_c(T)$ reflects an elastic coupling of the ordered CDW segments in the chains and sheets.

Acknowledgements

The authors thank the NSF, the TCSUH, Houston, TX, and the Robert A. Welch Foundation, Houston, TX, for financial support.

References

1. Sato H., Naito M. (1997) Physica C 274:221
2. Takagi H., Batlogg B., Kao H.L., Kwo J., Cava R.J. Krajewski J.J., Peck W.F.,Jr. (1992) Phys Rev Lett 69:2975
3. Fisk Z., Webb G.W. (1976) Phys Rev Lett 36:1084
4. Zhou J.-S., Goodenough J.B. (1996) Phys Rev B 54:12 488
5. Bersuker G.I., Goodenough J.B. (1997) Physica C 274:167
6. Grenier J.-C. et al. (1992) Physica C 202:209
7. Zhou J.-S., Chen H., Goodenough J.B. (1994) Phys Rev B 50:4168
8. Goodenough J.B., Zhou J.-S., Chan J. (1993) Phys Rev B 47:5275
9. Shengelaya A. (1999) this Proceedings, 91
10. Zhou J.-S., Goodenough J.B. (1995) Phys Rev B 51:3104
    Zhou J.-S., Goodenough J.B. (1996) Phys Rev B 53:R11967
    Zhou J.-S., Goodenough J.B. (1996) Phys Rev Lett:77:151
11. Axe J.D., et al. (1989) Phys Rev Lett 62:3058
12. Tranquada J.M., et al. (1995) Nature (London) 375:561
13. Zhou J.-S., Goodenough J.B. (1997) Phys Rev B 56:6288
14. Zhou J.-S., Goodenough J.B., Asamitsu A., Tokura Y. (1997) Phys Rev Lett 79:3234
15. Karpinski J. et al. (1998) Nature (Lond) 366:660
16. Zhou J.-S., Goodenough J.B., Dabrowski B., Rogacki K. (1996) Phys Rev Lett 77:4253
17. Bucher B. et al. (1993) Phys Rev Lett 70:2012
18. Zhou J.S., Goodenough J.B., Dabrowski B. (1998) Phys Rev B 58:R2956
19. Sendyka T.R. et al. (1995) Phys Rev B 51:6747
20. Zhou J.-S., Chen H., Goodenough J.B. (1994) Phys Rev B 49:9084
21. Yamada N., Ido M. (1992) Physica C 203:240
22. Dabrowski B. et al., (1996) Phys Rev Lett 76:1348
23. Mori N. et al. (1991) Physica C 185–189:40
24. Locquet J.P. et al. (1998) Nature (Lond) 394:453
25. Shen Z.-X. et al. (1998) Science 280:259
26. Norman M.R. et al. (1998) Nature (Lond) 392:157
27. Goodenough J.B., Zhou J.-S. (1999) J Supercond to be published [Proc. Stripes and High-T_c Superconductivity, Rome, 2–6 June, 1998]

List of Unpublished Lectures

List of Lectures read but not included in the present Proceedings

1. **A Decade of HTS Materials and Beyond,** by C. W. Chu *(Texas Center for Superconductivity and Dept. of Physics, Univ. of Houston, TX, USA)*

2. **Structural Response to an Electronic Singularity at the Maximum T_c ,** by J. Jorgensen *(Materials Sci. Div., Argonne Natl. Lab., Argonne, IL, USA)*

3. **HTS Materials for Levitator and Magnet Applications – A Study of Single Crystals and Textured Superconductors,** by B.W. Veal *(Mater. Sci. Div., Argonne Natl. Lab., IL, USA)*

4. **Far-Infrared Ellipsometry of the Out–of–Plane Response in Cuprate Superconductors,** by A. Wittlin *(Max–Planck–Inst. für Festkörperforschung, Stuttgart, Germany)*

5. **Anisotropy and Geometry Effects in the High-T_c Superconductors,** by John R. Clem *(Ames Laboratory and Dept. of Phys. & Astron., Iowa State Univ., Ames, IA, USA)*

6. **Luttinger–Liquid Phenomenology of High-T_c Superconductivity,** by K. Byczuk, J. Spałek, and W. Wójcik *(Inst. of Physics, Jagellonian University, Cracow, Poland)*

7. **Grain Boundary Properties of HTS,** by D. Larbalestier *(Dept. of Mater. Sci. & Engng, and Dept. of Phys., Appl. Superconductivity Cent., Univ. of Wisconsin, Madison, WI, USA)*

8. **Superconductivity in Intercalated Graphite and Fullerene,** by J. Stankowski, W. Kempiński, and L. Piekara–Sady *(Institute of Molecular Physics, Polish Ac. Sci., Poznań, Poland)*

9. **Magnetic Instabilities in High-Temperature Superconductors,** by Z. Tarnawski *(Fac. of Phys. and Nucl. Techn., Acad. of Mining & Metallurgy, Cracow, Poland)*, A. Gerber *(Fac. of Exact Sci., Univ. of Tel Aviv, Israel)*, and J. J. M. Franse *(Van der Vaals – Zeeman Lab., Univ. v. Amsterdam, The Netherlands)*

List of Posters

Posters presented at the Conference

1. **Pulsed Laser Deposited Amorphous and Crystalline YBaCuO Thin Films: Structural, Optical and Transport Properties,** by A.Abal'oshev, E. Dynowska, P. Gierłowski, A.Klimov, S.J. Lewandowski *(Inst. of Physics, Polish Ac. Sci., Warsaw, Poland)*, D. Okunev, Z.A.Samoilenko, and V.M. Svistunov *(Inst. for Physics and Engineering, Natl. Ac. Sci., Donetsk, Ukraine)*

2. **Superconducting Properties of Superconductor–Insulator Composites,** by B.Andrzejewski, J. Stankowski, A.Kaczmarek, B. Hilczer *(Inst. of Molec. Phys., Polish Ac. Sci., Poznań, Poland)*, and J. Marfaing *(Facult. des Scis. et Techniques, Lab. Matér., Organis. et Propért., CNRS, Marseille, France)*

3. **Superconductivity in the Pair–Tunneling Model Close to the Metal–Insulator Transition,** by J. Biesiada and J. Zieliński *(Inst. of Physics, Univ. of Silesia, Katowice, Poland)*

4. **Influence of Al_2O_3 or $BaZrO_3$ Substrates on Transport Properties of Bi2223 Superconductors,** by H. Bougrine, M. Ausloos, B. Robertz, R. Cloots *(Phys. Inst. and Chem. Inst., Univ. de Liège, Belgium)*, and J. Mucha *(Inst. of Low Temperature and Structure Research, Wrocław, Poland)*

5. **Thermodynamics of the BCS Model at Broken Particle–Hole Symmetry,** by J. Czerwonko *(Inst. of Physics, Univ. of Technology, Wrocław, Poland)*

6. **An EPR Measurements of the Spatial Distribution of Magnetic Field Near the Nb Superconducting Samples. The Object's Shape Effect,** by H. Drulis, L.Folcik, M. Drulis *(Inst. of Low Temperature and Structure Research, Wrocław, Poland)*, and N.M. Suleimanov *(Kazan' Physicotechnical Inst., Russian Ac. Sci., Kazan', Russian Federation)*

7. **Comparative Analysis of Inelastic Properties of 214 Lanthanum Perovskites,** by M. Gazda, B. Kusz, and K. Pieniaszek *(Fac. of Appl. Phys. and Mathem., Techn. Univ. of Gdańsk, Poland)*

8. **On Thermodynamic Properties of d–Wave Paired Superconductors,** by R.Gonczarek and M. Mulak *(Inst. of Phys., Techn. Univ. of Wrocław, Poland)*

9. **The Exchange of Conducting Electrons with Antiferromagnetic Chains,** by E. Hankiewicz, R. Buczko, and Z. Wilamowski *(Inst. of Phys. of Polish Ac. Sci., Warsaw, Poland)*

10. **Studies of Twinned Surface of Superconducting $YBa_2Cu_3O_{7-\delta}$ Single Crystals,** by E. Hankiewicz *(Inst. of Phys. of Polish Ac. Sci.,Warsaw, Poland)*, W. Sadowski, and T. Klimczuk *(Fac. of Appl. Phys. and Mathem., Techn. Univ. of Gdańsk, Poland)*

11. **Role of Anisotropy of the Impurity Scattering Potential in High-T_c Superconductors,** by G. Harań *(Inst. of Phys., Techn. Univ. of Wrocław, Poland)*

12. **The Superconducting Transition in $Pb(Sc_{1/2}Ta_{1/2})O_3$ – $YBa_2Cu_3O_{7-\delta}$ Composites,** by A. Kaczmarek, B. Andrzejewski, J. Stankowski, B. Hilczer *(Inst. of Molec. Phys., Polish Ac. Sci., Poznań, Poland)*, C. Caranoni, and J. Marfaing *(Fac. des Scis. et Techniq., Lab. Matér., Organis. et Propért., CNRS, Marseille, France)*

13. **Study of Thermal Stability of $Nd_{2-x}Ce_xCuO_{4-y}$ Single Crystals,** by T. Klimczuk and W. Sadowski *(Fac. of Appl. Phys. and Mathem., Techn. Univ. of Gdańsk, Poland)*

14. **Scaling Near the Zero–Temperature Critical Point in the Quantum Two–Dimensional Josephson Junction Arrays,** by T.K. Kopeć *(Inst. of Low Temperature and Structure Research, Wrocław, Poland)* and J.V. José *(Phys. Dept. and Center for Interdisc. Res. on Complex Systems, Northeast. Univ., Boston, MA, USA)*

15. **Electronic Tunneling into the Vortex Lattice States in Superconductors in High Magnetic Field,** by L. Kowalewski, M. Nogala, M. Thomas, and R.J. Wojciechowski *(Inst. of Phys., A. Mickiewicz Univ., Poznań, Poland)*

16. **Superconductivity in Correlated Systems: Modified Slave Boson Study,** by M. Krawiec, T. Domański, and K.I. Wysokiński *(Inst. of Phys., M. Curie-Skłodowska Univ., Lublin, Poland)*

17. **Photoinduced Nonlinear Optics Effects in Y–Ba–Cu–O Films,** by A. Kryza *(Inst. of Phys., Pedagogical Univ., Częstochowa, Poland)*

18. **Intrinsic Pinning in Layered Antiferromagnetic Superconductor,** by T. Krzysztoń *(Inst. of Low Temperature and Structure Research, Wrocław, Poland)*

19. **Percolation of Superconductivity,** by G. Litak *(Dept. of Mechan., Techn. Univ. of Lublin, Poland)* and B.L. Gyorffy *(H.H. Wills Phys. Lab., Univ. of Bristol, Engl., UK)*

20. **Van Hove Singularity and d–Wave Pairing in Disordered Superconductors,** by G. Litak *(Dept. of Mechan., Techn. Univ. of Lublin, Poland)*, A. Martin, B.L. Gyorffy, J.F. Annett *(H.H. Wills Phys. Lab., Univ. of Bristol, Engl., UK)*, and K.I. Wysokiński *(Inst. of Phys., M. Curie-Skłodowska Univ., Lublin, Poland)*

21. **Microwave Investigations of Py/Cu Multilayers with Various Magnetic Layers Coupling,** by M. Maciąg and J. Stankowski *(Inst. of Molec. Phys., Polish Ac. Sci., Poznań, Poland)*

22. **Pairing Symmetry in the Anisotropic BCS Model,** by M. Maska *(Dept. of Theor. Phys., Silesian Univ., Katowice, Poland)*

23. **Phonon–Induced and Phonon–Free Superconductivity in the Two–Dimensional Hubbard Model. A Strong Coupling Description,** by M. Mierzejewski, J. Zieliński, and A. Cebula *(Inst. of Phys., Silesian Univ., Katowice, Poland)*

24. **Hg–Based Superconducting Thin Films by Laser Ablation with High–Pressure, High–Temperature Treatment,** by A. Morawski, A. Paszewin, T. Łada, A. Presz *(High Press. Res. Center "Unipress", Polish Ac. Sci., Warsaw, Poland)*, K. Przybylski *(Fac. of Mater. Sci. and Ceramics, Acad. of Mining and Metallurgy, Cracow, Poland)*, P. Gierłowski *(Inst. of Phys., Polish Ac. Sci., Warsaw, Poland)*, and R. Gatt *(Synchrotr. Radiat. Cent., Univ. of Wisconsin, Stoughton, WI, USA)*

25. **Theoretical and Experimental Studies of the Magnetostriction Induced by the Pinning Forces in High Temperature Superconductors,** by A. Nabiałek and H. Szymczak *(Inst. of Phys., Polish Ac. Sci., Warsaw, Poland)*

26. **Principles of the Photoinduced Optical Detection of the Disordered Superconductors,** by J. Napieralski, A. Kryza, I.V. Kityk, and J. Kasperczyk *(Inst. of Phys., Pedagogical Univ., Częstochowa, Poland)*

27. **Measurements of Resistivity of HTS Materials,** by W. Nawrocki, B. Susła, and M. Wawrzyniak *(Inst. of Phys., Techn. Univ. of Poznań, Poland)*

28. **Effects of Diagonal Disorder on Charge Density Wave and Superconductivity in Local Pair Systems,** by G. Pawłowski and S. Robaszkiewicz *(Inst. of Phys., A. Mickiewicz Univ., Poznań, Poland)*

29. **Microstructure and Physicochemical Properties of (Hg,Pb)–Ba–Ca–Cu–0 (1223) Superconductor Prepared by High Pressure Gas Method,** by K. Przybylski *(Fac. of Mater. Sci. and Ceramics, Acad. of Mining and Metallurgy, Cracow, Poland)*,

A. Morawski, T. Łada, A. Paszewin *(High Press. Res. Center "Unipress", Polish Ac. Sci., Warsaw, Poland)*, and T. Brylewski *(Fac. of Mater. Sci. and Ceramics, Acad. of Mining and Metallurgy, Cracow, Poland)*

30. **Magnetism and Fine Electronic Structure of Mott Insulators: La_2CuO_4 and Other 3D Systems,** by R.J. Radwański, Z. Ropka, and R. Michalski *(Cent. for Solid State Physics, Cracow, Poland)*

31. **Thermal Treatment Study and Transport Properties of $PrBa_2Cu_3O_{7-\delta}$ Crystals,** by W. Sadowski, M. Łuszczek *(Fac. of Appl. Phys. and Mathem., Techn. Univ. of Gdańsk, Poland)*, and J. Olchowik *(Inst. of Phys., Dept. of Phys., Techn. Univ. of Lublin, Poland)*

32. **Selected Investigations of Applied Electromagnetism in High–Temperature Oxide Superconductors,** by J. Sosnowski *(Inst. of Electrotechnol., Warsaw, Poland)*

33. **The Insulator–to–Metal Transition in the Deoxygenated $Y_{1-x}Ca_xBa_2Cu_3O_{6+\delta}$ System,** by P. Starowicz, B. Penc, J. Sokołowski, and A. Szytuła *(Inst. of Phys., Jagellonian Univ., Cracow, Poland)*

34. **Thermopower Anisotropy for $YBa_2Cu_3O_{6+x}$ Single Crystals with $T_c = 0 - 50\,K$,** by Cz. Sułkowski, T. Plackowski *(Inst. of Low Temperature and Structure Research, Wrocław, Poland)*, and W. Sadowski *(Fac. of Appl. Phys. and Mathem., Techn. Univ. of Gdańsk, Poland)*

35. **Transport Properties and Time Relaxation of Oxygen Sublattice in $Nd_2CuO_{4-\delta}$,** by Cz. Sułkowski, A. Sikora, P.W. Klamut, R. Horyń, and M. Wołcyrz *(Inst. of Low Temperature and Structure Research, Wrocław, Poland)*

36. **Point Contact Measurements in Superconductors,** by B. Susła *(Poznań, Poland)*, W. Sadowski *(Fac. of Appl. Phys. and Mathem., Techn. Univ. of Gdańsk, Poland)*, and M. Kamiński *(Inst. of Phys., Techn. Univ. of Poznań, Poland)*

37. **Quantum Creep in "Dense" Mixed State of Type–II Superconductors,** by P. Tekiel *(Inst. of Low Temperature and Structure Research, Wrocław, Poland)*

38. **Crystal Field Effects on Magnetic Properties of R^{3+} Rare Earth Ions in R123 Compounds,** by J. Typek *(Inst. of Phys, Techn. Univ. of Szczecin, Poland)*, N. Guskos *(Dept. of Phys., Univ. of Athens, Greece)*, M. Wabia *(Inst. of Phys, Techn. Univ. of Szczecin, Poland)*, and V. Likodimos *(Dept. of Phys., Univ. of Athens, Greece)*

39. **Magnetic Studies of Flux Pinning in Hg–Based Single Crystals,** by A. Wiśniewski, R. Puźniak, R. Szymczak, M. Baran *(Inst. of Phys., Polish Ac. Sci., Warsaw, Poland)*, and J. Karpiński *(Lab. für Festk.physik d. ETH Zürich, Switzerland)*

40. **Superconducting Materials Based on the Binary Compound Mo$_6$Se$_8$,** by A. Wojakowski, R. Horyń *(Inst. of Low Temperature and Structure Research, Wrocław, Poland)*, O. Peña, C. Hamard, and F. Le Berre *(Chimie du Solide et Inorg. Moléc., Univ. de Rennes, France)*

41. **Symmetry of Binding in Doped Antiferromagnets,** by P. Wróbel *(Inst. of Low Temperature and Structure Research, Wrocław, Poland)* and Robert Eder *(Inst. für Theoretische Physik, Univ. Würzburg, Germany)*

Index of Contributors

Lecture Notes in Physics

For information about Vols. 1–508
please contact your bookseller or Springer-Verlag

Vol. 509: J. Wess, V. P. Akulov (Eds.), Supersymmetry and Quantum Field Theory. Proceedings, 1997. XV, 405 pages. 1998.

Vol. 510: J. Navarro, A. Polls (Eds.), Microscopic Quantum Many-Body Theories and Their Applications. Proceedings, 1997. XIII, 379 pages. 1998.

Vol. 511: S. Benkadda, G. M. Zaslavsky (Eds.), Chaos, Kinetics and Nonlinear Dynamics in Fluids and Plasmas. Proceedings, 1997. VIII, 438 pages. 1998.

Vol. 512: H. Gausterer, C. Lang (Eds.), Computing Particle Properties. Proceedings, 1997. VII, 335 pages. 1998.

Vol. 513: A. Bernstein, D. Drechsel, T. Walcher (Eds.), Chiral Dynamics: Theory and Experiment. Proceedings, 1997. IX, 394 pages. 1998.

Vol. 514: F. W. Hehl, C. Kiefer, R. J. K. Metzler, Black Holes: Theory and Observation. Proceedings, 1997. XV, 519 pages. 1998.

Vol. 515: C.-H. Bruneau (Ed.), Sixteenth International Conference on Numerical Methods in Fluid Dynamics. Proceedings. XV, 568 pages. 1998.

Vol. 516: J. Cleymans, H. B. Geyer, F. G. Scholtz (Eds.), Hadrons in Dense Matter and Hadrosynthesis. Proceedings, 1998. XII, 253 pages. 1999.

Vol. 517: Ph. Blanchard, A. Jadczyk (Eds.), Quantum Future. Proceedings, 1997. X, 244 pages. 1999.

Vol. 518: P. G. L. Leach, S. E. Bouquet, J.-L. Rouet, E. Fijalkow (Eds.), Dynamical Systems, Plasmas and Gravitation. Proceedings, 1997. XII, 397 pages. 1999.

Vol. 519: R. Kutner, A. Pękalski, K. Sznajd-Weron (Eds.), Anomalous Diffusion. From Basics to Applications. Proceedings, 1998. XVIII, 378 pages. 1999.

Vol. 520: J. A. van Paradijs, J. A. M. Bleeker (Eds.), X-Ray Spectroscopy in Astrophysics. EADN School X. Proceedings, 1997. XV, 530 pages. 1999.

Vol. 521: L. Mathelitsch, W. Plessas (Eds.), Broken Symmetries. Proceedings, 1998. VII, 299 pages. 1999.

Vol. 522: J. W. Clark, T. Lindenau, M. L. Ristig (Eds.), Scientific Applications of Neural Nets. Proceedings, 1998. XIII, 288 pages. 1999.

Vol. 523: B. Wolf, O. Stahl, A. W. Fullerton (Eds.), Variable and Non-spherical Stellar Winds in Luminous Hot Stars. Proceedings, 1998. XX, 424 pages. 1999.

Vol. 524: J. Wess, E. A. Ivanov (Eds.), Supersymmetries and Quantum Symmetries. Proceedings, 1997. XX, 442 pages. 1999.

Vol. 525: A. Ceresole, C. Kounnas, D. Lüst, S. Theisen (Eds.), Quantum Aspects of Gauge Theories, Supersymmetry and Unification. Proceedings, 1998. X, 511 pages. 1999.

Vol. 526: H.-P. Breuer, F. Petruccione (Eds.), Open Systems and Measurement in Relativistic Quantum Theory. Proceedings, 1998. VIII, 240 pages. 1999.

Vol. 527: D. Reguera, J. M. G. Vilar, J. M. Rubí (Eds.), Statistical Mechanics of Biocomplexity. Proceedings, 1998. XI, 318 pages. 1999.

Vol. 528: I. Peschel, X. Wang, M. Kaulke, K. Hallberg (Eds.), Density-Matrix Renormalization. Proceedings, 1998. XVI, 355 pages. 1999.

Vol. 529: S. Biringen, H. Örs, A. Tezel, J.H. Ferziger (Eds.), Industrial and Environmental Applications of Direct and Large-Eddy Simulation. Proceedings, 1998. XVI, 301 pages. 1999.

Vol. 530: H.-J. Röser, K. Meisenheimer (Eds.), The Radio Galaxy Messier 87. Proceedings, 1997. XIII, 342 pages. 1999.

Vol. 531: H. Benisty, J.-M. Gérard, R. Houdré, J. Rarity, C. Weisbuch (Eds.), Confined Photon Systems. Proceedings, 1998. X, 496 pages. 1999.

Vol. 532: S. C. Müller, J. Parisi, W. Zimmermann (Eds.), Transport and Structure. Their Competitive Roles in Biophysics and Chemistry. XII, 400 pages. 1999.

Vol. 533: K. Hutter, Y. Wang, H. Beer (Eds.), Advances in Cold-Region Thermal Engineering and Sciences. Proceedings, 1999. XIV, 608 pages. 1999.

Vol. 534: F. Moreno, F. González (Eds.), Light Scattering from Microstructures. Proceedings, 1998. XII, 300 pages. 2000.

Vol. 535: H. Dreyssé (Ed.), Electronic Structure and Physical Properties of Solids: The Uses of the LMTO Method. Proceedings, 1998. XIV, 458 pages. 2000.

Vol. 536: T. Passot, P.-L. Sulem (Eds.), Nonlinear MHD Waves and Turbulence. Proceedings, 1998. X, 385 pages. 1999.

Vol. 537: S. Cotsakis, G. W. Gibbons (Eds.), Mathematical and Quantum Aspects of Relativity and Cosmology. Proceedings, 1998. XII, 251 pages. 1999.

Vol. 538: Ph. Blanchard, D. Giulini, E. Joos, C. Kiefer, I.-O. Stamatescu (Eds.), Decoherence: Theoretical, Experimental, and Conceptual Problems. Proceedings, 1998. XII, 345 pages. 2000.

Vol. 539: A. Borowiec, W. Cegła, B. Jancewicz, W. Karowski (Eds.), Theoretical Physics. Fin de Siècle. Proceedings, 1998. XX, 319 pages. 2000.

Vol. 540: B. G. Schmidt (Ed.), Einstein's Field Equations and Their Physical Implications. Selected Essays. 1999. XIII, 429 pages. 2000

Vol. 541: J. Kowalski-Glikman (Ed.), Towards Quantum Gravity. Proceedings, 1999. XII, 376 pages. 2000.

Vol. 542: P. L. Christiansen, M. P. Sørensen, A. C. Scott (Eds.), Nonlinear Science at the Dawn of the 21st Century. Proceedings, 1998. XXVI, 458 pages. 2000.

Vol. 543: H. Gausterer, H. Grosse, L. Pittner (Eds.), Geometry and Quantum Physics. Proceedings, 1999. VIII, 408 pages. 2000.

Vol. 545: J. Klamut, B. W. Veal, B. M. Dabrowski, P. W. Klamut, M. Kazimierski (Eds.), New Developments in High-Temperature Superconductivity. Proceedings, 1998. VIII, 275 pages. 2000.

Vol. 546: G. Grindhammer, B. A. Kniehl, G. Kramer (Eds.), New Trends in HERA Physics 1999. Proceedings, 1999. XIV, 460 pages. 2000.

Monographs

For information about Vols. 1–20
please contact your bookseller or Springer-Verlag

Printing: Weihert-Druck GmbH, Darmstadt
Binding: Buchbinderei Schäffer, Grünstadt